CHEMICAL THERMODYNAMICS
With Examples for Nonequilibrium Processes

CHEMICAL THERMODYNAMICS

With Examples for Nonequilibrium Processes

Byung Chan Eu
McGill University, Canada

Mazen Al-Ghoul
American University of Beirut, Lebanon

 **World Scientific**

NEW JERSEY · LONDON · SINGAPORE · BEIJING · SHANGHAI · HONG KONG · TAIPEI · CHENNAI

Published by

World Scientific Publishing Co. Pte. Ltd.

5 Toh Tuck Link, Singapore 596224

USA office: 27 Warren Street, Suite 401-402, Hackensack, NJ 07601

UK office: 57 Shelton Street, Covent Garden, London WC2H 9HE

British Library Cataloguing-in-Publication Data
A catalogue record for this book is available from the British Library.

CHEMICAL THERMODYNAMICS
With Examples for Nonequilibrium Processes

ISBN-13 978-981-4295-11-6
ISBN-10 981-4295-11-6

Typeset by Stallion Press
Email: enquiries@stallionpress.com

Printed in Singapore.

*To students aspiring to attain a liberated perspective
on thermodynamics*

Preface

After a quick glance at it, one is liable to pass the judgment that thermodynamics appears to be an old subject in science. Yet it is hardly true, since, when closely examined, it is still a developing subject. It is regarded by many in science as an old subject because its basic principles were enunciated almost 160 years ago and the theory of reversible processes was basically completed in the hands of J. W. Gibbs within the span of 25 years from the time when the thermodynamic laws were stated by Clausius and Kelvin. It thus appears that there is nothing new to add to the subject. Nevertheless, the general theory of thermodynamic processes that include irreversible processes has been in an arrested state of evolution and has remained incomplete until L. Onsager, J. Meixner, and I. Prigogine formulated a theory of linear irreversible processes. This theory is still incomplete because irreversible processes about which the second law of thermodynamics is basically concerned have not received a fully satisfactory theoretical treatment. Fuller treatments of the subject are given serious attention in recent years. Therefore thermodynamics in the generalized sense is still worth serious study, if possible, from the viewpoint more general than that taken in the traditional approach to the subject and, in particular, equilibrium thermodynamics where only reversible processes are studied.

Since reversible processes are the idealized limits of irreversible processes observed in nature, the thermodynamics of reversible processes, namely, equilibrium thermodynamics, must be the limiting form of a more general theory of thermodynamic processes, and as such, it is worth having a fresh examination in what manner it is a limiting theory. We examine some basic aspects of equilibrium thermodynamics under such a motivation in this textbook, which also contains traditional treatments of various topics taught in courses in thermodynamics at an advanced undergraduate level and graduate level. The treatments given to the second law of thermodynamics and related topics, such as equilibrium conditions and stability of equilibrium in this work are different from those in conventional textbooks on equilibrium thermodynamics. In fact these are extended versions of those found in conventional textbooks. Examples of application of the extended treatment of the second law is discussed in the last chapter to show how one might apply the concepts to study irreversible processes far from equilibrium.

The materials on the conventional topics of thermodynamics in this book, excluding those related to the second law, have been taught by B.C.E. over many years in the courses of thermodynamics at McGill University. The treatments given to the second law of thermodynamics and related topics in this book, being new and more recent fruits of labor on the part of M.A.-G. have not been exposed to the classes in the past. However, we believe that they should be an integral part of a course on equilibrium thermodynamics. The reason is that not only the new results of research add to the science of thermodynamics, the mathematical representation of the second law of thermodynamics that adequately covers irreversible processes, and provides clarifications of various related topics about which the conventional treatment of the subjects leaves us uncomfortable, but also the equilibrium thermodynamics of reversible processes emerges as the limiting case of a more general theory of thermodynamic processes as it should be. Furthermore, the generalized mathematical representation of the second law of thermodynamics removes some nagging conceptual features that arise from the fact that the entropy was originally defined for reversible processes only. Yet one still thinks of the entropy as if it is a nonequilibrium quantity, when it comes to systems in the vicinity of an equilibrium state, as in the case of equilibrium conditions and stability of equilibrium states. Therefore, when equilibrium thermodynamics is considered from the generalized theory, namely, generalized thermodynamics, we can form more harmonized viewpoints towards the subject and contemplate formulation of a more comprehensive theory of thermodynamic processes, reversible or irreversible. In this potential lies our desire to examine equilibrium thermodynamics from an angle different from the traditional viewpoint. The examples discussed in the last chapter, albeit brief, illustrate the utility of the concept of calortropy in study of irreversible processes. The reader interested in more involved discussions on irreversible phenomena and hydrodynamics in the nonlinear regime is referred to monographs dealing with the subjects. Such monographs are available at present. It is the hope of the authors that this work will act as a stepping stone towards a more complete theory of irreversible phenomena.

B.C.E. & M.A.-G.
September 2009

Contents

Chapter 1

Introduction

Thermodynamics in the generalized sense is a branch of natural science in which we study heat, work, energy, their interrelationships, and the modes by which systems exchange heat, matter, and energy with each other and with the surroundings, by converting heat into work, and vice versa. Since all human activities and natural phenomena involve matter and energy of one form or another, the importance of such science is obvious, and for that reason it is in the foundations of physical, biological, and engineering sciences. As a matter of fact, thermodynamics owes its genesis to the urgent need at the early stage of the Industrial Revolution in the first half of the 19th century, to understand how steam engines work and improve their efficiencies, since the efficiencies of such engines had significant economic implications then. Such questions are still relevant even to this day in our everyday economic and industrial activities. On one hand, such a need motivated scientists and engineers to study the properties of steam in particular and gases in general to construct, for example, the steam table. On the other hand, it culminated in the idealization of engines with a reversible cycle by S. Carnot, who made a lasting contribution through his penetrating analysis of how heat engines operate, and his study resulted in his famous principle now known as Carnot's theorem (although his analysis was based on the caloric theory of heat which was proven to be an incorrect notion of heat). Later pioneers such as R. Clausius and W. Thomson (Lord Kelvin) adopted the correct notion of heat — in which heat is regarded as a form of energy — and developed a theory by retaining truthful features in Carnot's exposition and expanding on it. Through their efforts and the works by subsequent researchers, the science of thermodynamics was born in the second half of the 19th century. The subject was refined, especially in an important way, by J. W. Gibbs through his well-known work on heterogeneous equilibria. The modern form of the science of thermodynamics, laid on the foundations shaped by efforts by S. Carnot, Count Rumford (Benjamin Thompson), J. R. Mayer, R. Clausius, W. Thomson, H. Helmholtz, and J. W. Gibbs among others, has been developed by numerous other researchers, but its applications to chemical and chemical engineering problems owe a great deal to the works by M. Planck, W. Nernst, F. Haber, and G. N. Lewis and his school to name a few. The works of Max Born and C. Caratheodory have given equilibrium thermodynamics another mathematical aspect through Caratheodory's theorem, which opens up a geometrical viewpoint to thermodynamics. However, we will not discuss this line of thoughts in this work.

It is now generally believed that all natural macroscopic phenomena occur in full conformation to the laws governing thermodynamics. Although the subject of thermodynamics has been around over 160 years by now, it is not closed, since the pioneers in thermodynamics limited the development to reversible processes and thus to systems in equilibrium, it is still an open, developing science. It is therefore worth a serious study, especially since irreversible phenomena are not sufficiently well understood as yet from the standpoint of the laws of thermodynamics, especially if irreversible processes occur far removed from equilibrium.

From the viewpoint of thermodynamics of irreversible processes, equilibrium thermodynamics — which, more precisely, should be called thermostatics and we are going to study it here — is merely dealing with systems at a singular state of thermodynamic equilibrium. Since there are no macroscopically discernable processes occurring in equilibrium systems, equilibrium thermodynamics deals with idealized reversible processes and thus is not capable of describing what is happening in the real system over a finite time span and over space; rather, it is only able to tell us the possibilities of that particular event as far as the laws of thermodynamics are concerned. The description of the process over a finite span of time and space is in the realm of irreversible thermodynamics, which is still in the developing stage at present. The basic reason that the subject of equilibrium thermodynamics is useful and powerful despite the idealized reversible processes studied in it is that some macroscopic thermodynamic properties of a system which is going through irreversible processes can be related to the complementary quantities computed from the reversible processes, as will be shown later when we delve into the subject. This is because there is a state function of thermodynamic state variables for the system even if the processes are irreversible. This state function extends the notion of the equilibrium entropy into the domain of irreversible processes. The existence of such a state function — called calortropy — lifts thermodynamics from the level of studying only idealized reversible processes, as in equilibrium thermodynamics, to a level of using a more insightful and powerful mathematical tool for studying various aspects of irreversible behavior of the system. We will elaborate on this point in the chapters dealing with the second law of thermodynamics.

Since we are going to use various terms in our study, we fix their meanings by introducing the following system of terminology. A *system* is that part of the physical world which is under consideration, and the rest of the physical world is called the *surroundings*. When a system exchanges mass, heat, work, and any other forms of energy with the surroundings, it is called an *open system* (c). When a system does not exchange matter but energies with the surroundings, it is said to be a *closed system* (b). If a system has no interaction whatsoever with the surroundings, it is called an *isolated system* (a) (Fig. 1.1). It is possible to regard the system and the surroundings together as an isolated system. We will find it convenient to do so for some cases.

Thermodynamics is concerned with gross observables of a macroscopic system and their interrelationships. Since according to the atomic theory of matter a macroscopic system consists of an enormous number of atoms and molecules not counting elementary subnuclear particles, a microscopic description of the macroscopic system would entail knowledge of an enormous number of microscopic variables. However, a large number of molecules in an assembly exhibit as a rule a collective behavior, which may be described by a small number of variables called *macroscopic variables* or *macroscopic coordinates*. A macroscopic variable (coordinate) is an observable whose determination requires only measurements, over long time spans compared with periods of thermal motion, that take averages of microscopic variables

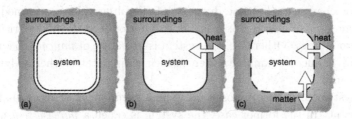

Fig. 1.1 The system and surroundings — universe. Panel (a) is an isolated system; (b) is a closed system; (c) is an open system.

over regions containing a large number of molecules, and involving large energies compared with individual energies of atoms and molecules. Pressure, volume, temperature, and internal energy, are examples of such macroscopic variables.

We often speak of *thermodynamic properties*. These are termed as the properties of the system which describe its macroscopic coordinates. Thermodynamic properties are classified into two categories. If a thermodynamic property is independent of the mass of the system, then it is called an *intensive property*. Examples are pressure, temperature, concentrations, and molar properties. If a thermodynamic property depends on the mass of the system, it is called an *extensive property*. Examples are the volume, energy, and entropy of a system which increase in proportion to the mass of the system. Intensive and extensive properties (variables) often appear as conjugate pairs of variables in thermodynamics. We may take the examples of pressure and volume, and temperature and entropy for such conjugate pairs of thermodynamic variables. Intensive properties, however, may vary with position in the space as do the extensive variables, if the system is not homogeneous. If the intensive properties are continuous functions of position throughout the system, then the system is called *homogeneous*, and if they are not continuous functions of position throughout the system, then the system is called *heterogeneous*. For example, a system is heterogeneous if the density changes discontinuously across the boundary of two homogeneous regions of the system. Such homogeneous regions of a heterogeneous system are called the *phases* of the system. An example for heterogeneous systems is a system of water and ice, and in this particular case there are two phases in the system.

We will often speak of a *thermodynamic state*. The thermodynamic state of a system is defined by its intensive properties. They may or may not change over space and time. A system is said to be in a state of *thermodynamic equilibrium* if (a) the thermodynamic state of the system does not change in the time duration of the observation performed; and (b) no material or energy flux exists in its interior or at the boundaries with the surroundings. Otherwise, the system is in a state of *nonequilibrium*. In thermodynamics we examine systems which are in a state of thermodynamic equilibrium.

A system may be composed of a single component or more than one component. In the former case, the system is called a *pure system* and in the latter case, it is called a *mixed system* or simply a *mixture*. In a mixture there may arise the question as to the number of *independent components* of the system: it is defined by the minimum number of chemical species from which the system can be prepared in each phase of the system by a specified set of physicochemical procedures. A practical way of determining the number of independent components is the total number of components minus the number of distinct chemical and other restrictive conditions such as chemical reactions and charge neutrality conditions. For example, consider a system formed when NaCl, KBr, and H_2O are mixed. If KCl, NaBr, NaBr·H_2O, KBr·H_2O, and NaCl·H_2O are isolated on chemical analysis, the distinct chemical reactions are

$$NaCl + KBr \rightarrow NaBr + KCl,$$
$$NaCl + H_2O \rightarrow NaCl·H_2O,$$
$$KBr + H_2O \rightarrow KBr·H_2O,$$
$$NaBr + H_2O \rightarrow NaBr·H_2O.$$

If the concentration of species i is denoted by c_i, and since the relations for the concentrations are

$$c_{NaCl}^{initial} = c_{NaBr} + c_{NaCl·H_2O} + c_{NaBr·H_2O}$$

and

$$c_{KCl} = c_{NaCl}^{initial} - c_{NaCl·H_2O},$$

where $c_{NaCl}^{initial}$ is the initial concentration of NaCl, we find

$$c_{KCl} = c_{NaBr} + c_{NaBr·H_2O}.$$

This means that the number of independent components is $(8 - 1) - 4 = 3$ in this case.

In Chapter 2 of this work, the notions of temperature, work, and heat are discussed. In Chapter 3, the first law of thermodynamics is discussed together with thermochemistry, which deals with measurements of heat released or absorbed by the system. In Chapter 4, the second law of thermodynamics is discussed. This important principle, which was literally enunciated by Lord Kelvin and R. Clausius, was given a mathematical representation in the form of inequality now known as the Clausius inequality. In this work, the Clausius inequality will be replaced by an equation as a general mathematical representation of the second law of thermodynamics, which remains valid even if there are irreversible processes present in the system. The said mathematical representation of the second law of thermodynamics permits us to develop the thermodynamics of irreversible processes in a general context. For this purpose we introduce the notion

of calortropy. This part of the treatment of the second law of thermodynamics sets the present work apart from the conventional methods used in
other works available in the literature on thermodynamics. The distinctive
point of the new quantity is that it is the extension to nonequilibrium of
the notion of equilibrium entropy that was originally introduced by Clausius for reversible processes only. By the accomplished extension, we are
now provided with the starting point of a theory of irreversible processes
in a general form, and even the equilibrium thermodynamics of Clausius
is provided with a window through which we can glimpse into the world
of irreversible phenomena, even if one studies just the reversible process
associated with the irreversible process in question. The nonequilibrium
extension of the Clausius entropy is given the new term *calortropy*, which
means heat evolution. This extension frees us from the shackles of entropy
defined for equilibrium only, and equilibrium thermodynamics consequently
becomes easier to comprehend than otherwise.

In the rest of the book, we treat the conventional subjects of equilibrium thermodynamics, which are commonly discussed in courses on thermodynamics. The subjects covered are thermodynamics of gases, liquids,
and solutions; heterogeneous equilibria; chemical equilibria; strong electrolytes; galvanic cells; and the Debye–Hückel theory of strong electrolytes,
which is the only concession we make to discuss a statistical treatment
of macroscopic phenomena. We will also discuss the thermodynamics of
systems subject to electromagnetic fields and the thermodynamics of interfacial phenomena. In the last chapter, we discuss a couple of examples of
application of irreversible processes, which are treated in the formalism developed for the notion of calortropy and the attendant macroscopic theory
of irreversible processes. This chapter is meant to introduce the reader to
the subject of irreversible phenomena.

We believe that thermodynamics is a subject that should be studied
without intrusion by a molecular theory approach, because it is a subject
that allows us to make deductions with regard to macroscopic properties of
matter without reference to the molecular picture of matter, and eventually serves as the ultimate aim of molecular theory of macroscopic matter,
which is developed by means of statistical mechanics. Mixing thermodynamics and statistical mechanics tends to confuse important issues *vis à vis*
thermodynamics, which are involved in the statistical mechanical treatment
of macroscopic properties of matter, and such possible confusion would hin-
der a logical development of the theory of irreversible processes. However,
a concession is made for the pedagogical importance of the Debye–Hückel

theory indispensable in the physical chemistry of electrolytes and plasmas, including the theory of electrolytic conductance.

A major portion of this book is based on the materials taught by B. C. Eu in the courses on thermodynamics, off and on, over a period of over two decades at McGill University. The same materials have been taught by M. Al-Ghoul at American University of Beirut for a couple of terms.

Chapter 2

Temperature, Work, and Heat

Temperature, work, and heat are three basic concepts underlying the science of thermodynamics, the scientific quantification of which traces back to the very beginning of thermodynamics. Feeling hot or cold is a physiological sensation that we have when we touch an object, but such sensation was not given an objective measure until the times of Galileo. Furthermore, the relation of temperature to heat and the relation of work to heat have evolved through the history of science, their evolution embodying our struggle to understand their nature. In this chapter, we discuss their quantification, so that they can be made use of in the subsequent study of thermodynamics in a scientific and logical manner.

2.1 Temperature

Temperature is not only one of the most important quantities in thermodynamics but also one of the basic state variables, in terms of which thermal properties of matter can be characterized and reckoned with. Its quantification had taken a long winding process of evolution in thoughts before the concept took the form currently in universal use. It is introduced as a quantifiable quantity by the zeroth law of thermodynamics stated below.

If two bodies of different degrees of hotness are put into contact, the difference in hotness eventually disappears between the two bodies, and we say that they are in thermal equilibrium. What in fact happens when such two bodies are put into thermal contact is that heat flows from the hotter body to the colder one until there is no difference in hotness or coldness. Such a phenomenon is experienced in our everyday life. In order to develop the science of thermodynamics it is necessary to quantify the

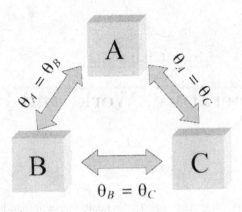

Fig. 2.1   Thermal equilibrium and the definition of temperature. Pairs of systems A, B, and C in thermal equilibrium no longer exchange heat or energy in another form, and the property characterizing the thermal equilibrium is the temperature of the systems.

sense of hotness and coldness. The desired quantification is essentially achieved by the following law:

The zeroth law of thermodynamics: *If two systems A and B are in thermal equilibrium with system C, then the systems A and B are also in thermal equilibrium.*

In other words, if there exists thermal equilibrium between A and C, there is a property (parameter of state) called temperature θ such that

$$\theta_A = \theta_C$$

and similarly for systems B and C

$$\theta_B = \theta_C.$$

Therefore there follows the equality

$$\theta_A = \theta_B. \tag{2.1}$$

These relations are illustrated in Fig. 2.1. They supply a scientific means to quantify temperature.

Let us examine how we might achieve the desired quantification of temperature. Imagine a balloon containing air is heated. The air inside the balloon then gets hotter than before heating and as the volume of the air in the balloon increases, the volume of the balloon increases. We thus see that the measure of hotness, that is, the temperature of the air inside the balloon, may be quantified in terms of the volume of the balloon. In other words, the physical quality of the air, temperature, is a function of volume, and vice versa. Variables such as volume or pressure, which can be used for measuring temperature, are called the *thermometric properties* of the system, and the instruments to quantify temperature are commonly called thermometers. There are other useful thermometric properties, such as electrical resistance, magnetic susceptibility, wavelengths of radiation, sound wave velocity, dielectric constants, and so on, that may be used to quantify the notion of temperature.[1]

Let us call P a thermometric property. Then, by the consideration made earlier, we may assume θ is a function of P:

$$\theta = \theta(P). \tag{2.2}$$

In order for the quantity P to qualify as a useful and practical thermometric property, the function $\theta(P)$ however, *is preferable to be monotonic and linear with respect to* P *in the temperature range where the property is utilized for measuring temperature*[2]:

$$\theta = a + bP, \tag{2.3}$$

where a and b are constants, which generally depend on the thermometric material used. The linearity and monotonicity of the relation are preferable for one-to-one correspondence between the temperature and the thermometric property as indicated in Fig. 2.2, because this relation is the simplest possible. Obviously, nonlinear relations such as that in Fig. 2.3

---

[1]See, for example, J. F. Schooley, ed., *Temperature* (American Institute of Physics, New York, 1982) for methods of measurement of temperature.

[2]The concept of temperature has taken a long time to acquire its present form. Galileo Galilei is credited to have devised for the first time in history (ca. 1592–1603) a thermometer, which was then called a thermoscope. The instrument was merely a glass tube with a bulb attached at one end, which contained air, and the other end of the tube immersed in water. As the temperature of the air in the bulb changes, the water level moved up or down as the air inside either contracted or expanded. In fact, it was a "barothermoscope" since the water level also depended on the pressure, although the pressure effect was not recognized until much later. His thermoscope had no fixed point. Thermometers with one fixed point were proposed in 1665 by Robert Boyle, Robert Hooke, and Christiaan Huygens. Thermometers with two fixed points were made in 1669 by Honoré Fabri, who adopted the melting point of snow for the lower fixed point, but rather vague "greatest summer heat" for the upper fixed point. In 1694, Carlo Renaldini proposed to take the freezing and boiling temperatures of water as the two fixed point temperatures. The present day centigrade scale is credited to Anders Celsius (1742), although it is believed to have been first suggested in 1710 by a Swede named Elvius.

Chemical Thermodynamics

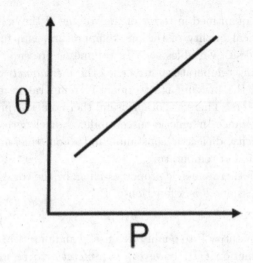

Fig. 2.2 Temperature is linear with respect to thermometric property P.

are not useful, since there is no one-to-one correspondence between θ and P in the interval $P_1 \leq P \leq P_2$, and hence the temperature cannot be uniquely determined in the interval. Since the linear relationship (Eq. (2.3)) between the thermometric property and temperature contains two parameters a and b, which must be obtained empirically, they are determined in reference to two typical points of θ. Since choice of such points can be arbitrary, there are various temperature scales possible.

2.1.1 *Centigrade (Celsius) Scale*

In the centigrade (Celsius) scale, two reference points of temperature are the freezing point and the boiling point of pure water at 1 atm pressure, respectively. Let us denote the values of thermometric property P at the freezing and boiling points by P_0 and P_{100}, respectively. The corresponding values of θ at the freezing and boiling points are taken to be equal to 0 and 100. Thus there are 100 degrees between the two reference points. We then find from Eq. (2.3)

$$0 = a + bP_0,$$
$$100 = a + bP_{100}.$$

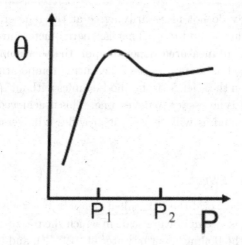

Fig. 2.3 Temperature is nonlinear with respect to thermometric property P.

These two equations are solved for a and b to obtain the relations

$$a = -\frac{100P_0}{P_{100} - P_0},$$

$$b = \frac{100}{P_{100} - P_0}.$$

Substituting back in Eq. (2.3), we obtain

$$\theta = \frac{1}{\alpha} \cdot \frac{(P - P_0)}{P_0}, \qquad (2.4)$$

where α is defined by

$$\alpha = \frac{P_{100} - P_0}{100P_0}. \qquad (2.5)$$

Therefore α is the mean relative change in P per degree between θ_0 and θ_{100}. For example, if the volume of a gas is taken for P then α represents a mean expansion coefficient of the gas (thermometric substance) between the two reference points of temperature. We therefore see that temperature can be measured by observing the variation of P relative to its value at one of the reference points. Well-known examples of this include alcohol and mercury thermometers. The temperatures measured by two different thermometric materials generally do not necessarily agree at the temperatures between two reference points θ_0 and θ_{100}. That is, thermometers are not universal and the temperature measured depends upon the thermometric materials and properties used for the purpose of measuring temperature.[3] Since it is difficult to carry on the science of thermodynamics without a universal scale of temperature, it is necessary to devise one. This is achieved with the ideal gas thermometer, and as will be seen, it provides a universal temperature scale.

2.1.2 *Fahrenheit Scale*

In some parts of the world the Fahrenheit temperature scale is used in daily life by custom. It is a temperature scale in which the freezing point of water is set at 32° (F), the boiling point of water at 212° (F), and the difference is divided by 180 for the measure of a degree. In science, the Fahrenheit scale of temperature is never used. We simply note the conversion relationship between the t degree (C) in the centigrade scale of temperature and t' degree (F) in the Fahrenheit scale of temperature is given by the equation

$$t°C = \frac{5}{9}[t'° F - 32].\qquad(2.6)$$

2.1.3 *Absolute Temperature Scale and Ideal Gas Thermometer*

To devise a universal temperature scale, the ideal gas thermometer employs the ideal gas as the thermometric material. The absolute temperature scale is obtained therewith.

If volume V is taken for the thermometric property P then α represents a mean expansion coefficient of the thermometric substance between the two reference points of temperature. The study of behaviors of gases under the influence of heat and pressure traces back to the early stages in the development of modern science, and the names like R. Boyle, J. Gay-Lussac, and H. V. Regnault were associated with it. It was found through extensive experiments that gases show certain universal behaviors towards heat and pressure as they become sufficiently diluted so that the pressure is low. The state of such gases is then called *ideal*. According to the modern molecular theory interpretation of the ideal gas behavior there are no interactions

---

[3]L. A. Guildner, *Physics Today*, p. 24, December, 1982.

between the molecules in ideal gases. Such a universal behavior indepen-dent of materials is obviously a desired property to exploit in devising a temperature scale.

If the volume is taken as the thermometric property for the ideal gas thermometer then the volume V at temperature θ is known to be given by

$$V = V_0(1 + \alpha\theta), \tag{2.7}$$

where, if the Celsius scale of temperature is used, α is given by

$$\alpha = \frac{V_{100} - V_0}{100V_0}. \tag{2.8}$$

Equation (2.7) is the result of rearranging Eq. (2.4) after setting the ther-mometric property P $= V$, and α is the mean expansion coefficient of the gas. It is empirically found that as the gas pressure p decreases to a sufficiently low value, α approaches a limit, that is,

$$\lim_{p \to 0} \alpha = \alpha^*, \tag{2.9}$$

where α^* is a universal constant independent of gases and has the value[4]

$$\alpha^* = \frac{1}{273.15}.$$

We will put

$$\alpha^* = \frac{1}{T_0}, \tag{2.10}$$

i.e., $T_0 = (273.15 \pm 0.02)°C$ for all ideal gases.

Equation (2.9) together with Eq. (2.4) implies that

$$\lim_{p \to 0} \theta = \theta^* \tag{2.11}$$

and thus in the low pressure limit

$$V = V_0(1 + \alpha^*\theta^*). \tag{2.12}$$

Equation (2.10) therefore indicates that there exists a temperature scale which does not depend on the material employed and consequently is *uni-versal*. Such a thermometer is called the ideal gas thermometer, and the scale of temperature based thereon the ideal gas temperature scale.

---

[4]Gay-Lussac found $\alpha = 1/267$ approximately, but Regnault in 1847 obtained $\alpha = 1/273$ by using an improved experimental procedure.

By using Eq. (2.10) we may rewrite Eq. (2.12) in the form

$$T = T_0 \frac{V}{V_0}, \tag{2.13}$$

where

$$T = T_0 + \theta^*. \tag{2.14}$$

Since the value of T_0 is subject to experimental errors, to fix T_0 universally and unequivocally it is agreed at the Tenth Conference of the International Committee on Weights and Measures that at the triple point of water the temperature is

$$T = 273.16 \text{ K exactly.}$$

The temperature scale so determined puts the temperature of freezing point of water at 1 atm pressure as

$$T_0 = 273.15. \tag{2.15}$$

This temperature scale is called the absolute temperature scale and the temperature in such a scale is expressed with the unit K, meaning Kelvin.

Let us now consider an immediate implication of the thermodynamic temperature scale devised with an ideal gas. It is empirically established through investigations by Boyle, Gay-Lussac, and others that for ideal gases there holds the relation

$$\lim_{p \to 0} pV = \beta(\theta) \tag{2.16}$$

and

$$\frac{\beta(\theta_1)}{\beta(\theta_2)} = \text{a universal constant independent of gases,} \tag{2.17}$$

where p is the pressure and β is a function of θ, which is independent of gases. Since as $p \to 0$

$$V = \frac{\beta(\theta)}{p},$$

it follows from Eq. (2.16) and Eq. (2.17)

$$\frac{T}{T_0} = \frac{\beta(\theta)}{\beta(\theta_0)} = c, \tag{2.18}$$

where c is a universal constant according to Eq. (2.17). Since Eq. (2.18) implies that

$$T \propto \beta(\theta),$$

we may write

$$\beta(\theta) = nRT, \tag{2.19}$$

where n denotes the number of moles of the gas contained in V and R is a universal constant, which is called the gas constant and its value is $8.3143 \, \mathrm{J\,K^{-1}\,mol^{-1}}$. Therefore Eq. (2.16) may now be written as[5]

$$\lim_{p \to 0} pV = nRT \tag{2.20}$$

or, more simply,

$$pV = nRT. \tag{2.21}$$

For a mole of an ideal gas it may be written as

$$pv = RT,$$

where $v = V/n$ denotes the molar volume. This is the ideal gas equation of state. When the equation of state is written in the form of Eq. (2.21), it must be understood that it holds only when p is sufficiently low so that the gas behaves as an ideal gas. The formulation of the ideal gas equation of state as in Eq. (2.21) indicates that the ideal gas equation of state is intimately tied up with the ideal gas temperature scale and, as will be seen later, all equations of state and thermodynamic properties for real substances are expressed in terms of the temperature scale devised on the basis of the ideal gas and reckoned in the absolute temperature scale.

---

[5]This limiting formula may be written in another form

$$T = \lim_{n \to 0} \frac{PV}{nR} = \lim_{n \to 0} \frac{pv}{R}.$$

Since the pressure–volume work per mole, pv, on the right-hand side may be calculated by using its molecular theory representation, i.e., statistical mechanical formula, this limiting form may be taken as the statistical mechanical formula for the temperature of fluids. Since the pressure–volume work in question is also extendable to nonequilibrium fluids, the formula makes it possible to extend the notion of temperature to the case of nonequilibrium fluids. If we denote the measure of nonequilibrium by ϵ, from the mathematical standpoint the extension of T mentioned may be regarded as an analytic continuation of T to $\epsilon > 0$.

Having elucidated the existence of a quantity called temperature and devised the absolute temperature scale based on the ideal gas, we now come to the question of whether it is possible to speak of temperature if a system is not in equilibrium internally. The zeroth law requires that two bodies in thermal contact be in thermal equilibrium for the concept of temperature to be meaningful. This does not mean that the two bodies have to be internally in equilibrium. The only condition required for parameter θ to exist is that there is no heat exchange between the two bodies over a characteristic span of observation time. If one of the two bodies is a thermometer calibrated against the ideal gas thermometer then the characteristic span of time influences the resolution power of the thermometer, which is the measure of how fast the thermometer responds to a heat transfer between the body and the thermometer. Even in such a case one still speaks of the temperature of the body. For example, if the body is animate like the human body, which is obviously undergoing complex irreversible processes within itself, we routinely speak of the temperature of the body, and it is recorded by the point of thermal equilibrium between the body and the thermometer in contact with the body. Similarly, one may insert a thermometer in a stream of a liquid which is obviously not in equilibrium, yet speaks of the temperature of the liquid. Depending on the processes that are going on in the liquid the temperature may be at a constant value over time or may be varying in time. The temperature scale devised based on the ideal gas thermometer introduced earlier can be still used to record the thermal state of systems where some irreversible processes are in progress. However, it should be kept in mind that the temperature so measured of a system in a nonequilibrium state may be only the temperature of the local volume sampled, but not the temperature of the global system. The latter would be the case if the system is globally maintained at a fixed temperature even if it is in a nonequilibrium state, as is the case for human or animate bodies. See footnote 5.

2.2   Pressure

Pressure is a mechanical quantity relevant to and, in fact, indispensable in developing thermodynamics of matter. It is defined as the force exerted on a unit area of a surface:

$$\text{pressure} = \frac{\text{force}}{\text{area}} = \frac{F}{A}.$$

Therefore the dimension of pressure is $N\,m^{-2} = kg\,m/s^2$, where N is the unit of force, Newton, in the SI system of units. This unit of pressure is called Pascal and denoted by Pa. In practice, since this SI unit is too small, other more practical units are used. For example,

$$1\,bar = 10^5\,N\,m^{-2} = 10^5\,Pa$$

and alternatively

$$1\,atm = 1.01325\,bar$$

with 1 atm of pressure defined as the pressure exerted by 0.760 m column of mercury at 0°C at the sea level. The standard value of the gravitational acceleration (g) is used for this calculation:

$$g = 9.80665\,m\,s^{-2}.$$

Pressure is occasionally expressed in the units of Torr, meaning Torricelli. The conversion factor of the units of Torr to the units of atm is

$$1\,atm = 760\,Torr.$$

The definition of pressure given earlier serves to provide a means to measure it. The simplest way to measure pressure is the use of a manometer. The pressure of a body (e.g., a fluid) is recorded, as mechanical equilibrium is established between the body and the manometer. It is important to recognize that as in the case of temperature, which is measured when the system is in thermal equilibrium with the thermometer, the measurement of pressure requires mechanical equilibrium between the system and the manometer — the measuring device. Therefore equilibrium is the prerequisite for measuring temperature and pressure of a system.

2.3   Work

The thermodynamic concept of work is similar to that in mechanics. When an object is displaced from one point in space to another under the influence of a force $\mathbf{F}$, we say that a work is done on the system, and for an infinitesimal displacement $d\mathbf{s}$ the differential work is

$$dW = \mathbf{F} \cdot d\mathbf{s}. \tag{2.22}$$

The work done to move the object (system) from point 1 to point 2 along the curve $\mathbf{s}$ is then obtained by integrating Eq. (2.22) over the path along

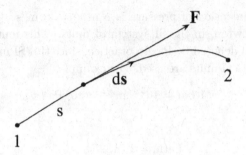

Fig. 2.4   Work by a force along path s.

which the work is performed (Fig. 2.4):

$$W_{12} = \int_1^2 \mathbf{F} \cdot d\mathbf{s}. \tag{2.23}$$

It should be noted that the integration is a line integral. This integral may be put in the form indicating the path-dependence of the value of the integral

$$W_{12} = \int_{1(\text{path})}^2 đW \tag{2.24}$$

with the subscript "path" meaning that the integration is path-dependent, that is, it is a line integral. We use a slash on the differential symbol đ to denote that, for example, đW is not an exact differential of W. The notation đW simply means that it is an infinitesimal quantity, namely, an infinitesimal work in this case, which arises from an infinitesimal displacement of a body by ds under the action of a force. This non-exactness property will be further elaborated shortly. It is important to note that work depends on the path and therefore the works done from point 1 to point 2 along two different paths are not generally equal, as we empirically well-recognize the truthfulness of this statement from our daily life experience. In mechanics, we know that they are equal, only if the force field has a potential energy, i.e., it is conservative. In thermodynamics, forces usually do not have a potential.

Since it is necessary to distinguish the work done on the system from the work done by the system and they are opposite in sign, we introduce a sign convention.

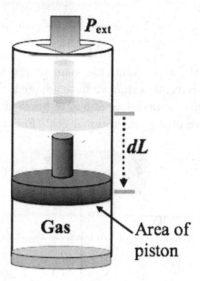

Fig. 2.5 Work by a gas in the piston under an external pressure. The gas in the piston is compressed.

Sign convention: *The work done on the system by the surroundings is taken as positive whereas the work done on the surroundings by the system is taken as negative.*

For example, if an external pressure P_{ext} is applied on area A of the container of a gas, which is capable of displacement and as a consequence the surface A is displaced by an infinitesimal distance dL in the negative direction (see Fig. 2.5), then the work done on the system is

$$\text{d}W = -P_{ext}\, dL \times A = -P_{ext}\, dV, \qquad (2.25)$$

where $dV = A\, dL$. The total work done on the system arising from the volume change from V_1 to V_2 is then

$$W = -\int_{V_1}^{V_2} P_{ext}\, dV. \qquad (2.26)$$

The integral is clearly path-dependent. The knowledge of P_{ext} as a function of V therefore will be necessary to perform the integration for W.

If the pressure remains constant then the integration is immediate, and the work is given by

$$W = P_{\text{ext}}(V_1 - V_2). \tag{2.27}$$

Since $V_1 - V_2 > 0$, the right-hand side, namely, the work, is positive in agreement with the convention that we have adopted.

As another example, suppose an object of mass M falls by a distance dh under the influence of the gravitational field (Fig. 2.6). Then the work

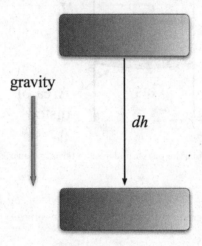

Fig. 2.6   Work by gravitational force.

done by the system on the surroundings (Earth) is given by

$$\text{d}W = -gM\,dh. \tag{2.28}$$

For another example, consider a force exerting on a rubber band (Fig. 2.7). If a rubber band of length L is stretched by dL by a tensile force f then the work done on the system is

$$\text{d}W = f\,dL. \tag{2.29}$$

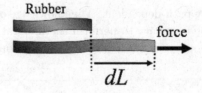

Fig. 2.7   Work by an extensional force.

Table 2.1   Other types of work

Type	đW
Ohmic	$\psi dq = \psi I dt$
Surface	γdA
Electric	$\mathbf{E} \cdot d\mathbf{D}$
Magnetic	$\mathbf{H} \cdot d\mathbf{M}$

Other kinds of work relevant in thermodynamics are listed in Table 2.1. The symbols in Table 2.1 are as follows: ϕ is the electric potential, q is the charge, I is the current (i.e., $I = dq/dt$), γ is the surface tension, A is the area, $\mathbf{E}$ is the electric field, $\mathbf{D}$ is the electric displacement, $\mathbf{H}$ is the magnetic field, and $\mathbf{M}$ is the magnetization.

If we denote by X_i the force and by x_i the conjugate variable, namely, displacement, the work done on the system by a set of forces is given by

$$\text{đ}W = \sum_i X_i \, dx_i, \tag{2.30}$$

where the sum is over all types of work performed. The total work corresponding to finite displacements from x_i^1 to x_i^2 is then given by

$$W = \sum_i \int_{x_i^1}^{x_i^2} X_i \, dx_i, \tag{2.31}$$

where the integrals are line integrals along the paths of work performed since $X_i \, dx_i$ is not an exact differential (see Fig. 2.8.).

In connection with the line integrals like work defined in Eq. (2.31), it is useful to add a mathematical note on nonexact differentials for which đW is an example. If a differential dz is exact then the value of the integral depends only on the values of z at the endpoints as in the following expression

$$\int_1^2 dz = z(2) - z(1).$$

However, if it is not exact, then

$$\int_1^2 \text{đ}z \neq z(2) - z(1),$$

and the value of the integral depends on the line along which the integration is performed. The differential work $đW$ has the latter property, and it is important to understand the basic difference between an exact differential, say, dz and a nonexact differential such as $đW$. For example, when integrated along the two different paths as in Fig. 2.8, its integral may be

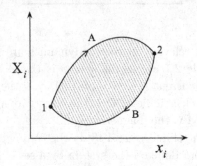

Fig. 2.8 Work performed under a force X_i in a cyclic process. The shaded region denotes the net work performed in the process.

broken up into two parts and written in the form

$$\int_{1(A)}^{2} đW - \int_{1(B)}^{2} đW \neq 0. \tag{2.32}$$

Since a line integral is the area under line (curve) defined by the path of integration between two endpoints, the left-hand side of Eq. (2.32) corresponds to the net work (the shaded area in Fig. 2.8) performed in the cyclic process.

This net work would have been equal to zero, if $đW$ were an exact differential and hence there existed a potential. In mechanics of a conservative field, the work over a cycle is equal to zero, since the force has a potential energy.[6] As an illustration of this point, let us consider Eq. (2.30) and assume that there exists a function $\varphi\,(x_1, x_2, \ldots, x_n)$ such that the force is given by the gradient of φ

$$X_i = -\frac{\partial \varphi}{\partial x_i} \quad (i = 1, 2, \ldots, n). \tag{2.33}$$

---

[6]In mathematical terminology it is said that the force is *monogenic* if it is derivable from a scalar function, whereas it is said to be *polygenic* if it is given by a differential form that is not expressible as a derivative of a scalar function. A typical example is the differential form (Eq. (2.30)), which is not a derivative of a scalar function.

Then Eq. (2.30) can be written as an exact differential of φ:

$$dW = -\sum_i \frac{\partial \varphi}{\partial x_i} dx_i = -d\varphi \tag{2.34}$$

and

$$W = \int_1^2 dW = -\int_1^2 d\varphi = \varphi(2) - \varphi(1). \tag{2.35}$$

The differential work dW in this case becomes an exact differential, and the function $\varphi(x_1, x_2, \ldots, x_n)$ is a potential energy mentioned earlier. In this case, in which the force is conservative, the differential form dW is monogenic in the mathematical terminology. In thermodynamics, the differential form for work is not monogenic and the work over a cycle is not equal to zero for the reason to be elaborated later when the notion of energy is discussed.

2.4 Heat

The concept of heat has evolved throughout the scientific history. Before Galilei (1623) and Francis Bacon (1620) it was believed that heat was a kind of matter, and a transfer of such a matter between bodies was the cause for a body being heated up by another. Thus it was believed that a body gets warmer when it receives a certain amount of such matter and gets colder when it gives it up. This concept was put on a scientific basis by Joseph Black (1760) in the late 18th century, who developed the caloric theory of heat. Galilei appears to be the first to realize that heat is not a matter, but a manifestation of the motion of particles and therefore a form of energy.[7] As was evident by Black's caloric theory of heat contrary to Galilei's idea, the revolutionary concept of Galilei and Bacon, however, had been laid dormant for over 200 years[8] until the observations made

---

[7]According to Joseph Black in his Lectures in the Elements of Chemistry (reproduced in *The Early Development of the Concepts of Temperature and Heat* by D. Roller (Harvard University Press, 1950), Francis Bacon (1620) proposed, on the basis of the consideration of means by which heat is produced, or made to appear, in bodies, such as the percussion of iron, the friction of solid bodies, the collision of flint and steel, that heat is motion. Considering this, it appears that Francis Bacon's notion of heat had more to do with the cause and generation of heat than the notion of heat itself, unlike the case with Galilei's.

[8]The idea that heat is motion was used by Daniel Bernoulli in ca. 1780 when he derived an ideal gas equation of state in his book *Hydrodynamica*, but this also did not draw attention.

by Count Rumford, Humphrey Davy, Robert Mayer, and James P. Joule, which indicated that heat cannot be a matter, but is a form of energy. Joule finally established through his famous experiments (e.g., paddle wheel experiment; see Fig. 3.1 in Chapter 3) that heat is a form of energy, namely, the mechanical equivalence of heat.

We have seen that there is a parameter that can be used as a gauge for measuring the degree of hotness of a body (system). We have called that measure the temperature of the body. Now, we can make use of this parameter to devise a way of quantifying heat.

When two systems A and B at different temperatures are put into thermal contact, there is a transfer between them of energy in the form of heat. The amount of heat transfer for a given temperature difference between A and B depends on the substances making up the systems, and we define an extensive property called the heat capacity of a substance. Heat was used to be measured in units of calorie. It was initially defined by two different ways:

Mean gram calorie: The mean gram calorie q_m is defined as *the mean amount of heat required to raise by* $1°C$ *the temperature of* 1 *gram of water at* 1 *atm pressure between* 0 *and* $100°C$, namely,

$$q_m = \frac{Q(0°\,\mathrm{C} \to 100°\,\mathrm{C}) \text{ of water at 1 atm pressure}}{100\,w\,\mathrm{g}}.$$

15° gram calorie: The 15° gram calorie $q_{15°}$ is defined as *the amount of heat required to raise by* $1°C$ *the temperature of* 1 *gram of water at* $14.5°C$ *and at* 1 *atm pressure*, namely,

$$q_{15°} = \frac{Q(14.5°\mathrm{C} \to 15.5°\mathrm{C}) \text{ of water at 1 atm pressure}}{w\,\mathrm{g}}.$$

Here w is the mass of the substance in gram units. The values of q_m and $q_{15°}$ do not exactly agree with each other, but form a certain ratio:

$$q_{15°} = 1.00024\,q_\mathrm{m}. \tag{2.36}$$

For this reason the Ninth International Conference on Weights and Measures recommended that the Joule be used as the unit of heat. Therefore thermo-mechanical calorie or simply calorie is defined as follows:

$$1 \text{ thermo-mechanical calorie} = 1 \,\text{cal}$$
$$= 4.184 \,\text{Joules}$$
$$= 4.184 \,\text{N m}.$$

The change in the heat content of a system after a process of change in state depends on the manner in which the change has occurred; in other words, it depends on the path (mode) along which the change has been effected on the system. Suppose, for example, a mass of a liquid is heated from temperature T_A to temperature T_B under constant pressure in one case, and under constant volume in another, it is experimentally known that the amounts of heat required for the same change in temperature are different for the two cases. These two modes of change in temperature are examples of different paths of change referred to earlier. If an infinitesimal heat change is denoted by đQ then the total heat change arising in a change in the state of the system from A to B along the path Γ is (Fig. 2.9)

$$Q = \int_{A(\Gamma)}^{B} đQ. \tag{2.37}$$

Note that đQ is not an exact differential and therefore the value of Q is path-dependent.

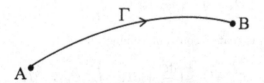

Fig. 2.9   Path of heat change from state A to state B along path Γ.

Before proceeding further on the discussion of heat, it is useful to introduce the sign convention for heat, since it is necessary to distinguish the heat absorbed from the surroundings by the system from the heat given up to the surroundings by the system.

Sign convention on heat: *Heat given up by the system to the surroundings is counted negative whereas heat absorbed by the system from the surrounding is counted positive.*

Clearly, because of the equivalence of heat to work and energy, this sign convention is consistent with the sign convention adopted earlier for work.

The heat capacity of a system is the amount of heat required to raise its temperature by $1°C$. The mean heat capacity may be defined by

$$\bar{C} = \frac{Q}{T_2 - T_1}, \tag{2.38}$$

where Q is the amount of heat absorbed by the system to raise its temperature from T_1 to T_2. The more useful parameter is the differential heat capacity, which is defined by

$$C = \frac{dQ}{dT} = \lim_{\Delta T \to 0} \frac{\Delta Q}{\Delta T}. \tag{2.39}$$

The heat capacities measured at constant pressure and at constant volume are different. In order to distinguish them we will affix a subscript p or v to C:

$$C_p = \left(\frac{dQ}{dT}\right)_p, \quad C_v = \left(\frac{dQ}{dT}\right)_v. \tag{2.40}$$

The integral in Eq. (2.37) may be rewritten in terms of heat capacity. If the path is that of constant pressure then

$$Q_p = \int_{T_1}^{T_2} \left(\frac{dQ}{dT}\right)_p dT = \int_{T_1}^{T_2} C_p\left(T\right) dT, \tag{2.41}$$

and if the path is that of constant volume, then

$$Q_v = \int_{T_1}^{T_2} \left(\frac{dQ}{dT}\right)_v dT = \int_{T_1}^{T_2} C_v\left(T\right) dT. \tag{2.42}$$

We have attached a subscript p or v to Q to distinguish the paths, that is, the modes of heat transfer. Since $C_p \neq C_v$, the heats Q_p and Q_v are not the same, indicating the path-dependence of the integral of the inexact differential dQ.

2.5   Reversible Processes and Reversible Work

Processes in real systems occur at a finite rate over a finite duration of time. Therefore work and heat transfer are generally performed irreversibly. The study of such irreversible processes is not a subject of equilibrium thermo-dynamics, which we wish to carry out in this book. In equilibrium thermo-dynamics a rather special class of processes is studied. They are reversible processes. A process is called *reversible*, only if there is continuous equilib-rium between the system and its surroundings throughout the process. We speak of a *reversible change* when the system passes through a sequence of internal equilibrium states without necessarily being in equilibrium, but in quasi-equilibrium with its surroundings.

Therefore it follows that a reversible process is an idealized limit process, which can be achieved by making the actual process infinitesimally slow and thus taking an infinite duration of time. This way, the causes making a pro-cess irreversible have vanished in the time span of interest, and equilibrium is established continuously between the system and its surroundings over the course of the entire process. Such processes obviously are not realiz-able in practice, since they are usually completed in a finite duration of time in real systems. Nevertheless, this idealization provides us with a very useful conceptual device with which to carry out calculations in thermody-namics, unencumbered by the knowledge required of the energy dissipation accompanying an irreversible process. In equilibrium thermodynamics of reversible processes, thermodynamic potentials can be obtained from dif-ferential forms which admit integrals, namely, potentials. This feature can be useful for studying thermodynamics of irreversible processes, since a re-versible process can be imagined to complete a cycle with an irreversible process, and the change in a thermodynamic potential over an irreversible process can be computed in terms of the thermodynamic potential over the reversible process complementary to the irreversible process making up the cycle. In this sense, the thermodynamics of reversible processes becomes a powerful theoretical tool for understanding some aspects of irreversible processes. It is then no longer a subject studying idealized situations of physical systems, but a theoretical tool to study some aspects of real pro-cesses in physical systems undergoing irreversible processes.

As is for reversible processes, we call a work reversible if it is performed through a reversible process which maintains the system in continuous equilibrium with the surroundings over the entire process of work. *The reversible work is a maximum work.* This statement will be proved later when we are mathematically better equipped in Chapter 4. What the state-ment implies is that there is an amount of unavailable work for the given task of work if the process is performed irreversibly, and this unavailable work is regarded as energy dissipation for the given task, which the work in question aims to achieve. Any irreversible process is accompanied by some sort of energy dissipation, and reversible processes may be defined as those in which energy dissipation is absent for the task in question. Therefore a reversible work is seen to be a maximum work for the given task.

The meaning of reversible processes can be made more precise in mathematical terms if the concept of thermodynamic forces is taken into account. If a system is not in thermodynamic equilibrium, there are intensive variables which are inhomogeneous in space or time, or both. That is, in a nonequilibrium system the intensive variables may vary over a distance or in time. The measure of spatial variations in intensive variables are given by their gradients, and *thermodynamic forces are defined as spatial gradients of intensive variables.* These forces are considered to be driving macroscopic irreversible processes in the system and to cause the latter to dissipate energy. In fact, the sum of squares of the thermodynamic forces is related to the energy dissipation associated with the irreversible processes. As these driving forces are spent in the course of time, the system reaches thermodynamic equilibrium under the given constraint and the energy dissipation arising from the driving forces vanishes. The reversible processes may then be defined as the limiting processes that occur under vanishingly small thermodynamic forces. They are, therefore, processes in which energy dissipation is absent. As a matter of fact, this was basically the definition used by Clausius when he introduced the notion of reversible processes in his formulation of equilibrium thermodynamics. However, this notion of Clausius has not been used in most textbooks on equilibrium thermodynamics to the detriment of a clearer understanding of the true meaning of reversible processes. We prefer this definition to the definition of reversible processes made with quasi-equilibrium processes maintained throughout the processes performed over an infinite duration of time. This aspect will be discussed in more detail in Chapter 4.

Chapter 3

The First Law of Thermodynamics

3.1 Equivalence of Heat and Energy

A scientific concept often trails a long tortuous path of evolution to attain the shape it currently assumes. Such a path of evolution necessarily reflects the evolution of our own thinking toward the phenomena underlying the concept, and an evolution means a modification of what is currently prevalent to a new and better suited form. Such a modification is usually prompted by our inherent desire to come up with a more encompassing, comprehensive viewpoint and theory, when faced with new empirical evidence which renders invalid or inappropriate the concept that has so far served us well and thus has been universally accepted as truthful.

In physical science, we empirically observe the states of mechanical objects and their changes and discern that their states change in rather intricate but seemingly haphazard manners. We then look for rules and laws governing the manners in which the states change, so as to find an order in the state of affair that appears complicated and complex. *Constancy* is a quality that stands out when contrasted to qualities that change. The concept of energy was born out of our desire to find a constant quality of mechanical systems that is preserved over the course of time irrespective of some changes of state that the systems have gone through.

In 1669, Christiaan Huygens discovered that the mass times the velocity squared, mv^2, was conserved during elastic collisions of mechanical bodies. Leibnitz called it *vis viva*. Much later, W. Thomson (Lord Kelvin) called $mv^2/2$ the kinetic energy. It was found that the kinetic energy was not conserved in inelastic collision even if friction was absent, but it was found that if the potential energy was added to the kinetic energy the sum was conserved. It is well known in mechanics that the energy of a system is conserved, if the force is conservative. Thus the kinetic energy of two elastically colliding billiard balls is conserved. But it is no longer conserved if an inelastic collision or friction is allowed between the two balls, since then the kinetic energy is transformed into another form of energy associated with the inelasticity of collision. In the case of inelastic collision, the total energy is conserved if that part of the energy associated with the inelastic excitation of internal states is added to the energy of the relative motion. However, if friction takes effect, heat is generated and the mechanical energy is proportionately diminished. Therefore the energy conservation law that holds for the elastic and inelastic collisions appears to be no longer valid.

Count Rumford (Benjamin Thompson[1]) observed that the amount of heat generated during the boring of a cannon is proportional to the mechanical energy expended. This observation was further developed conceptually and put on a firm scientific basis by J. P. Joule, who performed the first determination of the mechanical equivalent of heat. It was observed that heat generated by mechanical work is independent of the materials involved and strictly proportional to the work done, when the system is subjected to a cyclic process whose net result is merely the conversion of work into heat with the system returning to its initial state. This experimental discovery, coupled with the realization through Joule's experiments that *heat is another form of energy*, led to the first law of thermodynamics as a broad, generalized enunciation of the mechanical energy conservation law.

Joule found in his famous paddle wheel experiment (Fig. 3.1) that mechanical energy is converted into heat at a universal ratio independent of the substance and the processes of conversion. He was thereby able to determine the mechanical equivalent of heat:

$$W = JQ, \tag{3.1}$$

where W is the work, Q is the heat, and J is a constant, called the mechanical equivalent of heat. Its unit is called Joule, and 1 cal = 4.1840 J. Its value may be determined, for example, by measuring the amount of heat generated owing to the work done when a body (e.g., a ball) falls from a distance h under the influence of the gravitational field:

$$J = \frac{Mgh}{Q}.$$

---

[1]An American, who was in service to Elector of Bavaria, Ludwig. He was a colorful figure who was also a scientist, statesman, and businessman in addition to being briefly the husband to Madame Lavoisier, the widow of Antoine Laurent Lavoisier.

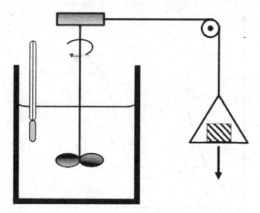

Fig. 3.1 Joule's paddle wheel experiment. As the weight on the right falls from a height h, the paddle wheel turns performing a work, which is converted into heat and thus raises the temperature of the water. The temperature change is then recorded by the thermometer on the left.

Joule's experiment implies that heat must be regarded as a form of energy. Therefore the energy conservation law must be considered with mechanical energy and heat together. Computed in the units of Joule, the total internal energy change dE due to a process must be the sum of the change in work $đW$ and the change in heat $đQ$:

$$dE = đQ + đW, \tag{3.2}$$

where we call E the internal energy of the system. This way, the energy conservation law is generalized so that thermal processes as well as mechanical motions are brought under its aegis in a unified form.

3.2 The First Law of Thermodynamics

Rudolph Clausius in 1850 stated the first law of thermodynamics as follows:

"The energy of an isolated system (universe) is constant."

In other words, the internal energy of an isolated system is conserved. This may be put in another equivalent statement as made by M. Planck:

"It is impossible to construct a perpetual machine of the first kind—a machine that, working in a cycle, expends no heat to produce an equivalent work."

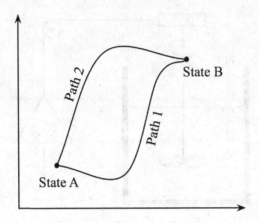

Fig. 3.2   A cyclic process from state A to state B consisting of two segments.

Let us examine a consequence of this statement. Suppose there is a cycle of process $A \xrightarrow{1} B \xrightarrow{2} A$, which takes the system from state A to state B through path 1 and then returns it back to state A through path 2 as schematically depicted in Fig. 3.2.

For the process $A \xrightarrow{1} B$ the internal energy change is given by the expression

$$\int_{A(1)}^{B} dE = \int_{A(1)}^{B} \dbar Q + \int_{A(1)}^{B} \dbar W.$$

Similarly, for the process $B \xrightarrow{2} A$ it is given by

$$\int_{B(2)}^{A} dE = \int_{B(2)}^{A} \dbar Q + \int_{B(2)}^{A} \dbar W.$$

Addition of these two equations yields a relationship between circular integrals as follows:

$$\oint dE = \oint \dbar Q + \oint \dbar W, \tag{3.3}$$

where the circle on the integral sign means an integration over the cycle performed in the space of state variables. According to the first law of thermodynamics, energy cannot be created over the cycle since, upon the system's returning to its original state through a cyclic process, the internal energy of the system must assume the same value.

Therefore, expressed mathematically, the cyclic integral of the internal energy must vanish:

$$\oint dE = 0. \tag{3.4}$$

This implies that

$$Q + W = 0. \tag{3.5}$$

Here

$$Q = \oint dQ, \quad W = \oint dW. \tag{3.6}$$

Equation (3.4) embodies in a mathematical form the experimental fact that heat is an equivalent form of work or energy as was enunciated by Joule. It follows from Eq. (3.4) that the energy change ΔE_{AB} along path 1 must equal the energy change ΔE_{AB} along path 2:

$$\Delta E_{AB(1)} = \Delta E_{AB(2)}, \tag{3.7}$$

where

$$\Delta E_{AB(i)} = \int_{A(i)}^{B} dE = E(B) - E(A) \quad (i = 1, 2),$$

that is, the change in E is independent of the path along which the change of state is effected; it depends only on the initial and the final state.

Such a function is called a *state function*. The differential of a state function is said to be exact in the macroscopic state space of the system: dE is an *exact differential in such a space*. We see that the exactness of dE is demanded by the first law of thermodynamics as a general energy conservation law. Equation (3.2) is a mathematical statement of the first law of thermodynamics in a differential form for an infinitesimal process. A rigorous formulation of the first law in local form is given in Appendix A at the end of the book.

In obtaining Relation (3.7) there was no mention of whether the process is reversible or not. If a process is not reversible, there can be dissipative effects on work or heat transfer that we will not be able to deduce if the irreversible process is not explicitly examined. Suppose the process 2 is performed reversibly whereas the process 1 is irreversible. In this case, Relation (3.7) makes it possible to deduce the effects of the irreversible process on the internal energy change simply in terms of $\Delta E_{AB(2)}$, which is the internal energy change accompanying the reversible segment of the cycle. And this $\Delta E_{AB(2)}$ is rather simple to compute by means of equilibrium thermodynamics as will be shown. This is the point alluded to earlier in Chapter 2 with regard to the theoretical utility of equilibrium thermodynamics in connection with irreversible processes. This aspect will receive further examination in the course of development of the theory.

For a closed system with no external forces except for a uniform normal pressure $p = p(V, T)$ on the system, the differential work is given by the pressure–volume work

$$\dbar W = -p\,dV \qquad (3.8)$$

and Eq. (3.2) takes the form

$$dE = \dbar Q - p\,dV. \qquad (3.9)$$

This is the differential form of the first law of thermodynamics in the case of pressure–volume work alone, when there is no dissipative internal work, which we do not assume to be present in equilibrium thermodynamics, the subject of interest in this work. If the pressure–volume work and the heat transfer are performed reversibly, the expression (3.9) is for a reversible process. In this case, the calculation of changes in the internal energy is fairly simple to perform, as will be seen.

3.3 Enthalpy

The first law of thermodynamics has been shown to give rise to the conclusion that the internal energy E of the system is a state function. There is another function related to the internal energy which is also a state function depending only on the initial and final states of a process. It sometimes is more convenient to use this new function than the internal energy in thermodynamic considerations. We introduce the new function below.

Suppose the volume of a system is changed at constant pressure from V_1 to V_2. We shall designate the initial and the final states 1 and 2, respectively. The heat change accompanying the process is

$$Q = \int_1^2 \dbar Q. \qquad (3.10)$$

By using Eq. (3.9) in the integral, if the pressure–volume work is the only work, we find

$$Q = \int_1^2 dE + \int_{V_1}^{V_2} p\,dV$$
$$= E_2 + pV_2 - (E_1 + pV_1), \qquad (3.11)$$

where we have made use of the fact that p is constant throughout the change. It is convenient to define a new state function

$$H = E + pV, \tag{3.12}$$

which is called the enthalpy. It represents the heat content of the system. With this definition the heat change Q at constant p is given by

$$Q = \int_1^2 dH = H_2 - H_1 = \Delta H, \tag{3.13}$$

which confirms the notion that enthalpy is a measure of the heat content of the system. Equation (3.13) appears to suggest that Q is a state function, but it must be noted that Q is not necessarily equal to ΔH, if p is not kept constant. This point becomes evident if we examine the differential form for enthalpy.

3.3.1 *Differential Forms for Enthalpy*

Since the first law of thermodynamics is expressible in the differential form (Eq. (3.2)), an equivalent differential form useful for the enthalpy defined can be obtained. It is easy to deduce the desired form

$$dH = \dj Q + \dj W + d(pV). \tag{3.14}$$

If the work term is split into the pressure–volume work and the rest as in

$$\dj W = -p\,dV + \dj W', \tag{3.15}$$

where $\dj W'$ represents the totality of the other kinds of work than the pressure–volume work, then dH is given by

$$dH = \dj Q + V\,dp + \dj W'. \tag{3.16}$$

If $\dj W' = 0$, then

$$dH = \dj Q + V\,dp. \tag{3.17}$$

These differential forms for enthalpy may be regarded as expressions equivalent to the first law of thermodynamics, which was expressed in the differential form of internal energy

$$dE = \dj Q + p\,dV. \tag{3.18}$$

Having introduced state functions, internal energy (E) and enthalpy (H), it is useful to reexamine heat capacities in terms of either the internal energy or the enthalpy just defined. If we assume that there is a pressure–volume work only, then from Eq. (3.18) the constant volume heat capacity defined in Eq. (2.40) may be written as

$$C_v = \left(\frac{\partial E}{\partial T}\right)_V \qquad (3.19)$$

and, similarly, the constant pressure heat capacity defined in Eq. (2.40) as

$$C_p = \left(\frac{\partial E}{\partial T}\right)_p + p\left(\frac{\partial V}{\partial T}\right)_p .$$

By the definition of enthalpy, we find

$$\left(\frac{\partial E}{\partial T}\right)_p = \left(\frac{\partial H}{\partial T}\right)_p - p\left(\frac{\partial V}{\partial T}\right)_p . \qquad (3.20)$$

Substitution of this into the equation above yields

$$C_p = \left(\frac{\partial H}{\partial T}\right)_p . \qquad (3.21)$$

This also follows more directly from Eq. (3.17). These two heat capacities are clearly not equal to each other. The difference of the two heat capacities may be computed as follows:

$$C_p - C_v = \left(\frac{\partial H}{\partial T}\right)_p - \left(\frac{\partial E}{\partial T}\right)_V$$

$$= p\left(\frac{\partial V}{\partial T}\right)_p + \left(\frac{\partial E}{\partial T}\right)_p - \left(\frac{\partial E}{\partial T}\right)_V . \qquad (3.22)$$

Since the internal energy may be regarded as a function of T and V — in other words, if the state space is assumed to be spanned by T and V — we obtain the relation

$$dE = \left(\frac{\partial E}{\partial T}\right)_V dT + \left(\frac{\partial E}{\partial V}\right)_T dV. \qquad (3.23)$$

Differentiation of Eq. (3.23) with respect to T at constant p yields

$$\left(\frac{\partial E}{\partial T}\right)_p = \left(\frac{\partial E}{\partial T}\right)_V + \left(\frac{\partial E}{\partial V}\right)_T \left(\frac{\partial V}{\partial T}\right)_p . \qquad (3.24)$$

When this is substituted into the right hand side of Eq. (3.22), there follows the equation

$$C_p - C_v = \left[p + \left(\frac{\partial E}{\partial V}\right)_T\right]\left(\frac{\partial V}{\partial T}\right)_p . \qquad (3.25)$$

Since the right-hand side of Eq. (3.25) is not equal to zero, the two specific heats are clearly not identical. Mathematically speaking, this nonvanishing difference is a consequence of the non-exactness of dQ since heat transfer depends on the path along which it is performed, and a constant pressure path is certainly different from a constant volume path. We will see later that the right-hand side of Eq. (3.25) can be expressed in another form more convenient for measurement. Since, as will be shown later,

$$\left(\frac{\partial E}{\partial V}\right)_T = -p + T\left(\frac{\partial p}{\partial T}\right)_V, \tag{3.26}$$

we find

$$C_p - C_v = T\left(\frac{\partial p}{\partial T}\right)_V \left(\frac{\partial V}{\partial T}\right)_p, \tag{3.27}$$

which can be calculated if the equation of state is given. We will return to this quantity in a later chapter.

3.3.2 The Difference in Isobaric and Isochoric Heat Capacities for Ideal Gases

As was established in Chapter 2, the ideal gas is an idealized substance, which obeys the equation of state

$$pV = nRT. \tag{3.28}$$

This was also seen to be intimately connected to the ideal gas temperature scale. From the viewpoint of energy, ideal gases are defined as those that do not cost energy as the gases are isothermally expanded. This is mathematically expressed in the form

$$\left(\frac{\partial E}{\partial V}\right)_T = 0. \tag{3.29}$$

Alternatively, this second condition means that the internal energy of an ideal gas is independent of its volume. This is a consequence of the absence of interactions between the molecules in the ideal gas, which is the behavior attained by the gas as the pressure is diminished to a vanishingly small value.

When Eqs. (3.28) and (3.29) are used in Eq. (3.25), there follows the relation

$$C_p - C_v = nR, \tag{3.30}$$

and the relation of C_p and C_v has become particularly simple and independent of substances. This is another aspect of the universality of the behavior of ideal gases. Recall that the equation of state (Eq. (3.28)) is also independent of substances since there is no material parameter present in it.

3.4 Work and Heat of Isothermal Reversible Expansion

Suppose a vessel containing an ideal gas is maintained at thermal equilibrium with a thermostat at temperature T. Then the gas is reversibly expanded. Since Eq. (3.29) holds for the ideal gas by definition and since $dT = 0$ by the experimental condition, the internal energy of the ideal gas is conserved as the gas is reversibly but isothermally expanded. This conclusion follows from Eq. (3.23) since

$$dE = 0. \tag{3.31}$$

It therefore follows from Eq. (3.9) that the heat change then must be balanced by the work:

$$đQ = p\,dV.$$

This means that if an ideal gas is isothermally expanded, heat Q, equal in amount to the work W done by the system, must be absorbed from the heat reservoir (surroundings). For a reversible isothermal expansion from V_1 to V_2 the heat absorbed therefore is expressible as

$$Q = -W = \int_{V_1}^{V_2} p\,dV. \tag{3.32}$$

By using the ideal gas equation of state (Eq. (3.28)), we calculate the integral in Eq. (3.32) to find that the isothermal reversible heat change is

$$Q = -W = nRT \ln\left(\frac{V_2}{V_1}\right) = -nRT \ln\left(\frac{p_2}{p_1}\right). \tag{3.33}$$

In the case of reversible isothermal compression the sign in Eq. (3.33) is reversed since the work is done on the system.

3.5 Work of Adiabatic Expansion

Suppose a vessel containing an ideal gas is thermally isolated so that the gas neither receives heat from, nor gives it up to, its surroundings. (A thermal insulator may be used to isolate the vessel thermally.) Then the gas is adiabatically and reversibly expanded from V_1 to V_2. This change is accompanied by a temperature decrease from T_1 to T_2. Since $dQ = 0$ owing to the process being adiabatic, we have

$$dE = \left(\frac{\partial E}{\partial T}\right)_V dT = -p\,dV. \tag{3.34}$$

Therefore the reversible adiabatic work is

$$W = -\int_{V_1}^{V_2} p\,dV = \int_{T_1}^{T_2} C_v(T)\,dT. \tag{3.35}$$

The pressure and volume of a gas obey a certain relation during an adiabatic expansion or compression. Let us find the relation. Substitution of Eq. (3.28) into Eq. (3.34) and use of the definition of C_v puts the equation in the form

$$T^{-1}C_v(T)\,dT = -RV^{-1}\,dV.$$

Integration of this equation under the assumption of constant C_v yields

$$T_1 V_1^{\gamma-1} = T_2 V_2^{\gamma-1}, \tag{3.36}$$

where γ is the polytropic ratio of the gas:

$$\gamma = \frac{C_p}{C_v}. \tag{3.37}$$

We have made use of Eq. (3.30) for Eq. (3.36). It must be emphasized that Eq. (3.36) holds if the equation of state is given by Eq. (3.28) and C_v is independent of temperature. When the equation of state (Eq. (3.28)) is made use of in Eq. (3.36), it may be written as

$$p_1 V_1^{\gamma} = p_2 V_2^{\gamma}$$

or

$$pV^{\gamma} = \text{constant}. \tag{3.38}$$

The case of real gases will be considered for a relation equivalent to Eq. (3.36) when the thermodynamics of real gases is studied later.

3.6 Heat Capacity and Heat Change

The heat change accompanying a reversible process in a substance at constant pressure can be computed from the heat capacity data of the substance. Heat capacities are measured in the laboratory by means of calorimetry, that is, by using a calorimeter. Calorimetry is a well-developed experimental discipline widely practiced in thermodynamics and in physical chemistry as a means of studying thermal and molecular properties of substances. Heat capacity data so measured are usually expressed as a power series in T at different pressures. It is found generally useful to summarize heat capacity data in the form of series in T:

$$C_p(T) = a_{-2}T^{-2} + a_0 + a_1 T + \cdots, \qquad (3.39)$$

where a_{-2}, a_0, and a_1 are empirical coefficients determined by calorimetry. They are generally dependent on pressure. Such coefficients are tabulated and available in the literature.[2] Since the pressure dependence of heat capacities can be computed if the equation of state is known for the substance, it is useful to tabulate them in a condition at which the heat capacities are independent of pressure. Since it is known that the heat capacities of ideal gases are independent of pressure, the coefficients must tend to limits

$$\lim_{p \to 0} a_{-2} = a^0_{-2},$$

$$\lim_{p \to 0} a_0 = a^0_0,$$

$$\lim_{p \to 0} a_1 = a^0_1,$$

and therefore

$$C_p^0 \equiv \lim_{p \to 0} C_p = a^0_{-2}T^{-2} + a^0_0 + a^0_1 T + \cdots, \qquad (3.40)$$

where a^0_{-2}, a^0_0, and a^0_1 are material constants independent of p. For this reason the heat capacities of gases are usually tabulated for ideal gases in the literature. When the temperature expansion (Eq. (3.39)) is used for ideal gases, Eq. (3.40) must be understood. In the case of solids and liquids they are tabulated at 1 atm. Some typical values for the coefficients in Eq. (3.39) are listed in Table 3.1. The heat capacities at other pressure values can then be computed, based on the tabulated values, by means of the thermodynamic procedure that will be developed later in this work.

The heat capacity data can be used to compute other thermodynamic quantities. Here, we consider an example. By using Eq. (3.39), we can compute ΔH for an isobaric process where the temperature of the system at pressure p changes from T_1 to T_2:

$$\Delta H = \int_1^2 dH = \int_{T_1}^{T_2} dT\, C_p(T)$$

$$= a_{-2}(T_1^{-1} - T_2^{-1}) + a_0(T_2 - T_1) + \frac{1}{2}a_1(T_2^2 - T_1^2) + \cdots. \qquad (3.41)$$

---

[2]See the NIST Standard Reference Database: http://webbook.nist.gov/chemistry/fluid.

Table 3.1   Heat capacities for some substances.

Substance	a_{-2} (cal deg^2/mole)	a_0 (cal/mole)	a_1 (cal/mole deg)
Gases (298–2000 K)			
S	0.36×10^5	5.26	-0.10×10^{-3}
H_2	0.12×10^5	6.52	0.78×10^{-3}
O_2	-0.40×10^5	7.16	1.00×10^{-3}
N_2	-0.12×10^5	6.83	0.90×10^{-3}
CO	-0.11×10^5	6.79	0.98×10^{-3}
CO_2	-2.06×10^5	10.57	2.10×10^{-3}
H_2O	—	7.30	2.46×10^{-3}
NH_3	-0.37×10^5	7.11	6.00×10^{-3}
Liquids			
H_2O	—	18.04	—
Solids (298 to T_m or 2000 K)			
C (graphite)	-2.04×10^5	4.03	1.14×10^{-3}
Al	—	4.94	2.96×10^{-3}
Cu	—	5.14	1.50×10^{-3}

The arabic numerals for the endpoints in the first integral represent the initial and final states (T_1, p) and (T_2, p), respectively.

Joseph Black of Edinburgh in the 18th century discovered that there is a characteristic amount of heat absorbed or released, for example, when a given amount of a liquid is vaporized or condensed from its vapor. He termed it the *latent heat*. The latent heat is a characteristic property of a substance and differs from substance to substance. Generally, when there is a phase transformation in a substance, there is a latent heat associated with the transformation. The latent heats associated with vaporization, melting, sublimation, and so on are called, respectively, the heat of vaporization, the heat of melting, the heat of sublimation, and so on.

Suppose the temperature of a substance is increased from T_1 to T_p, at which it undergoes a phase transition to another phase, and then the temperature of the substance in the new phase is further increased from T_p to T_2. The enthalpy change for this reversible change consists of three parts:

$$\Delta H = \int_{T_1}^{T_p} dT \; C_p^{(1)}(T) + \Delta H_p + \int_{T_p}^{T_2} dT \; C_p^{(2)}(T), \qquad (3.42)$$

where $C_p^{(1)}$ and $C_p^{(2)}$ are the heat capacities of the low- and high-temperature phases 1 and 2, and ΔH_p is the latent heat for the phase transition. Note that the reversible change is at constant pressure. We will discuss in a later chapter the pressure dependence of ΔH for a process in which pressure is changed under constant temperature condition. If the temperature dependence is known for the heat capacities in Eq. (3.42), the integrals may be evaluated more explicitly as in Eq. (3.41).

3.7 Thermochemistry

Heat changes involved in chemical systems and, in particular, chemical reactions are important quantities, which can be used to deduce some thermodynamic properties of the reacting systems. Here we consider some of such quantities, which will be useful for computing other thermodynamic quantities, such as equilibrium constants and free energy changes, discussed in subsequent chapters.

3.7.1 *Heat of Reaction*

Changes in heat content accompany chemical reactions occurring at constant pressure and temperature. Such a heat content change is referred to as the heat of reaction. Since a change in heat content at constant pressure is found to be equal to a change in enthalpy, it is sufficient to find the latter. More precisely, for the chemical reaction written in a general form

$$\sum_{i \in R} \nu_i^{(R)} X_i = \sum_{i \in P} \nu_i^{(P)} Y_i,$$

where X_i and Y_i are the reactants and products, the superscripts R and P stand for the reactants and products, respectively, and $\nu_i^{(R)}$ and $\nu_i^{(P)}$ are stoichiometric coefficients, the heat of reaction is defined as the difference between the total enthalpies of the products and reactants of the chemical reaction at constant pressure and temperature:

$$\Delta H = \sum_{i \in P} h_{Pi} - \sum_{i \in R} h_{Ri}. \tag{3.43}$$

Here h_{Pi} and h_{Ri} are the molar enthalpies of products and reactants, respectively. They are available in the literature, generally tabulated at 298.15 K and 1 atm pressure. When ΔH is evaluated according to Eq. (3.43), an assumption is implicit that the reactants are completely converted into the products.

Since heats of reaction can differ if the states of aggregation of the constituents are different even if the reaction involves the same constituents, it is customary to indicate the states of aggregation of the constituents. For example, we write

$$C(s) + O_2(g) = CO_2(g) \tag{3.44}$$

and

$$CH_4(g) + 2O_2(g) = CO_2(g) + 2H_2O(l), \qquad (3.45)$$

where s, l, and g, respectively, refer to the solid, liquid, and gaseous state of the constituents. Thus, if we take the first of the aforementioned reactions as an example for illustration of Eq. (3.43), its heat of reaction is

$$\Delta H = h(CO_2, g) - h(C, s) - h(O_2, g)$$
$$= -393.41 \text{ kJ mol}^{-1} \text{ (at 298.15 K and 1 atm)}. \qquad (3.46)$$

In the case of the second reaction

$$\Delta H = h(CO_2, g) + 2h(H_2O, l) - h(CH_4, g) - h(O_2, g)$$
$$= -890.36 \text{ kJ mol}^{-1} \text{ (at 298.15 K and 1 atm)}.$$

The IUPAC convention used in expressing ΔH for chemical reactions is that ΔH refers to the enthalpy change for a mole of the chemical reaction as expressed.

3.7.2 Standard States and Heat of Formation

It is of particular interest to find the heats of reaction for chemical reactions in which compounds are formed from their constituent elements, since such information can be used to deduce heats of reaction for other related chemical reactions as will be shown shortly. Since heats of reaction are the differences in enthalpies, it is possible to fix heats of reaction in reference to the standard state where the enthalpies for elements are set equal to zero. It is customary to choose the state of aggregation as that form of an element which is most stable at the temperature under consideration.

Particularly, *for gaseous elements the standard state is chosen as the state of the hypothetical ideal gas at* 1 atm *pressure and* 298.15 K. *For solid and liquid elements the standard state is chosen at* 1 atm *pressure and* 298.15 K. With the standard states so chosen and the enthalpies of stable elements set equal to zero, the heat of reaction for a mole of a compound formed from its elements in the standard state is simply the standard heat of formation, which will be indicated by ΔH_f^0. To illustrate this we consider Reaction (3.44). Since $h(C, s) = h(O_2, g) = 0$ at the standard states, we find

$$\Delta H_f^0 = h(CO_2, g) \text{ (at 298.15 K and 1 atm)},$$

which is now the standard heat of formation of CO_2. A few values are listed in Tables 3.2–3.4 for standard heats of formation.

Table 3.2   Standard heats of formation.

Compound	State	ΔH^0_{298} (kJ mol^{-1})
H_2O	g	-239.75
H_2O	l	-283.38
H_2O_2	g	-132.03
HF	g	-258.01
HCl	g	-91.52
CO	g	-109.57
HBr	g	-35.92
HI	g	-25.72
HIO_3	c	-236.56
H_2S	g	-19.97
H_2SO_4	l	-804.34
SO_2	g	-294.34
SO_3	g	-391.78
CO_2	g	-390.13
$SOCl_3$	l	-204.08
S_2Cl_2	g	-23.64

Table 3.3   Standard heats of formation.

Substance	Formula	ΔH^0_{298} (kJ mol^{-1})
Paraffins		
Methane	CH_4	-74.10
Ethane	C_2H_6	-83.75
Propane	C_3H_5	-102.66
N-Butane	C_4H_{10}	-123.26
Isobutane	C_4H_{10}	-130.04
N-Pentane	C_5H_{12}	-140.10
2-Methylbutane	C_5H_{12}	-152.11
Tetramethylmethane	C_5H_{12}	-163.46
Monolefines		
Ethylene	C_2H_4	52.08
Propylene	C_3H_6	20.56
1-Butene	C_4H_5	1.59
cis-2-Butene	C_4H_3	-5.76
trans-2-Butene	C_4H_5	-9.70
2-Methylpropene	C_4H_5	-13.29
1-Pentene	C_5H_{10}	-19.26

3.7.3 *Hess's Law*

Heats of reaction can be determined by direct measurement of changes in heat content. However, it is not necessary to measure the heat of reaction for every chemical reaction that we encounter, since it is often possible to deduce it from the enthalpy changes of other reactions. Such deductions are possible because enthalpy is a state function which is determined by

Table 3.4 Standard heats of formation.

Substance	Formula	ΔH_{298}^0 (kJ mol^{-1})
Allenes	C_3H_4	191.00
1,3-Butadiene	C_4H_6	111.44
1,3-Pentadiene	C_5H_5	78.33
1,4-Pentadiene	C_5H_5	106.04
Acetylenes		
Acetylene	C_2H_2	224.94
Methylacetylene	C_3H_4	183.79
Dimethylacetylene	C_4H_6	146.10

its values at the initial and final states of the system and that the mass as well as the energy is conserved in a chemical reaction. Suppose ΔH is the enthalpy change for a reversible process from state a to state z. It is then possible to write it in terms of ΔH's for other reversible processes:

$$\Delta H = H_z - H_a$$
$$= (H_z - H_y) + (H_y - H_x) + (H_x - H_b) + \cdots + H_b - H_a$$
$$= \sum_i \Delta H_i, \tag{3.47}$$

where ΔH_i stands for $H_z - H_y$, $H_y - H_x$, and so on. This is a mathematical consequence of the exactness of the differential dH since it is possible to write

$$\Delta H = \int_a^z dH = \int_y^z dH + \int_x^y dH + \int_b^x dH + \cdots + \int_a^b dH$$
$$= \sum_i \Delta H_i.$$

In the case of chemical reactions this decomposability of ΔH is known as Hess's law. We now translate this mathematical expression into the case of chemical reactions. Suppose there are m chemical reactions. We compactly express them by the equation

$$\sum_{i=1}^r \nu_i^{(\sigma)} X_i = 0 \quad (\sigma = 1, 2, \ldots, m), \tag{3.48}$$

where X_i denotes species (compound) i and $\nu_i^{(\sigma)}$ is the stoichiometric coefficient for species i of reaction σ in a set of chemical reactions. It is counted positive for the products, and negative for the reactants. For example, if the reaction

$$\tfrac{1}{2}H_2 + \tfrac{1}{2}Cl_2 = HCl$$

is designated the σth reaction of the set, then $\nu_{\mathrm{HCl}}^{(\sigma)} = 1$, $\nu_{\mathrm{Cl_2}}^{(\sigma)} = -\frac{1}{2}$, and $\nu_{\mathrm{H_2}}^{(\sigma)} = -\frac{1}{2}$. The enthalpy change for reaction σ may be expressed in the form

$$\Delta H^{(\sigma)} = \sum_{i=1}^{r} \nu_i^{(\sigma)} h_i, \tag{3.49}$$

where h_i is the molar enthalpy of substance i. We have earlier discussed a method of calculating ΔH when pressure is kept constant. We will discuss how to compute them in general from the basic thermodynamic data in a later chapter.

Suppose there is a chemical reaction that can be composed by taking a linear combination of the m reactions in the set of chemical reactions in Eq. (3.48). We will denote the reaction by

$$\sum_{i=1}^{r} \nu_i X_i = 0. \tag{3.50}$$

The enthalpy change for this reaction may be given in the form

$$\Delta H = \sum_{i=1}^{r} \nu_i h_i. \tag{3.51}$$

Since Reaction (3.50) is a linear combination of Reactions (3.48), there exists a set of constants ω_σ such that

$$\sum_{\sigma=1}^{m} \sum_{i=1}^{r} \omega_\sigma \nu_i^{(\sigma)} X_i = \sum_{i=1}^{r} \nu_i X_i.$$

By comparing both sides of this equation, we find the relation of the stoichiometric coefficients ν_i in Reaction (3.50) to those in the set of Reactions (3.48):

$$\nu_i = \sum_{\sigma=1}^{m} \omega_\sigma \nu_i^{(\sigma)}. \tag{3.52}$$

Inserting it into Eq. (3.51) and using Eq. (3.49), we obtain for Reaction (3.50) the enthalpy change ΔH in terms of $\Delta H^{(\sigma)}$ for the set of reactions Eq. (3.48):

$$\Delta H = \sum_{\sigma=1}^{m} \omega_\sigma \Delta H^{(\sigma)}. \tag{3.53}$$

This result embodies Hess's law. Note that ΔH here has exactly the same coefficients ω_σ of the linear combination as in the linear combination of the stoichiometric coefficients in Eq. (3.52). Equation (3.53) shows how the enthalpy change of a reaction might be obtained from the sources of information on $\Delta H^{(\sigma)}$ for the chemical reactions making up the set in Eq. (3.48). Hess's law holds at any temperature and pressure as long as the reactions involved are for the substances in the same conditions, namely, if the temperature, pressure, and the states of aggregation are the same for the compounds in Reaction (3.50) and those in the set of Reactions (3.48). If the conditions are altered at the end of a process, a proper account should be made of the changes in the heat content arising from the altered conditions. A partial account is already given for such a correction in the previous section. See Eq. (3.42).

We consider an example for application of Hess's law below. Consider the reaction

$$C_2H_4(y) + H_2(y) - C_2H_6(g). \tag{3.54}$$

The heats of reaction for the following reactions are known in the units of $kJ\,mol^{-1}$:

(1) $C_2H_4(g) + 3O_2 = 2CO_2(g) + 2H_2O(l)$ $\Delta H^{(1)} = -1411.26$
(2) $2H_2(g) + O_2(g) = 2H_2O(l)$ $\Delta H^{(2)} = -571.70$
(3) $2C_2H_6(g) + 7O_2(g) = 4CO_2 + 6H_2O(l)$ $\Delta H^{(3)} = -3119.59$

Take $\omega_1 = 1$, $\omega_2 = 1/2$, and $\omega_3 = -1/2$. Then Reaction (3.54) is obtained from Reactions (1)–(3) above and the heat of reaction for Reaction (3.54) is accordingly obtained as follows:

$$\Delta H = \Delta H^{(1)} + \tfrac{1}{2}\Delta H^{(2)} - \tfrac{1}{2}\Delta H^{(3)}$$
$$= -137.31\,kJ\,mol^{-1}.$$

If $\Delta H^{(\sigma)}$ are taken for the standard state heats of reaction then ΔH will be the standard heat of reaction for the reaction of interest.

3.7.4 Kirchhoff's Equation

The heat of reaction varies with temperature and pressure, and there often arise cases where the heat of reaction at other states than the standard state is of interest. Especially, the variation of ΔH with temperature can be desired. It can be derived in the following manner. Differentiating Eq. (3.43) with respect to T at constant p, we obtain

$$\left(\frac{\partial}{\partial T}\Delta H\right)_p = \sum_{i\in P}\left(\frac{\partial}{\partial T}H_{Pi}\right)_p - \sum_{i\in R}\left(\frac{\partial}{\partial T}H_{Ri}\right)_p, \tag{3.55}$$

but since

$$C_{pPi} = \left(\frac{\partial}{\partial T} H_{Pi}\right)_p, \quad C_{pRi} = \left(\frac{\partial}{\partial T} H_{Ri}\right)_p,$$

the specific heat change for the reaction is

$$\Delta C_p = \sum_{i \in P} C_{pPi} - \sum_{i \in R} C_{pRi}. \tag{3.56}$$

If we make use of Eq. (3.53) instead of Eq. (3.43), then the specific heat change is

$$\Delta C_p = \sum_{\sigma=1}^{m} \omega_\sigma c_p^{(\sigma)}, \tag{3.57}$$

where $c_p^{(\sigma)}$ is a molar heat capacity. This formula may be considered as Hess's law applied to specific heats. From Eqs. (3.55) and (3.57) follows the equation

$$\left(\frac{\partial \Delta H}{\partial T}\right)_p = \Delta C_p. \tag{3.58}$$

This is called the Kirchhoff equation. If ΔH is for a single chemical reaction, then ΔC_p is for the reaction, but if ΔH is for the set of chemical reactions defined by Eq. (3.43), then ΔC_p is given by Eq. (3.57). In either case, the Kirchhoff equation formally remains the same. By integrating the Kirchhoff equation and using calorimetric data on C_p it is possible to calculate ΔH at temperature T in terms of the enthalpy change at T_0:

$$\Delta H(T, p) = \Delta H(T_0, p) + \int_{T_0}^{T} dT' \, \Delta C_p(T', p). \tag{3.59}$$

If the calorimetric data are represented as in Eq. (3.39), the integration can be easily performed, and the result is similar to that in Eq. (3.42). The integration is left to the reader as an exercise.

3.8   Mathematical Notes

Here we discuss some mathematical notions and relations which are handy in studying, and performing calculations in, thermodynamics. They are quite well known in calculus and the reader versed in calculus may skip this section.

3.8.1 *Exact Differentials*

We elaborate a little more on exact differentials in the case where the number of independent variables is 2. If it is larger than 2, the conditions for the differentials to be exact differentials, namely, the integrability conditions, are rather complicated. The reader is referred to theory of differential forms and differential manifolds[3] for this topic. Consider a differential dz of a function $z(x, y)$:

$$dz(x, y) = M(x, y) \, dx + N(x, y) \, dy. \tag{3.60}$$

The necessary and sufficient condition for dz to be an exact differential is

$$\frac{\partial M}{\partial y} = \frac{\partial N}{\partial x}. \tag{3.61}$$

Since

$$M(x, y) = \frac{\partial z}{\partial x}, \quad N(x, y) = \frac{\partial z}{\partial y}, \tag{3.62}$$

Condition (3.61) implies that

$$\frac{\partial^2 z}{\partial y \partial x} = \frac{\partial^2 z}{\partial x \partial y}. \tag{3.63}$$

Thus, if a function of two variables $z(x, y)$ satisfies this condition, its differential is exact.

With this mathematical preparation let us examine if $đQ$ is indeed inexact. Since the internal energy E may be regarded as a function of T and V, we may write

$$dE = \left(\frac{\partial E}{\partial T}\right)_V dT + \left(\frac{\partial E}{\partial V}\right)_T dV. \tag{3.64}$$

By substituting it into Eq. (3.9) and rearranging the terms we find

$$đQ = \left(\frac{\partial E}{\partial T}\right)_V dT + \left[p + \left(\frac{\partial E}{\partial V}\right)_T\right] dV. \tag{3.65}$$

Since

$$\frac{\partial^2 E}{\partial x \partial y} = \frac{\partial^2 E}{\partial y \partial x},$$

---

[3]See, for example, I. N. Sneddon, *Elements of Partial Differential Equations* (McGraw-Hill, New York, 1957); H. Flanders, *Differential Forms* (Academic, New York, 1963).

by the exactness of dE, for $đQ$ to be exact it is necessary that

$$\left(\frac{\partial p}{\partial T}\right)_V = 0,$$

but this is not true in general. Therefore it is concluded that $đQ$ is not an exact differential.

3.8.2 *Chain Relations*

A functional relation between two variables x and y may be put in the implicit form

$$f(x, y) = C, \tag{3.66}$$

where C is a constant and $f(x, y)$ is a differentiable function of x and y. For example, consider the relation for a circle of radius R

$$x^2 + y^2 = R^2.$$

We wish to calculate the derivative dy/dx with the implicit function. Taking the differential of f, we obtain

$$df = \left(\frac{\partial f}{\partial x}\right)_y dx + \left(\frac{\partial f}{\partial y}\right)_x dy = 0.$$

If f is kept constant and the equation is differentiated with dx, we obtain

$$\left(\frac{\partial f}{\partial x}\right)_y + \left(\frac{\partial f}{\partial y}\right)_x \left(\frac{\partial y}{\partial x}\right)_f = 0, \tag{3.67}$$

which may be put in the form[4]

$$\left(\frac{\partial y}{\partial x}\right)_f = -\frac{\left(\frac{\partial f}{\partial x}\right)_y}{\left(\frac{\partial f}{\partial y}\right)_x}. \tag{3.68}$$

---

[4]It is useful to devise a mnemonic scheme for this useful relation. To construct the right-hand side of Eq. (3.68) we can proceed as follows: for the numerator we construct a derivative from the derivative on the left-hand side, starting from the subscript and moving clockwise as the derivative is formed with the third variable in the chain $f \rightarrow x \rightarrow y$ becoming the subscript in the new derivative formed. That is, from the chain $f \rightarrow x \rightarrow y$, forms the derivative $\left(\frac{\partial f}{\partial x}\right)_y$. For the denominator we start from f in the counterclockwise chain $f \rightarrow y \rightarrow x$ and form the derivative $\left(\frac{\partial f}{\partial y}\right)_x$ with the third member of the chain becoming the subscript in the derivative formed, and finally put a negative sign on the ratio of the derivatives. This procedure mechanically produces the right-hand side of Eq. (3.68) from the derivative on the left-hand side without having to go through the steps used for obtaining it from Eq. (3.67).

This relation can be generalized to a situation of three variables x, y, and z. Let there be a function f such that

$$f(x, y, z) = C, \qquad (3.69)$$

where C is a constant. One variable then may be considered a function of the other two. Take the differential of f:

$$df = \left(\frac{\partial f}{\partial x}\right)_{y,z} dx + \left(\frac{\partial f}{\partial y}\right)_{x,z} dy + \left(\frac{\partial f}{\partial z}\right)_{x,y} dz = 0. \qquad (3.70)$$

If one variable, say, z is kept constant in addition to f, then this differential may be written as

$$\left(\frac{\partial y}{\partial x}\right)_{z,f} = -\frac{\left(\frac{\partial f}{\partial x}\right)_{z,y}}{\left(\frac{\partial f}{\partial y}\right)_{x,z}}, \qquad (3.71)$$

which is in the same form as Eq. (3.68). If y is kept constant instead of z, then there follows the relation

$$\left(\frac{\partial z}{\partial x}\right)_{y,f} = -\frac{\left(\frac{\partial f}{\partial x}\right)_{y,z}}{\left(\frac{\partial f}{\partial z}\right)_{y,x}}. \qquad (3.72)$$

The aforementioned mnemonic scheme also applies to these chain relations of derivatives if y is kept invariant. A similar relation can be obtained if variable x is kept constant. If we make the functional relation explicit and write

$$z = g(x, y), \qquad (3.73)$$

and take the differential of z, we obtain

$$dz = \left(\frac{\partial g}{\partial x}\right)_{y} dx + \left(\frac{\partial g}{\partial y}\right)_{x} dy. \qquad (3.74)$$

If z is kept constant then this differential form may be written as

$$\left(\frac{\partial y}{\partial x}\right)_{z} = -\frac{\left(\frac{\partial g}{\partial x}\right)_{y}}{\left(\frac{\partial g}{\partial y}\right)_{x}} = -\frac{\left(\frac{\partial z}{\partial x}\right)_{y}}{\left(\frac{\partial z}{\partial y}\right)_{x}}, \qquad (3.75)$$

which is the same as the relation in Eq. (3.68). This relation will be useful for some calculations in thermodynamics.

3.8.3 Jacobians

Suppose there is a transformation of a two-dimensional space to another: $(x, y) \to (u, v)$. This may be expressed by a pair of functional relations:

$$x = u(x, y), \quad y = v(x, y). \tag{3.76}$$

Under this transformation the elements of area in the two spaces are related to each other in the following manner:

$$dx \, dy = |J| \, du \, dv,$$

where J is called the Jacobian of transformation and defined by the determinant of partial derivatives

$$J(x, y, u, v) = \frac{\partial(x, y)}{\partial(u, v)} = \begin{vmatrix} \left(\frac{\partial x}{\partial u}\right)_v & \left(\frac{\partial x}{\partial v}\right)_u \\ \left(\frac{\partial y}{\partial u}\right)_v & \left(\frac{\partial y}{\partial v}\right)_v \end{vmatrix}. \tag{3.77}$$

The absolute value must be taken for the determinant. Since the value of the determinant is invariant to the interchange of rows and columns, the Jacobian may be written as

$$J = \begin{vmatrix} \left(\frac{\partial x}{\partial u}\right)_v & \left(\frac{\partial y}{\partial u}\right)_v \\ \left(\frac{\partial x}{\partial v}\right)_u & \left(\frac{\partial y}{\partial v}\right)_u \end{vmatrix}. \tag{3.78}$$

The Jacobian has the following properties which are derived from those of determinants:

$$\frac{\partial(x, y)}{\partial(u, v)} = -\frac{\partial(x, y)}{\partial(v, u)},$$

$$\frac{\partial(x, x)}{\partial(u, v)} = 0,$$

$$\frac{\partial(x, y)}{\partial(u, u)} = 0, \tag{3.79}$$

$$\frac{\partial(k, y)}{\partial(u, u)} = 0,$$

where k is a constant. The aforementioned identities can be obtained from the fact that a determinant changes its sign when two rows or columns are interchanged and a determinant vanishes if two rows or columns are equal. From the rule for the product of two determinants follows the identity

$$\frac{\partial(x, y)}{\partial(u, v)} \cdot \frac{\partial(u, v)}{\partial(t, s)} = \frac{\partial(x, y)}{\partial(t, s)}. \tag{3.80}$$

The definition of the Jacobian yields the identity

$$\left(\frac{\partial y}{\partial u}\right)_x = \frac{\partial(x,y)}{\partial(x,u)}. \tag{3.81}$$

We apply the aforementioned properties to the following example. It was previously shown that from the implicit relation (3.69) follows the chain relation

$$\left(\frac{\partial x}{\partial y}\right)_z = -\frac{\left(\frac{\partial z}{\partial y}\right)_x}{\left(\frac{\partial z}{\partial x}\right)_y}. \tag{3.82}$$

This relation also follows if the properties of Jacobians presented earlier are made use of:

$$\left(\frac{\partial x}{\partial y}\right)_z = \frac{\partial(x,z)}{\partial(y,z)} = \frac{\partial(x,z)}{\partial(x,y)}\frac{\partial(x,y)}{\partial(y,z)}$$

$$= -\left(\frac{\partial z}{\partial y}\right)_x \left(\frac{\partial x}{\partial z}\right)_y$$

$$= -\frac{\left(\frac{\partial z}{\partial y}\right)_x}{\left(\frac{\partial z}{\partial x}\right)_y}. \tag{3.83}$$

The properties in Eq. (3.79) can be generalized if the dimension of the space is larger than 2, and the relations similar to Eq. (3.83) can be easily obtained by taking advantage of the properties of Jacobians.

Problems

(1) Show that for the following reversible cycle performed by an ideal gas

$$(p_1, V_1, T_1) \to (p_1, V_2, T_1) \to (p_1, V_2, T_2) \to (p_2, V_1, T_2) \to (p_1, V_1, T_1)$$

the internal energy is conserved, namely,

$$\oint dE = 0,$$

whereas

$$\oint dW \neq 0, \quad \oint dQ \neq 0.$$

Show also that

$$\oint dH = 0.$$

(2) Prove that

$$C_p = C_v + \left[V - \left(\frac{\partial H}{\partial p}\right)_T\right]\left(\frac{\partial p}{\partial T}\right)_V.$$

(3) Show that for an ideal gas

$$\left(\frac{\partial C_v}{\partial V}\right)_T = 0, \quad \left(\frac{\partial C_p}{\partial V}\right)_T = 0.$$

(4) Show that

$$\left(\frac{\partial C_v}{\partial V}\right)_T = T\left(\frac{\partial^2 p}{\partial T^2}\right)_V.$$

(5) Calculate the change in ΔH for reaction

$$H_2(g) + \tfrac{1}{2}O_2(g) = H_2O(g)$$

for the temperature change from T to $2T$ at constant pressure. The following calorimetric data are available:

$$C_p(H_2O, g) = \left(30.20 + 9.933 \times 10^{-3}\, T + 1.117 \times 10^{-6}\, T^2\right)\, J\,mol^{-1},$$
$$C_p(H_2, g) = \left(29.07 - 0.837 \times 10^{-3}\, T + 2.012 \times 10^{-6}\, T^2\right)\, J\,mol^{-1},$$
$$C_p(O_2, g) = \left(25.50 + 1.361 \times 10^{-2}\, T - 4.255 \times 10^{-6}\, T^2\right)\, J\,mol^{-1}.$$

(6) Show that for an ideal gas

$$\left(\frac{\partial H}{\partial V}\right)_T = 0.$$

(7) The equation of state for gases at relatively low pressures may be expressed in the form

$$pV = RT + Bp,$$

where B is independent of p or V. By using this equation of state, derive an expression for the quantity

$$\left(\frac{\partial H}{\partial V}\right)_T - \left(\frac{\partial E}{\partial V}\right)_T.$$

(8) Let the equation of state be given by $pV = RT + Bp$ as in Problem 5. Obtain expressions for Q and W of isothermal reversible expansion by using the equation of state and compare them with those of an ideal gas.

(9) Use Eqs. (3.39) and (3.59) to compute ΔH at T and p.

Chapter 4

The Second Law of Thermodynamics

The first law of thermodynamics elucidates the nature of heat and what we mean by the energy of thermal systems as well as the mutual relationship between work and heat. It also requires that all natural processes conserve energy at the end of a cyclic change. This conservation law of energy then demands that the internal energy be a state function, and as a consequence its differential is an exact differential in thermodynamic state space. However, the first law does not tell us whether a natural process is possible to occur spontaneously or not. It merely imposes a restriction with regard to the energy of the system at the end of a process.

In nature, we see that some processes occur spontaneously whereas some others do not. It is an indisputable fact that the river flows downstream, not upstream, unless an external agency intervenes with it to make the reverse process possible by some sort of compensation in energy. Similarly, a vacuum is filled by air if it is connected with a container filled with air, but the vacuum is not created spontaneously in a container filled with gas without a compensation. There are numerous examples of this kind in nature that while some processes can happen spontaneously, their reverse processes are not spontaneously possible, even if they are energetically possible, not being in violation of the first law. Such a directional preference by natural phenomena is made an axiom in the form of the second law of thermodynamics. The second law of thermodynamics was preceded by Carnot's theorem enunciated by Sadi Carnot who, on examination of an idealized cyclic operation characteristic of steam engines, discovered a universal principle for thermal macroscopic processes in 1824. It is perhaps useful to observe that the study of cyclic processes is on the basis of the Carnot theorem and also the second law of thermodynamics. Therefore, it is indispensable to consider them to gain a proper understanding of the essence and evolution of the idea of the second law of thermodynamics.

Carnot originally formulated his theorem on the basis of the caloric theory of heat which was later discarded as an incorrect concept, but the essence of his theorem[1] was found to be correct if the notion of heat (distinguished as *chaleur* from *calorique* by him) was identified with a form of a mechanical energy, as enunciated by Joule, Davy, Count Rumford, and others mentioned in Chapter 3. After examination of Carnot's work, R. Clausius (1850) and W. Thomson (1851) independently were able to modify the Carnot theorem and enunciate the second principle of thermodynamics when systems undergo cyclic processes. Their universal principle is now known as the second law of thermodynamics, and it was stated by them as follows:

Clausius principle: *It is impossible to transfer heat from a colder to a hotter body without converting at the same time a certain amount of work into heat at the end of a cycle of changes.*

W. Thomson (Lord Kelvin) equivalently stated it as follows:

Kelvin principle: *In a cycle of processes it is impossible to transfer heat from a heat reservoir and convert it all into work, without transferring at the same time a certain amount of heat from a hotter to a colder body.*

These statements can be also put in another still equivalent form by M. Planck:

Planck principle: *It is impossible to construct a perpetual machine of the second kind which transfers heat from a colder to a hotter body without a compensation.*

These principles can be shown to be equivalent.[2] Schematically, the second law states that the following process is not possible unless some work is done: Let the temperatures of two systems be T_1 and T_2, respectively, and $T_2 < T_1$. Let the amount of heat given up by the system at T_1 to the system at T_2 be Q_1. Then it is impossible to return all of Q_1 to the system at T_1 without an equivalent amount of work. Similar examples can

---

[1]In Carnot's theory, there appears the notion of *calorique*, which is distinctive from *chaleur*. He did not clarify its true nature, but it is safe to assume it is a quantity associated with heat transfer and equivalent to entropy or, more precisely, calortropy in the present work.

[2]See, for example, J. G. Kirkwood and I. Oppenheim, *Chemical Thermodynamics* (McGraw-Hill, New York, 1961).

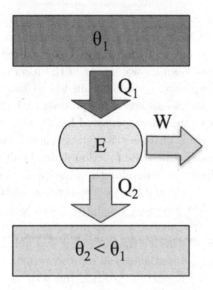

Fig. 4.1 A schematic rendering of the essence of the second law of thermodynamics, which states that Q_2 cannot be equal to zero for a realizable process (engine).

be constructed with the gravitational force and so on. Another schematic way of looking at the essence of the second law is to examine the essential feature of a heat engine depicted in Fig. 4.1. The engine (E) takes heat Q_1 from the high temperature heat reservoir at T_1 and performs work W returning some heat Q_2 to the lower temperature heat reservoir at T_2. It is impossible to make an engine which converts all of Q_1 into work W without some $Q_2 \neq 0$ emitted to the lower temperature reservoir. That is, it is not possible to have $Q_2 = 0$.

Since the efficiency of the engine in question may be defined by

$$\epsilon = -\frac{W}{Q_1} \tag{4.1}$$

for work done by the system and by the first law of thermodynamics

$$W = -(Q_1 - Q_2) \tag{4.2}$$

for the cyclic process, the efficiency will be unity if $Q_2 = 0$. Therefore the second law of thermodynamics implies that the efficiency of a cycle cannot be unity.

However universal and truthful a principle may be, it is not possible to formulate a physical theory with only a literal statement of it, and it is no exception for the second law of thermodynamics as stated earlier. One of the statements must be translated into an equivalent mathematical form. Since the two statements are equivalent, it is sufficient to have a mathematical representation for one of them. A mathematical representation of the second law of thermodynamics was made in the form of an inequality by R. Clausius. The inequality, nowadays called the Clausius inequality, was then specialized to the case of reversible processes which Clausius defined as the processes where the uncompensated heat vanishes. He showed that in such processes there exists a quantity (state function) called *entropy*.[3] It, however, is a thermodynamic state function defined only for the case of reversible processes. In order to develop this concept we now follow Clausius, who used the Carnot cycle and the Carnot theorem for the purpose.[4] The restriction of reversible processes can be removed, and the notion of entropy can be generalized to irreversible processes. This will be discussed later. It is interesting to notice that Clausius was able to give the second law of thermodynamics a mathematical representation by using the Carnot theorem. Therefore the important position that Carnot's theorem takes in thermodynamics cannot be overstated.

4.1 Carnot Cycle

In the early 19th century, the industrial revolution, powered by steam engines, was progressing in earnest in England, and it made England dominant, economically and politically, in the world at that time. In the book[5] entitled *"Réflexions sur la Puissance motrice du Feu et sur les Machines"* published in 1824 Sadi Carnot proposed to study an idealized cyclic process in order to understand steam engines which he called the English machines, since the English were using them to power the industrial revolution and

---

[3] *It was coined by Clausius from a Greek word root tropy to which prefix en was added so that it looks akin to the term energy. In Greek the term entropy means evolution.* See R. Clausius, *Ann. Phys.* (Leipzig) **125**, 313 (1865).

[4] Later, when statistical mechanics of gases was developed initially by J. C. Maxwell (1867) and L. Boltzmann (1872) followed by J. W. Gibbs (1902), the second law of thermodynamics was given statistical mechanical interpretations along with the statistical mechanical formula for the entropy and its inequality. In some textbooks on thermodynamics the statistical mechanical entropy inequality is used to the exclusion of adequate accounts of its evolution as a law that imposes a limitation upon macroscopic physical processes in nature. Such a treatment of this important physical law is not only unjustifiable, but also detrimental to the development of thermodynamic theory of irreversible processes.

[5] English translation: E. Mendoza, *Reflections on the Motor Power of Fire and on the Machines* (Peter Smith, Gloucester, MA, 1977).

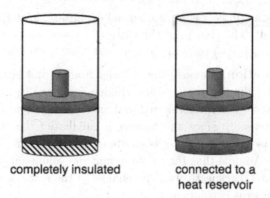

completely insulated   connected to a
heat reservoir

Fig. 4.2 The cylinder on the left is fitted with an insulator (adiabatic wall) at the bottom, and the bottom wall of the cylinder on the right is diathermal. Both pistons and the cylinder wall are adiabatic.

became powerful. This cyclic process, now called the Carnot cycle, does not really represent the working of real engines, but is contrived to capture the essence of their working. As mentioned earlier, Carnot used the notion of heat according to the caloric theory of heat, which was overturned about a quarter century later by Joule, Rumford, and others. Nevertheless, he captured the essential truth for the working of engines and macroscopic processes, physical and biological, in nature. We discuss the essence of his work in the following.

Imagine a working substance, for example, steam or air, is contained in a cylinder fitted at one end with a movable, adiabatic piston. The other end of the cylinder can be made perfectly diathermal or adiabatic at will, but the cylinder wall is adiabatic. The following sequence of operations is performed on the substance in the cylinder (see Fig. 4.3.):

First operation. Put the heat conducting wall in contact with the hot reservoir A at temperature[6] θ_1 and let the piston rise to volume V_2 from V_1 by reducing the external pressure from p_1 to p_2, see Fig. 4.3.

---

[6]In the present discussion it is not necessary to specify a particular temperature scale; it is sufficient to have two thermal reservoirs at given temperatures. However, the subsequent calculations using the ideal gas equation of state will make it abundantly clear that the absolute temperature scale is meant by θ. We use the symbol θ to distinguish it from the thermodynamic temperature T which will be introduced later on the basis of the Carnot theorem and the efficiency of a Carnot cycle. This thermodynamic temperature is eventually made coincident with the absolute temperature.

Since the process is an isothermal expansion, heat Q_1 is absorbed from A. The change in the state of the system in this process is from (p_1, V_1, θ_1) to (p_2, V_2, θ_1).

Second operation. Remove the cylinder from A and insulate the base of the cylinder so that the cylinder is thermally isolated. Then the pressure is reversibly reduced and as a consequence the volume is increased. Since the process is adiabatic, the temperature is decreased. This process is continued until the temperature drops to θ_2. Assume that the pressure and the volume at that point are p_3 and V_3, respectively. The change in the state of the system in this process is from (p_2, V_2, θ_1) to (p_3, V_3, θ_2).

Third operation. The insulating base of the cylinder is removed from the cylinder, and the cylinder is put in thermal contact with the cold reservoir B at temperature θ_2. Then the pressure is reversibly increased to p_4 and the volume is decreased to V_4 as a consequence. This pressure is taken such that the system returns to state (V_1, p_1, θ_1) at the end of the fourth operation. Since the process is isothermal and the volume is decreased, the substance gets hotter under compression and hence heat Q_2 is transferred to the cold reservoir B. The change in the state of the system in this process is from (p_3, V_3, θ_2) to (p_4, V_4, θ_2).

Fourth operation. The cylinder is removed from B and the base of the cylinder is thermally insulated, and then the substance is compressed reversibly and adiabatically until the temperature rises to θ_1. Since p_4 was so chosen as to return the system to its original state, the system will return to (V_1, p_1) at θ_1. The change in the state of the system in this process is from (p_4, V_4, θ_2) to (p_1, V_1, θ_1).

Thus the system is returned to the initial state at the end of a cyclic operation. This cyclic reversible process is depicted in four steps in Fig. 4.3. The diagram as in Fig. 4.3 was introduced for the first time by E. Clapeyron, who gave a mathematical analysis of Carnot's *Réflexion* in which mathematical analysis was avoided because the book was intended for popular consumption.

Let us calculate the net amount of work done by the cycle by using a mole of an ideal gas so as to make calculation simple. We will assume that the specific heat is independent of temperature.

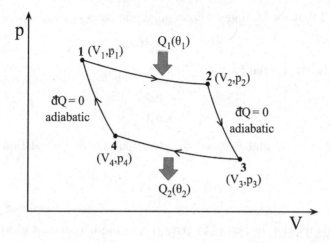

Fig. 4.3 A Carnot cycle in the p–V plane. The four operations are described in the text. The horizontal segments represent isothermal processes, and the vertical segments adiabatic processes.

The work done by the system in the four steps are as follows:

Step 1:

$$W_{12} = -R\theta_1 \ln\left(\frac{V_2}{V_1}\right),$$

Step 2:

$$W_{23} = \int_{\theta_1}^{\theta_2} d\theta\, c_v(\theta) = c_v(\theta_2 - \theta_1),$$

Step 3:

$$W_{34} = -R\theta_2 \ln\left(\frac{V_4}{V_3}\right),$$

Step 4:

$$W_{41} = \int_{\theta_2}^{\theta_1} d\theta\, c_v(\theta) = c_v(\theta_1 - \theta_2).$$

The c_v is the molar heat capacity of the working substance, namely, the ideal gas. Note that c_v for the ideal gas is independent of temperature θ. The total work is given by

$$W = W_{12} + W_{23} + W_{34} + W_{41}$$
$$= -R\theta_1 \ln\left(\frac{V_2}{V_1}\right) - R\theta_2 \ln\left(\frac{V_4}{V_3}\right). \tag{4.3}$$

Now recall that for ideal gases in an adiabatic process

$$\theta V^{\gamma - 1} = \text{constant}.$$

More explicitly written for Steps 2 and 4,

$$\theta_1 V_2^{\gamma - 1} = \theta_2 V_3^{\gamma - 1},$$
$$\theta_1 V_1^{\gamma - 1} = \theta_2 V_4^{\gamma - 1}.$$

By eliminating θ_1 and θ_2 from the equations we find a relation among volumes:

$$\left(\frac{V_1}{V_2}\right)^{\gamma - 1} = \left(\frac{V_4}{V_3}\right)^{\gamma - 1}. \tag{4.4}$$

Use of this relation in Eq. (4.3) gives rise to the net amount of work in a rather simple form:

$$W = -R(\theta_1 - \theta_2) \ln\left(\frac{V_2}{V_1}\right). \tag{4.5}$$

The heat transfers Q_1 and Q_2 during the isothermal processes (i.e., Steps 1 and 3) are easily calculated in the case of an ideal gas. According to the result obtained in Sec. 3.4 they are as follows:

$$Q_1 = -W_{12} = R\theta_1 \ln\left(\frac{V_2}{V_1}\right),$$
$$Q_2 = -W_{34} = R\theta_2 \ln\left(\frac{V_4}{V_3}\right) \tag{4.6}$$
$$= R\theta_2 \ln\left(\frac{V_1}{V_2}\right).$$

For the second equality of Q_2 Eq. (4.4) is used.

The efficiency of the Carnot cycle (engine) is defined by the ratio of the work W to the heat supplied Q_1:

$$\eta = \frac{-W}{Q_1} = \frac{|W|}{Q_1}. \tag{4.7}$$

Since for a cycle the internal energy is conserved, that is,

$$\oint dE = 0,$$

the work done by the reversible Carnot cycle is

$$-W = Q_1 - Q_2 \tag{4.8}$$

and hence we may express the efficiency in terms of heat alone:

$$\eta = \frac{Q_1 - Q_2}{Q_1} = 1 - \frac{Q_2}{Q_1}. \tag{4.9}$$

Upon use of Q_1 and Q_2 in Eq. (4.6) the efficiency for the reversible cycle may be expressed only in terms of temperatures of the reservoirs and no other parameters:

$$\eta = \frac{R(\theta_1 - \theta_2)\ln(V_2/V_1)}{R\theta_1 \ln(V_2/V_1)} = \frac{\theta_1 - \theta_2}{\theta_1} = 1 - \frac{\theta_2}{\theta_1}. \tag{4.10}$$

The temperatures in this expression is in the absolute temperature scale because of the ideal gas equation of state used for calculations involved. We emphasize that this efficiency formula is shown to be valid in the case of an ideal gas undergoing reversible processes. We will see in Sec. 4.4 that it holds in general for real substances, reinforcing the universality of the result as demanded by the Carnot theorem.

4.2 Carnot's Theorem

As a result of the analysis made of the working of a Carnot cycle as an idealization of real engines, Carnot obtained a theorem concerning the efficiency of reversible cycles and its relationship to the efficiency of irreversible cycles. This theorem is now known as Carnot's theorem.

Carnot's theorem: *The efficiency of reversible Carnot cycles is independent of materials and modes of operation, and is maximum. It depends only on the temperatures of the heat reservoirs. In other words, if we denote the efficiencies of reversible cycles by η_{rev}, η'_{rev}, and so on, and that of an irreversible cycle by η_{irr} then it may be stated that*

$$\eta_{rev} = \eta'_{rev} \quad and \quad \eta_{rev} \geq \eta_{irr}.$$

Moreover, η_{rev} depends only on the temperatures θ_1 and θ_2 of the two heat reservoirs, hot and cold.

Proof. In order to prove the first part of the theorem we first consider two reversible engines operating between two heat reservoirs at temperatures θ_1 and θ_2 ($\theta_1 \geq \theta_2$). Engine En' takes heat Q'_1 from the hotter reservoir and converts part of it into work W and emits heat Q'_2 to the colder reservoir. The other engine En takes heat Q_1 from the hotter reservoir and converts part of it into work W and emits heat Q_2 to the colder reservoir. The two engines are now coupled such that engine En' is a prime

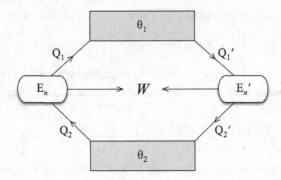

Fig. 4.4 Coupling of a prime mover and a refrigerator as a device to prove Carnot's theorem.

mover and the other engine En is a refrigerator; see Fig. 4.4. Engine En thus operates reversely to the mode by which the engine En' operates, pumping heat Q_2 from the colder reservoir and returning heat Q_1 to the hotter reservoir at the expense of work W. Now assume that

$$\eta'_{rev} > \eta_{rev}.$$

By the definition of efficiency, this may be written equivalently as

$$\frac{-W}{Q'_1} > \frac{-W}{Q_1},$$

which means the inequality

$$Q_1 - Q'_1 > 0.$$

This implies that heat $\Delta Q = Q_1 - Q'_1$ is transferred at the end of the cycle from the colder to the hotter reservoir without a compensation. This is against the second law of thermodynamics. Therefore the assumption is false.

Let us then assume that the reverse is true, that is,

$$\eta'_{rev} < \eta_{rev}.$$

By reversing the cycle, that is, by considering En as a prime mover and En' as a refrigerator, we reach the same conclusion as before. Therefore we must conclude that

$$\eta'_{rev} = \eta_{rev}. \qquad (4.11)$$

Since the same argument can be carried out with another engine En'' instead of En', the first part of the theorem is proved.

In order to prove the second part of the theorem we regard En' as an irreversible prime mover and En as a reversible refrigerator. Now suppose that

$$\eta_{irr} > \eta_{rev}.$$

Then, by the definition of efficiency

$$\frac{-W}{Q_1'} > \frac{-W}{Q_1},$$

which means

$$Q_1 - Q_1' > 0.$$

This again contradicts the second law of thermodynamics, since it implies that heat is transferred from the colder body to a hotter one at the end of a cycle without a compensation. Therefore, we must conclude that the assumption is false, and

$$\eta_{rev} \geq \eta_{irr}. \tag{4.12}$$

The reverse process is not possible, since it involves an irreversible engine.

We now prove[7] the last part of the theorem: *that the reversible efficiency depends only on the temperatures of the two heat reservoirs, hot and cold.*

The reversible Carnot cycle is entirely determined when the adiabatics and isotherms of the cycle are known. If the working substance is specified so that the constitutive relations of the substance are given, the isotherms are determined by the temperatures θ_1 and θ_2, and the adiabatics by some independent variables, say, v_1 and v_2 corresponding to the temperatures. The efficiency of the reversible Carnot cycle is then thought to be a function of θ_1, θ_2, v_1, and v_2 and of the working substance C. Let us therefore assume

$$\frac{-W}{Q_1} = f(\theta_1, \theta_2, v_1, v_2, C). \tag{4.13}$$

This is a continuous function of the variables. The Carnot theorem states that f depends only on θ_1 and θ_2. Consider two bodies C and C' which transform between the same heat reservoirs and describe cycles En and En', respectively, the former being a prime mover and the latter working in the reverse sense—a refrigerator. For this to be possible the temperatures must satisfy certain inequalities. Let θ_1 and θ_2 be the temperatures of two reservoirs, hot and cold; and θ_1' and θ_2' be the temperatures of the isotherms

---

[7]See H. Poincaré, *Thermodynamique* (Georges Carré, Paris, 1892).

of the cycle En; and θ_1'' and θ_2'' the temperatures of the isotherms of the cycle En'. Then for the coupled cycles to be possible the inequalities of temperatures must hold

$$\theta_1'' > \theta_1 > \theta_1' > \theta_2' > \theta_2 > \theta_2''.$$

Denote by W the work produced by the cycle En, Q_1 the heat that is taken from the hot reservoir, Q_2 the heat that it cedes to the cold reservoir. And by W', Q_1', and Q_2' similar quantities corresponding to cycle En'. Then

$$\frac{-W}{Q_1} \le \frac{-W'}{Q_1'}. \tag{4.14}$$

This is proved as follows. If m and m' are the masses of bodies C and C', which are transformed in the cycles, we have for the heat taken from the hot heat reservoir by the combined cycles

$$mQ_1 - m'Q_1'.$$

Since Q_1 and Q_1' are positive, we can take for m and m' values such that

$$mQ_1 - m'Q_1' = 0. \tag{4.15}$$

But then the work

$$-mW + m'W'$$

produced by the two cycles cannot be positive because then we will have a production of work with only one source of heat, contrary to the second law of thermodynamics. Therefore

$$-mW + m'W' \le 0. \tag{4.16}$$

Replacing m and m' with Q_1^{-1} and $Q_1'^{-1}$ we obtain Inequality (4.14).

Consider now two Carnot cycles En and En' which are, respectively, defined by the variables $\theta_1', \theta_2', v_1'$, and v_2' for the former and by $\theta_1'', \theta_2'', v_1''$, and v_2'' for the latter. The cycle En runs in the direct sense and the cycle En' in the reverse sense between two same heat reservoirs of temperatures θ_1 and θ_2, respectively. For this to be possible we must have the inequalities

$$\theta_1' < \theta_1, \quad \theta_2' > \theta_2 \quad \theta_1'' > \theta_1, \quad \theta_2'' < \theta_2.$$

Then it follows from Inequality (4.14) that

$$f(\theta_1', \theta_2', v_1', v_2', C) \le f(\theta_1'', \theta_2'', v_1'', v_2'', C'). \tag{4.17}$$

If we suppose

$$\theta_1' > \theta_1, \quad \theta_2' < \theta_2 \quad \theta_1'' < \theta_1, \quad \theta_2'' > \theta_2,$$

we can describe the cycle En in the reverse direction and the cycle En' in the direct sense. Then we have the inequality

$$f(\theta_1', \theta_2', v_1', v_2', C) \geq f(\theta_1'', \theta_2'', v_1'', v_2'', C'). \tag{4.18}$$

Since f is continuous, we can make $\theta_1', \theta_1'' \to \theta_1$ and $\theta_2', \theta_2'' \to \theta_2$ without changing the sign of Inequalities (4.17) and (4.18). We then have

$$f(\theta_1, \theta_2, v_1', v_2', C) \leq f(\theta_1, \theta_2, v_1'', v_2'', C'), \tag{4.19}$$

$$f(\theta_1, \theta_2, v_1', v_2', C) \geq f(\theta_1, \theta_2, v_1'', v_2'', C'). \tag{4.20}$$

These inequalities cannot be satisfied simultaneously. Therefore we have

$$f(\theta_1, \theta_2, v_1', v_2', C) = f(\theta_1, \theta_2, v_1'', v_2'', C'), \tag{4.21}$$

which means that f is independent of variables v_1, v_2, and C. This proves the last part of the Carnot theorem. Therefore the Carnot theorem is completely proved.

The proof presented is in fact seen also as the proof of equivalence of the second law of thermodynamics and the Carnot theorem. The Carnot theorem provides not only the more transparent mathematical representation, but also an important concept that is not transparent in the Clausius and Kelvin principles, since it shows that *the efficiencies of irreversible cycles cannot exceed the efficiency of the reversible cycles.* Furthermore, we now see that the second law of thermodynamics, which is based on everyday experience and therefore less transparent mathematically than the Carnot theorem, is given a clearer mathematical representation by means of Carnot's theorem. We reiterate that both the second law of thermodynamics and the Carnot theorem are phrased about cyclic processes, since cyclic processes are the only kind of processes in which the system is assured of returning to the original state without fail and thus the existence of conserved quantities characteristic of the system is implied without recourse to any other measure to ascertain their conserved nature. And conserved quantities are certainly convenient to use in formulating a theory of macroscopic processes.

From the Carnot theorem it is possible to deduce that *a reversible work is a maximum work possible.* To show it let us assume that there are two engines operating between two heat reservoirs and both of them absorb heat Q_1 from the hotter reservoir and emit heat Q_2 to the colder reservoir. The work performed by the reversible engine is W and that by the irreversible engine is W'.

Then according to the Carnot theorem there holds the inequality

$$\eta_{rev} = \frac{-W}{Q_1} \geq \frac{-W'}{Q_1} = \eta_{irr}, \tag{4.22}$$

which means that

$$-W \geq -W'. \tag{4.23}$$

This proves the statement. It is seen as a corollary to the Carnot theorem. By the sign convention on work, $-W$ is positive since W is the work performed by the system, and similarly for $-W'$. Therefore, Inequality (4.23) implies that if the process is irreversible there is an amount of work unavailable to the given task of the process. This is related to the energy dissipation accompanying an irreversible process. It is convenient to introduce the unavailable work W_{ua} by the relation

$$W = W' + W_{ua}. \tag{4.24}$$

The precise nature of unavailable work depends on the irreversible process in question and the system as well as the device with which the work is performed. This unavailable work must be elucidated if the theory of irreversible processes is the aim.

4.3 Thermodynamic Temperature

The Carnot theorem makes it possible to give the second law of thermodynamics a mathematical representation. To achieve the desired aim it is necessary to lay foundations for it. First of all, let us consider implications of the Carnot theorem. One of the most important is the notion of thermodynamic temperature which was first deduced from the Carnot theorem by Lord Kelvin. We consider the gist of it here. Since the efficiency of reversible Carnot cycles is independent of materials and the modes of operation, but depends only on the temperatures θ_1 and θ_2 of the heat reservoirs according to Carnot's theorem, we see that the ratio of Q_2 to Q_1 is a function of θ_1 and θ_2 only:

$$\frac{Q_2}{Q_1} = f(\theta_2, \theta_1). \tag{4.25}$$

Since we may write the left-hand side in the form

$$\frac{Q_2}{Q_1} = \frac{Q_2}{Q_0} \cdot \frac{Q_0}{Q_1} = f(\theta_2, \theta_0) f(\theta_0, \theta_1),$$

we conclude that $f(\theta_2, \theta_1)$ is a function with the following property:

$$f(\theta_2, \theta_1) = f(\theta_2, \theta_0) f(\theta_0, \theta_1). \qquad (4.26)$$

Let $\theta_2 = \theta_1$. Then

$$f(\theta_1, \theta_0) = \frac{1}{f(\theta_0, \theta_1)}. \qquad (4.27)$$

The properties of f described in Eqs. (4.26) and (4.27) imply that there exists a function of θ, $\phi(\theta)$, such that

$$f(\theta_2, \theta_1) = \frac{\phi(\theta_2)}{\phi(\theta_1)} \qquad (4.28)$$

and consequently it is possible to write

$$\frac{Q_2}{Q_1} = \frac{\phi(\theta_2)}{\phi(\theta_1)}. \qquad (4.29)$$

The function $\phi(\theta)$ is arbitrary. It will be denoted by T:

$$T = \phi(\theta). \qquad (4.30)$$

This is called the thermodynamic temperature. The temperature based on the Carnot theorem must be universal since the Carnot theorem is universal owing to the fact that *the reversible efficiencies are independent of the modes of operation and the working substances.* In the manner shown, the Carnot theorem can serve as the basis of introducing a universal thermodynamic temperature.[8] Since the efficiency calculated for a reversible cycle in Eq. (4.10) is expressed in absolute temperature, it is appropriate to regard θ here as a temperature in the absolute scale. In this event, the efficiency formula for a reversible cycle in Eq. (4.10) suggests that $\phi(\theta)$ must be linearly proportional to θ, and the proportionality constant may be taken such that θ coincides with the thermodynamic temperature T:

$$T = \theta. \qquad (4.31)$$

In this manner, the thermodynamic temperature is made to coincide with the absolute temperature based on the ideal gas temperature, and they now can be interchangeably used. It then is possible to express the efficiency of

---

[8]The existence of universal thermodynamic temperature presupposes a reversible cycle or process. Without it the temperature defined is no longer universal and hence generally dependent on the substance or the system.

a reversible cycle in terms of the absolute temperature (or thermodynamic temperature):

$$\eta = \frac{Q_1 - Q_2}{Q_1} = \frac{T_1 - T_2}{T_1}. \tag{4.32}$$

It is important to recall that T_1 and T_2 are originally the temperatures of the heat reservoirs and, only in the case of a reversible cycle, can they be regarded as the temperatures of the system which is in thermal equilibrium with the heat reservoirs in various steps of operation in the cycle.

4.4 Entropy and Calortropy

4.4.1 *Clausius Inequality*

With the Carnot efficiency for reversible cycles expressed in terms of the thermodynamic temperature, the Carnot theorem now acquires a firmer mathematical grip on the second law of thermodynamics. To pursue this line further, which eventually gives rise to the Clausius inequality, it is convenient at this point to adopt a sign convention for heat transfer in a way consistent with the sign convention for work introduced earlier.

Sign convention: *If heat is transferred to the system, Q is taken positive whereas, if it is transferred out of the system, Q is taken negative.*

With this convention, we may write the efficiency of a reversible cycle in the form

$$\frac{Q_1 - (-Q_2)}{Q_1} = \frac{T_1 - T_2}{T_1}.$$

Rearranging the terms in this equation we find for the reversible cycle

$$\frac{Q_1}{T_1} + \frac{Q_2}{T_2} = 0. \tag{4.33}$$

If the amounts of heat injected into and rejected by or withdrawn from the system (cycle) are infinitesimal, we may write Eq. (4.33) in a differential form in terms of differential elements dQ_i:

$$\frac{dQ_1}{T_1} + \frac{dQ_2}{T_2} = 0. \tag{4.34}$$

This equation may be further generalized by the following mathematical device. A reversible cycle may be regarded as consisting of a large number

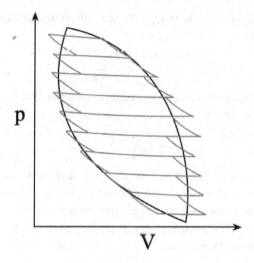

Fig. 4.5 Decomposition of a Carnot cycle into infinitesimal Carnot cycles $\{i\}$. The ith cycle takes heat dQ_i from the adjacent $(i-1)$th cycle and emits heat dQ_{i+1} to its immediate neighbor, the $(i+1)$th cycle in the sequence.

of infinitesimal reversible cycles as shown in Fig. 4.5. The horizontal lines represent the isotherms, and the vertical lines the adiabatics. The cycle i takes an infinitesimal amount of heat dQ_i from the $(i-1)$th cycle and emits an infinitesimal amount of heat dQ_{i+1} to the $(i+1)$th cycle. The temperatures of the isotherms are denoted by T_i. For the whole collection of the infinitesimal cycles we may then write

$$\sum_i \frac{dQ_i}{T_i} = 0, \tag{4.35}$$

which is a generalization of Eq. (4.34). If the number of infinitesimal cycles is taken to infinity the summation may be replaced with an integral over the whole cycle:

$$\oint \frac{dQ_{rev}}{T} = 0. \tag{4.36}$$

It must be emphasized that this equation holds for reversible cycles only and dQ_{rev} denotes a differential heat transfer in a reversible process. They are cycles in which processes are performed such that the system is in continuous equilibrium with the surroundings or there is no energy dissipation, the meaning of which will be elucidated later more precisely.

Let us now consider a more general case involving an irreversible cycle. In this case, the Carnot theorem reads

$$\eta_{irr} = \frac{Q_1 - Q_2}{Q_1} \leq \frac{T_1 - T_2}{T_1} = \eta_{rev}. \tag{4.37}$$

By applying the same method of reasoning as for that leading to Eq. (4.36) we obtain the inequality

$$-\oint \frac{dQ}{T} \geq 0. \tag{4.38}$$

This inequality, first obtained by Clausius, is called the Clausius inequality. It is important to remember that T is the temperature associated with the reversible cycle and thus equilibrium, and even if the process in question is irreversible the temperature factor T in the inequality is still that of the reversible process corresponding to the irreversible process of interest; see Eq. (4.37).

Clausius called dQ the compensated heat since it is a heat transfer between the system and the surroundings (heat reservoirs) that occurs irreversibly. He also identified another basic quantity that he called the uncompensated heat, and denoted it by N. This quantity, which is always positive and vanishes if the process is reversible, is equal in value to the closed contour integral (simply called the circular integral) in the inequality (Eq. (4.38)):

$$N = -\oint \frac{dQ}{T} \geq 0. \tag{4.39}$$

Before we examine this form of the Clausius inequality in detail, it is useful in historical deference to the development made by Clausius regarding a special case of his inequality. This special case gives rise to equilibrium thermodynamics.

4.4.2   *Entropy*

Let us return to Eq. (4.36), which holds for reversible processes. Although Clausius discovered the existence of uncompensated heat N, its precise nature was not known sufficiently well to him at that time.[9] To make some

---

[9] In connection with this line of work and much before he put forward the notion of entropy, Clausius had made use of the notion of *disgregation*, which appeared to be akin to a quantity he later called *entropy*. However, he never developed and quantified the notion beyond the entropy and into the regime of irreversible processes.

progress in the absence of the detailed knowledge of N Clausius considered the case of reversible processes[10] for which $N = 0$ identically. Since the line integral of the compensated heat thus obtained is nothing other than Eq. (4.36) and meets the mathematical criterion for exact differentials, it led him to conclude that there exists an exact differential dS such that

$$dS = \frac{dQ_{rev}}{T} \qquad (4.40)$$

in the space of thermodynamic variables characteristic of the system for the reversible process considered. Moreover, it can be shown that there exists such a function S for any reversible process since the Carnot theorem holds for all reversible processes. Clausius called S the entropy[11] of the system, and it is a state function in the space of thermodynamic variables for the system, which is spanned by the internal energy and volume and, in the case of a mixture, also by the concentrations of species.

The second law of thermodynamics, expressed by Eq. (4.36) for a reversible process, is now represented by the vanishing circular integral

$$\oint dS = 0. \qquad (4.41)$$

It should be emphasized that this vanishing integral is true only if the process is reversible. It must be also noted that dQ_{rev} denotes a non-exact differential for heat change although the symbol đ is not used in place of the symbol d. When dQ_{rev} is divided by the absolute temperature the result becomes an exact differential. Such factors which make non-exact differentials exact are called integrating factors, and $1/T$ is an integrating factor in the case of dQ_{rev}. Therefore the entropy change for a reversible process $A \to B$ is obtained on integration of Eq. (4.40):

$$\Delta S = S_B - S_A = \int_A^B dS = \int_A^B \frac{dQ_{rev}}{T}, \qquad (4.42)$$

where the integration is over the reversible path. For an isolated system there is no heat put in or taken out of the system and hence

$$dQ_{rev} = 0,$$

---

[10]Reversible processes may be defined as the processes in which the uncompensated heat N vanishes everywhere. This definition is preferable to the definition based on the process of continuous quasi-equilibrium introduced in Section 2.5.

[11]See Footnote 3 for the reference.

which then implies

$$\Delta S = 0.$$

In other words,

$$S_A = S_B$$

for isolated systems at equilibrium. It should be clearly understood that the entropy S is defined for reversible processes only.

The differential for S given in Eq. (4.40) can be combined with the first law of thermodynamics to yield the fundamental relation for thermodynamics of reversible processes or equilibrium

$$dE = T \, dS - p \, dV, \tag{4.43}$$

in the case of pressure–volume work alone. If there are other kinds of work besides the pressure–volume work then the differential form for energy may be written as

$$dE = TdS - pdV + \sum_i X_i \, dx_i, \tag{4.44}$$

where X_i are forces and x_i are displacements whose examples were given in Chapter 2. We will consider thermodynamics in the cases of such forms of work in later chapters. The differential forms (Eqs. (4.43) or (4.44)) are often referred to the mathematical expression for the second law of thermodynamics, but it is only for reversible processes or systems in equilibrium.

4.4.3 *Calortropy*

Before we discuss the application of the notion of entropy in thermodynamics it is useful to examine more closely the form of the Clausius inequality (Eq. (4.39)), to learn about the nature of the uncompensated heat N. It must be associated with the unavailable work appearing in Eq. (4.24) since the numerator in the right member of Eq. (4.14) is equal to $-W'$ which is less than the work $-W$ performed by the reversible cycle corresponding to the right member of Inequality (4.37). By adding a term $-W_{ua}/Q_1$ to the left member of Inequality (4.37), which thereby is made an equation, we can cast Inequality (4.37) in the form of an equation

$$-\frac{W_{ua}}{T_2} = -\oint \frac{dQ}{T}. \tag{4.45}$$

If the left-hand side is identified with the uncompensated heat

$$N \equiv -\frac{W_{ua}}{T_2} > 0, \tag{4.46}$$

then the Clausius inequality (Eq. (4.39)) is recovered and a little more definite meaning of the uncompensated heat is attained. Clearly, the uncompensated heat so identified is always positive in the sign convention adopted for work and vanishes only if the process is reversible. This consideration made here indicates that *the compensated and uncompensated heats are two independent physical quantities which exactly balance each other for the cycle.* In other words, we may write Eq. (4.45) in the form

$$\oint \frac{dQ}{T} + N = 0. \tag{4.47}$$

The compensated and uncompensated heats are characteristic of the system and the surroundings in a given heat transfer process involved between them. The symbol N clearly is not another way of writing the circular integral on the right-hand side of Eq. (4.45), which appears to be an identity at a quick glance. Neither is the circular integral on the right-hand side of Eq. (4.45) the definition of N, which should be elucidated on its own right if an irreversible process was an object of study. It is important to recognize that the Clausius inequality is derived from the Carnot theorem. Therefore the temperature T appearing in the Clausius inequality or Eq. (4.47) is the temperature expressing the efficiency of the reversible cycle with which the efficiency of an irreversible cycle is compared.

The uncompensated heat may be expressed as an integral over the path of the cycle

$$N = \oint dN \geq 0, \tag{4.48}$$

where dN is such that $(dN/dt) \geq 0$ everywhere in the cycle, since otherwise it is possible to construct a cycle, which contravenes the positivity of the integral and thus the second law of thermodynamics. Then the Clausius inequality can be written as

$$\oint \left(\frac{dQ}{T} + dN \right) = 0. \tag{4.49}$$

This is a mathematical representation for the second law of thermodynamics, which was literally stated in the forms of the Clausius and Kelvin principles. It now is written in a form of equation instead of an inequality initially obtained by Clausius, and such an equation is realized by treating the compensated and uncompensated heats as two independent physical quantities on the equal footing. Although it may look as a simple and trivial rewriting of the Clausius inequality, it provides *a fresh way of looking at the thermodynamics of irreversible processes and generalizing the classical equilibrium thermodynamics to nonequilibrium*. We now examine how the classical thermodynamics of reversible processes or equilibrium can be recovered and made free from the troublesome features, especially the one related to the concept of entropy, which is defined only when the system is in equilibrium, but, as will be shown below, has been nevertheless used in the context in which the system is away from equilibrium.

As in the case of Eq. (4.36) holding for reversible processes, Eq. (4.49) implies that even if the process is irreversible there still exists a state function Ψ in the space of thermodynamic variables whose differential form is given by

$$d\Psi = \frac{dQ}{T} + dN \qquad (4.50)$$

and vanishes on integration over the cycle:

$$\oint d\Psi = 0. \qquad (4.51)$$

This is reminiscent of the vanishing circular integral of dS in Eq. (4.41). The vanishing circular integral (Eq. (4.51)) implies that $d\Psi$ is an exact differential in the thermodynamic space which may be spanned by the internal energy, volume, concentrations of species, and macroscopic variables characteristic of the nonequilibrium system in which irreversible processes occur. The precise nature of the thermodynamic space should be elucidated upon detailed examination of the irreversible processes in hand, but it is clear that the space should be spanned by macroscopic variables describing the irreversible process in hand in addition to the usual variables characterizing the equilibrium system of temperature T, such as the internal energy, volume, and concentrations of species involved. This state function is called the calortropy.[12] Since the process is irreversible Ψ is a generalized form of the Clausius entropy S, which holds for reversible

---

[12]It is a composite word meaning heat (*calor*) evolution (*tropy*). It is akin to Carnot's *calorique*, which he used in his *Réflexion* to mean a quantity that is similar to heat, but not quite the same as heat. Since *calorie* is already in use, the term *calorique* is inappropriate to use. The term coined thus seems to be suitable. Some research workers in irreversible thermodynamics use the term nonequilibrium entropy to mean what is essentially the calortropy here, but the latter is simpler and more concise than the former.

processes only. The vanishing circular integral (Eq. (4.51)) is a mathematical representation of the second law of thermodynamics for the irreversible cyclic process under consideration. For the same cyclic process the first law of thermodynamics is also expressible as a vanishing circular integral

$$\oint dE = 0.$$ (4.52)

This pair of circular line integrals is over the same irreversible path in the thermodynamic space for the system. We recall that although dW and dQ are not exact differentials, their sum dE is an exact differential. It is then interesting to see that, just as in the case of the first law of thermodynamics, although dQ/T and dN are not exact differentials if the process is irreversible, their sum $d\Psi$ is an exact differential, and the first and second laws of thermodynamics are mathematically expressible as a pair of vanishing circular integrals if the process is cyclic. We summarize these results as a theorem:

Theorem. *In the case of a cyclic process, regardless of whether it is reversible or irreversible, the first and second laws of thermodynamics are, respectively, expressible as a pair of vanishing cyclic integrals in the thermodynamic space:*

$$\oint dE = 0,$$ (4.53)

$$\oint d\Psi = 0.$$ (4.54)

4.4.4 *Inequalities of Entropy and Calortropy*

Since $(dN/dt) \geq 0$ where dt is an infinitesimal time interval, for an infinitesimal interval of the cycle there holds the inequality

$$d\Psi \geq \frac{dQ}{T}.$$ (4.55)

Since $(dN/dt) = 0$ if the process is reversible, not only the equality holds in Eq. (4.55) but also $d\Psi$ becomes identical with dS, namely,

$$d\Psi_{rev} = \frac{dQ_{rev}}{T}$$
$$= dS,$$ (4.56)

where Ψ_{rev} and Q_{rev} are the calortropy and heat transfer accompanying the reversible process.

If Inequality (4.55) is integrated over a finite irreversible process from state 1 to state 2 there follows the inequality

$$\Delta\Psi \geq \int_{1(irr)}^{2} \frac{dQ}{T}, \tag{4.57}$$

where

$$\Delta\Psi = \int_{1(irr)}^{2} d\Psi. \tag{4.58}$$

Inequality (4.57) also follows directly from Eq. (4.50) if the latter is integrated over the same interval:

$$\Delta\Psi = \int_{1(irr)}^{2} \frac{dQ}{T} + \int_{1(irr)}^{2} dN$$

$$\geq \int_{1(irr)}^{2} \frac{dQ}{T}. \tag{4.59}$$

The equality holds, only if the process is reversible in which case we may write

$$\Delta\Psi_{rev} = \Delta S = \int_{1(rev)}^{2} \frac{dQ_{rev}}{T}. \tag{4.60}$$

If the system is isolated, then there is no heat exchange between the system and the surroundings. In this case, dQ vanishes identically over the process and Inequality (4.57) gives rise to the inequality for the calortropy

$$\Delta\Psi \geq 0, \tag{4.61}$$

where the equality holds, only if the process is reversible. This is a special case of Eq. (4.60) for which

$$\Delta\Psi_{rev} = \Delta S = 0. \tag{4.62}$$

For irreversible processes in an isolated system

$$\Delta\Psi > 0. \tag{4.63}$$

Therefore the calortropy tends toward a maximum and attains a maximum value identical with the entropy at equilibrium. This conclusion should be compared with the famous statement for the second law of thermodynamics made by Clausius in 1865: *The entropy of the world tends towards a maximum.* Since by definition the entropy is an attribute of the system at equilibrium, it makes sense that calortropy should replace the term entropy in the statement in our opinion. We elaborate on this assertion with the following examination.

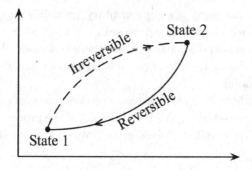

Fig. 4.6   A cyclic path consisting of an irreversible and a reversible segment.

Suppose a system has irreversibly and spontaneously reached a final state from an initial state as in Fig. 4.6. If the system is not closed, it is always possible to make the system return to the original state through a reversible path, provided that the system is properly compensated to do so. For such a system it is always possible to construct a cyclic process consisting of two segments, one irreversible and the other reversible, as depicted in Fig. 4.6, and find from Eq. (4.51)

$$\Delta \Psi = \int_{1(irr)}^{2} d\Psi = \int_{1(rev)}^{2} dS = \Delta S. \qquad (4.64)$$

This equation together with Inequality (4.57) implies the inequality

$$\Delta S \geq \int_{1(irr)}^{2} \frac{dQ}{T}. \qquad (4.65)$$

This is the Clausius–Duhem inequality known in thermodynamics as a mathematical form of the second law of thermodynamics. This means that Inequality (4.57) implies the Clausius–Duhem inequality. We emphasize that Eq. (4.64) holds only if there exists a reversible segment in the cycle, which in this particular case requires that the system be not closed since some sort of compensation is necessary for the system to return to the original state from which it irreversibly departed. An isolated system that has irreversibly and spontaneously reached a final state cannot spontaneously return to the initial state since the necessary compensation can only come from the surroundings for the system to do so. The Clausius–Duhem inequality (Eq. (4.65)) is mathematically and conceptually impeccable, but it does not mean that for an isolated system there holds the inequality

$$\Delta S \geq 0, \qquad (4.66)$$

since there does not exist a complementary reversible process Γ_{rev} that makes up a cycle with irreversible process Γ_{irr}, if the system is isolated. The existence of such a path Γ_{rev} implies that there is a mode of compensation from the surroundings. If so, the system is no longer isolated and inequality $\Delta S > 0$ is misleading.

Inequality (4.66), in fact, causes a considerable conceptual difficulty when closely examined, since ΔS is not defined if the process is irreversible, and $\Delta S = 0$ for a reversible process in an isolated system, since

$$\Delta S = \int_{1(rev)}^{2} \frac{dQ_{rev}}{T} = 0 \qquad (4.67)$$

identically because $dQ_{rev} = 0$ for an isolated system. Therefore it cannot be concluded that $\Delta S > 0$ if there is an irreversible process in an isolated system. The confusing part of Inequality (4.66) is the conclusion that $\Delta S > 0$ although the system is isolated so that $dQ_{rev} = 0$ and thus, by definition, $\Delta S = 0$. This difficulty disappears if ΔS is replaced by $\Delta \Psi$ so that Inequality (4.61) holds for the case of an isolated system undergoing an irreversible process; Inequality (4.61) has no such problem as for Inequality (4.66) for the reason that it is not equal to zero even if the system is isolated and thus $dQ_{rev} = 0$ because of a nonvanishing uncompensated heat for an irreversible process.

The laxity of the Clausius statement has been a source of debate[13] and confusion in the fine points of thermodynamics, but we now seem to have a way to avoid such confusions. The discussion presented here on the basis of Eq. (4.64) and Inequality (4.65) suggests that the conventional Clausius–Duhem inequality for the Clausius entropy should be phrased in terms of the calortropy over the irreversible segment as in Inequality (4.57) and, in the case of an isolated system, as in Inequality (4.61). Inequality (4.61) is easier to comprehend than the Clausius–Duhem inequality for an isolated system because it does not present a difficulty that there holds Eq. (4.67) for a reversible process in an isolated system. We further elaborate on this point.

Inequality (4.66) may be understood in the following sense. Suppose the irreversible process Γ_{irr} in Fig. 4.6 occurs when the system is isolated and the reversible process Γ_{rev} occurs when the system is not isolated. In other words, after the irreversible process Γ_{irr} is completed, the system

---

[13]See, for example, J. Kestin, *The Second Law of Thermodynamics* (Dowden, Hutchinson, & Ross, Stroudburg, PA, 1976).

is allowed to interact with its surroundings and made to return reversibly from state B to state A. In this case

$$\Delta S = \Delta \Psi \left(\Gamma_{irr} \right) = \int_{A(irr)}^{B} dN \geq 0,$$

where the equality holds if Γ_{irr} is reversible. This equation, as Eq. (4.64), allows to calculate $\Delta \Psi (\Gamma_{irr})$ in terms of ΔS over a complementary reversible process, but only implies that if an irreversible process occurs in an isolated system then the ΔS of a reversible process that is complementary to the irreversible process in question and occurs in interaction with the surroundings is positive. This is probably what Clausius should have meant by Inequality (4.66).

If the nature of uncompensated heat is elucidated in detail, it should be possible to calculate $\Delta \Psi$. We briefly comment on this aspect. Let us imagine that there is a finite segment of an irreversible process $1 \to 2$. Then, on integrating Eq. (4.50), there follows

$$\Delta \Psi = \int_{1(irr)}^{2} \frac{dQ}{T} + \Xi_g, \qquad (4.68)$$

where

$$\Xi_g = \int_{1(irr)}^{2} dN \geq 0. \qquad (4.69)$$

This inequality is a form of the second law of thermodynamics. If it is possible to take the system to the initial state 1 through a reversible process then the entropy change for the reversible process ΔS is given by the equation

$$\Delta S = \int_{1(irr)}^{2} \frac{dQ}{T} + \Xi_g \qquad (4.70)$$

as was originally suggested by Clausius. Provided we know how to compute Ξ_g for the process, it is possible to compute the integral in Eq. (4.70) by computing ΔS for a reversible process complementary to the irreversible process in question, or Ξ_g from the knowledge of the integral in Eq. (4.70) and ΔS.

Let us examine the significance of Eq. (4.70) with the following example. A metal bar at temperature T_2 is separated by an insulator C from another of the same metal at temperature T_1, which is lower than T_2. The insula-

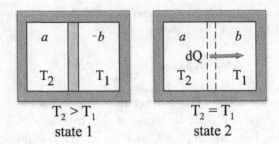

Fig. 4.7   Two metal bars in contact at two different temperatures which go through an irreversible process of heat transfer.

tor is then removed and the metal bars are allowed to interact thermally (Fig. 4.7). They will reach equilibrium at which point the temperature will be $\frac{1}{2}(T_1 + T_2)$. The systems are subsequently brought to the original states by a reversible process by making them interact with the surroundings. This reversible process is accompanied by an entropy change ΔS in the system. Since the irreversible process occurs in an isolated condition the integral in Eq. (4.70) is equal to zero and ΔS is given by

$$\Delta S = \Xi_g \geq 0. \qquad (4.71)$$

The entropy change ΔS for the reversible process can be easily calculated:

$$\Delta S = C_p \ln\left(\frac{T_1 + T_2}{2\sqrt{T_1 T_2}}\right), \qquad (4.72)$$

for which we have assumed, for the simplicity of the result, that the specific heat C_p is independent of temperature. Since there holds the inequality

$$\frac{1}{4}\left(T_1 + T_2\right)^2 \geq T_1 T_2,$$

we conclude that indeed

$$\Delta S \geq 0$$

and Inequality (4.71) is satisfied, indicating that the irreversible process in question is spontaneous with the value of the global uncompensated heat estimated for the irreversible process as in Eq. (4.72). Therefore, we see that Eq. (4.70) can be used for estimating the uncompensated heat Ξ_g from the knowledge of ΔS for a reversible process complementary to the irreversible process in question. We emphasize that ΔS in

Eq. (4.72) is for a reversible process in a system interacting with the surroundings.

4.5 Carnot's Theorem and Real Gases

We have calculated the efficiency of a reversible Carnot cycle when the working substance is ideal. However, the working substance is not ideal in practice but real, since the molecules do interact with each other and the ideal gas equation of state holds only in the low pressure limit. If a real gas is taken for the working substance, some steps in the calculation we have performed for the Carnot cycle are not valid. There then arises the question as to the generality of the result obtained for the efficiency. The result, however, holds in general for real working substances, and we would like to prove this statement for the Carnot cycle depicted in Fig. 4.4 in the following.

The equation of state for 1 mole of a real gas may be given in the form

$$p = \frac{RT}{V} + f(V, T), \qquad (4.73)$$

where $f(V, T)$ is a function of V and T which we do not have to specify except that it is a well-behaved function and piecewise integrable. Note that in the case of real gases

$$\left(\frac{\partial E}{\partial V}\right)_T \neq 0.$$

This requires a new mode of calculation for the heat change accompanying an isothermal expansion, which is different from that used for an ideal gas.

The work of isothermal expansion is

$$W_{ij} = -\int_{V_i}^{V_j} p \, dV \ (i = 1, 3; \ j = 2, 4), \qquad (4.74)$$

which may be explicitly calculated if the integration is performed with Eq. (4.73) substituted into it. Such calculation, however, is not necessary for our purpose here.

By regarding the internal energy as a function of V and T we may write

$$dQ = dE + p \, dV$$

$$= C_v \, dT + \left[T\left(\frac{\partial p}{\partial T}\right)_V - p\right] dV + p \, dV. \qquad (4.75)$$

Therefore, since $dQ = 0$ for an adiabatic process, the work of adiabatic expansion (or compression) may be written in the form

$$W_{23} = -\int_{V_2}^{V_3} p\,dV = -\int_{T_1}^{T_2} C_v\,dT + \int_{V_2}^{V_3} \left[p - T\left(\frac{\partial p}{\partial T}\right)_V \right]_{T_2} dV,$$

(4.76)

$$W_{41} = -\int_{V_4}^{V_1} p\,dV = \int_{T_1}^{T_2} C_v\,dT + \int_{V_4}^{V_1} \left[p - T\left(\frac{\partial p}{\partial T}\right)_V \right]_{T_1} dV.$$ (4.77)

The total work is then the sum of the four components calculated above:

$$\begin{aligned} W &= W_{12} + W_{23} + W_{34} + W_{41} \\ &= -\int_{V_1}^{V_2} p(T_1)\,dV - \int_{V_3}^{V_4} p(T_2)\,dV + \int_{V_2}^{V_3} \left[p - T\left(\frac{\partial p}{\partial T}\right)_V \right]_{T_2} dV \\ &\quad + \int_{V_4}^{V_1} \left[p - T\left(\frac{\partial p}{\partial T}\right)_V \right]_{T_1} dV. \end{aligned}$$ (4.78)

By the first law of thermodynamics

$$\oint dE = 0$$

for the cycle, and we obtain

$$-W = Q_1 - Q_2.$$

Therefore the efficiency is

$$\eta = \frac{-W}{Q_1} = 1 - \frac{Q_2}{Q_1},$$

which takes the same form as for the ideal working substance. We would like to show that, as is for the ideal gas, the efficiency for the real gas is still given by

$$\eta = 1 - \frac{T_2}{T_1}.$$

For the purpose we first observe that for an isothermal process

$$dQ = T\left(\frac{\partial p}{\partial T}\right)_V dV,$$

which follows from Eq. (4.75) if $dT = 0$. Therefore the heat changes accompanying the isothermal expansion and compression in the cycle are

$$Q_1 = \int_1^2 dQ = \int_{V_1}^{V_2} dV\, T_1 \left(\frac{\partial p}{\partial T} \right)_V (T_1),$$

$$Q_2 = \int_4^3 dQ = \int_{V_4}^{V_3} dV\, T_2 \left(\frac{\partial p}{\partial T} \right)_V (T_2),$$

(4.79)

where we have used the sign convention on heat emission and absorption. Along the adiabatics

$$\frac{C_v}{T} dT + \left(\frac{\partial p}{\partial T} \right)_V dV = 0,$$

(4.80)

which follows from Eq. (4.75) if $dQ = 0$. Moreover, since

$$dC_v(T,V) = \left(\frac{\partial C_v}{\partial V} \right)_T dV \quad \text{at } T = \text{constant},$$

we obtain

$$C_v(T,V) = C_v^0 + \int^V dV \left(\frac{\partial C_v}{\partial V} \right)_T$$

$$= C_v^0 + \int^V dV\, T \left(\frac{\partial^2 p}{\partial T^2} \right)_V$$

$$= C_v^0 + \int^V dV\, T \left(\frac{\partial^2 f}{\partial T^2} \right)_V \quad (T = \text{constant}),$$

(4.81)

where C_v^0 is generally a function of T, but we will consider the case of constant C_v^0. Substitution of Eq. (4.81) into Eq. (4.80) and use of the equation of state (Eq. (4.73)) yield

$$d \left[R \ln(T^\alpha V) + \int^V dV \left(\frac{\partial f}{\partial T} \right)_V \right] = 0,$$

(4.82)

where $\alpha = C_v/R$. Integrating it, we obtain

$$R \ln(T^\alpha V) + \int^V dV \left(\frac{\partial f}{\partial T} \right)_V = \text{constant}.$$

(4.83)

Since from the equation of state

$$\left(\frac{\partial p}{\partial T} \right)_V = \frac{R}{V} + \left(\frac{\partial f}{\partial T} \right)_V,$$

using Eqs. (4.79) and (4.83) we find

$$Q_1 = RT_1 \ln\left(\frac{T_2}{T_1}\right)^\alpha, \quad Q_2 = RT_2 \ln\left(\frac{T_2}{T_1}\right)^\alpha \qquad (4.84)$$

and consequently

$$\frac{Q_2}{Q_1} = \frac{T_2}{T_1}. \qquad (4.85)$$

Finally, we thus find

$$\eta = 1 - \frac{T_2}{T_1}, \qquad (4.86)$$

even if the working substance is real. This result also holds even if the assumption on constant C_v is removed. For the following holds true:

$$\int^T \frac{C_v^0}{T} dT + \int^V dV \left(\frac{\partial p}{\partial T}\right)_V (T) = \text{constant}. \qquad (4.87)$$

By the calculation performed above, we have proved that the efficiency of a reversible Carnot cycle of a real fluid for the working substance obeying the equation of state (Eq. (4.73)) still depends only on the temperatures of the heat reservoirs between which the cycle operates.

4.6   Examples of Other Cycles

Carnot's cycle was the precursor to other cycles invented later. Some examples will be shown in this section. They underlie some of practical engines used in machines including automobiles. The Diesel cycle, for example, is one typical example of application.

4.6.1   *Rankine Cycle*

The Scottish engineer W. J. M. Rankine, a contemporary of Kelvin and an important pioneer in thermodynamics, invented the Rankine cycle, which basically describes the working of steam engines. It may be schematically represented by the diagram in the $p-V$ plane in Fig. 4.8.

4.6.2   *Otto Cycle*

The German engineer N. Otto built an engine in 1876 invented by the Frenchman A. Beaude Rochas in 1862. Otto's engine became the prototype of automobile engines made in the United States. Gasoline engines may be

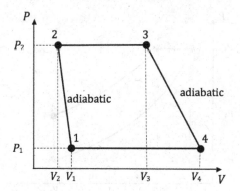

Fig. 4.8   Rankine cycle: step 2–3, isobaric expansion; step 3–4, adiabatic expansion; step 4–1, isobaric compression; step 1–2, adiabatic compression.

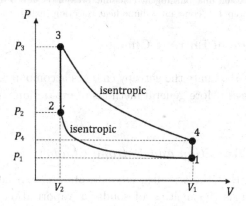

Fig. 4.9   Otto cycle: step 1–2, adiabatic compression (isentropic); step 2–3, constant volume heating; step 3–4, adiabatic expansion (isentropic); step 4–1, constant volume cooling.

represented by the Otto cycle which is described by the diagram in the p–V plane in Fig. 4.9.

4.6.3   *Diesel Cycle*

In about 1900, the German engineer R. Diesel built engines now known as Diesel engines. The Diesel engines may be described by the following p–V diagram depicted in Fig. 4.10. It is called the Diesel cycle and is currently in use for various automobiles and power generators. The efficiency of the Diesel cycle is greater than the efficiency of an Otto cycle at the same peak pressure.

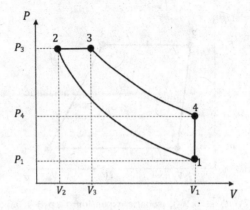

Fig. 4.10   Diesel cycle: step 1–2, adiabatic (isentropic) compression; step 2–3, isobaric expansion (fuel injection and burning at constant pressure); step 3–4, adiabatic (isentropic) expansion; step 4–1, constant volume heat rejection (exhaust and intake of air).

4.7   Calculation of Entropy Change

In this section, we evaluate the entropy changes accompanying some simple reversible processes. More general methods of evaluation will be discussed in later chapters.

4.7.1   *Phase Transition and Entropy Change*

It is well known that the so-called latent heats are associated with phase transitions, such as the melting of solids or vaporization of liquids and solids. Such heats are called more specifically the heat of fusion (melting), the heat of vaporization, and so on. There are entropy changes associated with such phase transitions. They are called the entropy of fusion, the entropy of vaporization, and so on, and are easily evaluated from the latent heats. Let us recall that since

$$dQ = T \, dS$$

and

$$H = E + pV,$$

if the process involved is reversible, combining them with the first law of thermodynamics, Eq. (3.9), yields

$$dH = T \, dS + V \, dp \tag{4.88}$$

for the enthalpy change.

Using this equation at $p = \text{constant}$, we can calculate the entropy change ΔS_{lat} corresponding to the latent heat ΔH_{lat} of a phase transition at constant p:

$$\Delta S_{lat} = \frac{\Delta H_{lat}}{T}, \tag{4.89}$$

where T is the phase transition temperature. Note that ΔS_{lat} and ΔH_{lat} are discontinuous changes between two phases at T.

4.7.2 Entropy Changes of an Ideal Gas

For reversible processes the entropy change may be written

$$dS = \frac{1}{T}(dE + p\, dV), \tag{4.90}$$

which results from Fundamental Relation (3.9). Since in the case of an ideal gas E is a function of T only, Eq. (4.90) may be cast in the form

$$dS = \frac{1}{T}c_v\, dT + RV^{-2}\, dV. \tag{4.91}$$

Here we are considering 1 mole of an ideal gas and c_v denotes the molar specific heat at constant volume. Suppose a reversible process takes the system from state (T_1, V_1) to state (T_2, V_2). The entropy change for this process may be evaluated as follows: The temperature of the system is increased from T_1 to T_2 at $V = V_1$ and then the volume is increased from V_1 to V_2 at $T = T_2$. This process is depicted in Fig. 4.11. The entropy change for the isochoric process $(T_1, V_1) \rightarrow (T_2, V_1)$ is

$$(\Delta S)_{isochoric} = \int_{T_1}^{T_2} dT\, \frac{c_v(T)}{T}$$

$$= c_v \ln\left(\frac{T_2}{T_1}\right), \tag{4.92}$$

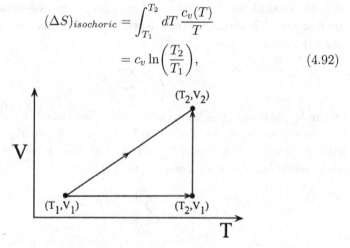

Fig. 4.11 A reversible path in $T-V$ plane and its distortion into an isochoric and an isothermal path.

where the second line holds if C_v is independent of T. The entropy change for the isothermal process $(T_2, V_1) \rightarrow (T_2, V_2)$ is

$$(\Delta S)_{isothermal} = R \int_{V_1}^{V_2} dV \frac{1}{V} = R \ln\left(\frac{V_2}{V_1}\right). \tag{4.93}$$

Combining Eqs. (4.92) and (4.93) yields the entropy change for the process $(T_1, V_1) \rightarrow (T_2, V_2)$:

$$\Delta S = (\Delta S)_{isochoric} + (\Delta S)_{isothermal}$$
$$= \int_{T_1}^{T_2} dT \frac{c_v(T)}{T} + R \ln\left(\frac{V_2}{V_1}\right)$$
$$= c_v \ln\left(\frac{T_2}{T_1}\right) + R \ln\left(\frac{V_2}{V_1}\right), \tag{4.94}$$

where the third line holds if c_v is independent of T. Since $c_p = c_v + R$ for a mole of an ideal gas, we may write Eq. (4.94) to obtain ΔS in the form

$$\Delta S = \int_{T_1}^{T_2} dT \frac{c_p(T)}{T} + R \ln\left(\frac{p_1}{p_2}\right)$$
$$= c_p \ln\left(\frac{T_2}{T_1}\right) + R \ln\left(\frac{p_1}{p_2}\right), \tag{4.95}$$

the second line holding for the case of c_p independent of T.

Let us now remove the assumption of a constant specific heat, since the specific heats for ideal diatomic and polyatomic gases generally depend on temperature. The temperature dependence of c_v may be represented by the expansion

$$c_v(T) = \sum_{n=-\infty}^{\infty} b_n T^n, \tag{4.96}$$

where b_n are coefficients independent of T. Note that if $b_n = 0$ for $n < -2$, we recover an expansion similar to Eq. (3.39). When Eq. (4.96) is substituted into the first line of Eq. (4.94) and the integration is performed, there follows the isochoric component of ΔS:

$$(\Delta S)_{isochoric} = \sum_{\substack{n=-\infty \\ (n \neq 0)}}^{\infty} \left[\frac{b_n}{n}T^n + b_0 \ln T\right]_{T_1}^{T_2}.$$

By combining this with the isothermal component of ΔS we obtain the overall entropy change for ideal diatomic or polyatomic gases.

4.8 Free Energies

4.8.1 Helmholtz Free Energy

The fundamental relations introduced earlier such as for dE and dH are differential forms in the space of variables (S, V) and (S, p), respectively. It is often convenient to use other sets of variables. For example, the variable set (T, V) may be used because it is more convenient than, for example, the variable pair (S, V). To express the fundamental relation in this set of variables, we introduce the Legendre transformation

$$A = E - TS. \tag{4.97}$$

Taking differential of this relation and using Fundamental Relation (4.43) we find the fundamental relation for Helmholtz free energy A:

$$dA = -S\ dT - p\ dV. \tag{4.98}$$

This is clearly a differential form in the space of variables (T, V). This is a measure of energy available for useful work for the task of interest.

4.8.2 Gibbs Free Energy

It is possible to have a Legendre transformation of H as follows:

$$G = H - TS. \tag{4.99}$$

On substitution of H from this relation into the fundamental relation for H in Eq. (4.88) there follows the fundamental relation for Gibbs free energy, G:

$$dG = -S\ dT + V\ dp, \tag{4.100}$$

which is a differential form in the space of variables (T, p). This variable set is the most convenient of the variable sets considered so far from the thermodynamic viewpoint, since temperature and pressure are easily amenable to measurement. This is also one of the reasons that the Gibbs free energy is one of the most convenient and commonly used among thermodynamic functions. It is also a measure of available energy for a given task.

4.9 Maxwell's Relations

The various forms of energy introduced in the previous section are all state functions and their differential forms are exact. They are combined expressions for the first and the second law in the case of a closed system undergoing a reversible process. We will collect them in one place:

$$dE = T\, dS - p\, dV, \tag{4.101}$$

$$dH = T\, dS + V\, dp, \tag{4.102}$$

$$dA = -S\, dT - p\, dV, \tag{4.103}$$

$$dG = -S\, dT + V\, dp. \tag{4.104}$$

Since these are all exact differentials, they satisfy the exactness conditions discussed in the mathematical notes in Chapter 2.

Equation (4.101) enables us to compute two derivatives of E:

$$\left(\frac{\partial E}{\partial S}\right)_V = T, \quad \left(\frac{\partial E}{\partial V}\right)_S = -p. \tag{4.105}$$

Differentiating the derivatives once again with V and S, respectively, and remembering that

$$\frac{\partial^2 E}{\partial S \partial V} = \frac{\partial^2 E}{\partial V \partial S},$$

we find the relation

$$\left(\frac{\partial T}{\partial V}\right)_S = -\left(\frac{\partial p}{\partial S}\right)_V. \tag{4.106}$$

Performing similar calculations with Eqs. (4.102)–(4.104) we derive the following relations:

$$\left(\frac{\partial T}{\partial p}\right)_S = \left(\frac{\partial V}{\partial S}\right)_p, \tag{4.107}$$

$$\left(\frac{\partial S}{\partial V}\right)_T = \left(\frac{\partial p}{\partial T}\right)_V, \tag{4.108}$$

$$\left(\frac{\partial S}{\partial p}\right)_T = -\left(\frac{\partial V}{\partial T}\right)_p. \tag{4.109}$$

These derivatives involving T, S, p, and V are called Maxwell's relations. They are equivalent to the integrability conditions for the differential forms (Eqs. (4.101)–(4.104)) and are extremely powerful and useful relations for various calculations in thermodynamics.

> *Mnemonics:* It is very useful for computational purposes to devise a mnemonic scheme to construct the Maxwell relations, particularly, since their similarity creates considerable difficulty to

remember them yet one does not wish to compute them from the fundamental relations whenever encountering them. The following rules for constructing them are found effective.

First, take the quadruplet of S, T, p, and V of variables and group them into conjugate pairs, (S, T) and (p, V). Then the products of differentials are balanced in the form[14]

$$dS \, dT = (-1)^{\alpha} \, dp \, dV. \tag{4.110}$$

Note that the conjugate variable pairs (S, T) and (p, V) are each made up of an intensive and an extensive variable. The value of exponent α will be determined by the rules stated below — it is the essence of the mnemonics presented here.

The Maxwell relations are of partial derivatives which are formed with intensive and extensive variables, one each, from the two conjugate variable pairs of the quadruplet. For example, to form Relation (4.109), we divide the equation with $dp \, dT$ in which p and T are the variables with which we differentiate S on the left and V on the right. Thus we obtain

$$\left(\frac{\partial S}{\partial p} \right)_l = (-1)^{\alpha} \left(\frac{\partial V}{\partial T} \right)_r .$$

Then the value of l on the left is $l = T$, the conjugate variable of S of the pair (S, T), whereas the value of r on the right is $r = p$, the conjugate variable of the pair (p, V). Thus we obtain

$$\left(\frac{\partial S}{\partial p} \right)_T = (-1)^{\alpha} \left(\frac{\partial V}{\partial T} \right)_p .$$

The value of α is determined by the following rule:

$\alpha = 1$ if both intensive variables (e.g., p and T) occur on both sides of the equation.

---

[14]The mathematical basis is as follows: Under the transformation $(S, T) \to (p, V)$

$$dS \, dT = \frac{\partial (S, T)}{\partial (p, V)} dp \, dV$$

and

$$\frac{\partial (S, T)}{\partial (p, V)} = \pm 1.$$

The Jacobian of transformation is usually taken 1, but here it is relaxed to take -1 as well.

$\alpha = 0$ if both intensive variables occur on one side only, left or right, of the equation.

Since $\alpha = 1$ for the present example, it follows

$$\left(\frac{\partial S}{\partial p}\right)_T = -\left(\frac{\partial V}{\partial T}\right)_p,$$

which is Relation (4.109).

Using these rules, other Maxwell relations can be easily constructed. For example, by dividing Eq. (4.110) with $dV\ dT$ we find the relation

$$\left(\frac{\partial S}{\partial V}\right)_l = (-1)^\alpha \left(\frac{\partial p}{\partial T}\right)_r$$

where by the rules stated earlier, $l = T$, $r = V$, and $\alpha = 0$. Hence

$$\left(\frac{\partial S}{\partial V}\right)_T = \left(\frac{\partial p}{\partial T}\right)_V$$

which is Relation (4.108). If we divide Eq. (4.110) with the product $dS\ dp$ and use the rules stated we find

$$\left(\frac{\partial T}{\partial p}\right)_S = \left(\frac{\partial V}{\partial S}\right)_p$$

and if we divide with $dV\ dS$ there follows relation

$$\left(\frac{\partial T}{\partial V}\right)_S = -\left(\frac{\partial p}{\partial S}\right)_V$$

which is Relation (4.106).

We have earlier utilized calorimetric data to compute the internal energy and the enthalpy as well as the entropy. In order to reinforce the method of computation we observe the following relations:

$$C_v = \left(\frac{\partial E}{\partial T}\right)_V = T\left(\frac{\partial S}{\partial T}\right)_V, \qquad (4.111)$$

$$C_p = \left(\frac{\partial H}{\partial T}\right)_p = T\left(\frac{\partial S}{\partial T}\right)_p. \qquad (4.112)$$

Notice in these relations that the entropy has the same dimension as the specific heats.

We illustrate applications of the Maxwell relations with a few examples in the following.

Example 1. $(\partial T/\partial p)_E$ may be expressed in terms of more easily measurable quantities. By the chain relation introduced in the mathematical notes in Chapter 3

$$\left(\frac{\partial T}{\partial p}\right)_E = -\frac{\left(\frac{\partial E}{\partial p}\right)_T}{\left(\frac{\partial E}{\partial T}\right)_p}.$$

Since

$$\left(\frac{\partial E}{\partial T}\right)_p = \left[\frac{\partial}{\partial T}(H - pV)\right]_p$$

$$= C_p - p\left(\frac{\partial V}{\partial T}\right)_p, \tag{4.113}$$

$$\left(\frac{\partial E}{\partial p}\right)_T = \left[\frac{\partial}{\partial p}(G - pV + TS)\right]_T$$

$$= -p\left(\frac{\partial V}{\partial p}\right)_T - T\left(\frac{\partial S}{\partial p}\right)_T$$

$$= -p\left(\frac{\partial V}{\partial p}\right)_T + T\left(\frac{\partial V}{\partial T}\right)_p, \tag{4.114}$$

by putting these equations together, we find

$$\left(\frac{\partial T}{\partial p}\right)_E = -\frac{V(p\kappa - T\alpha)}{C_p - pV\alpha}, \tag{4.115}$$

where α and κ are respectively the isobaric expansion coefficient and the isothermal compressibility:

$$\alpha = \frac{1}{V}\left(\frac{\partial V}{\partial T}\right)_p,$$

$$\kappa = -\frac{1}{V}\left(\frac{\partial V}{\partial p}\right)_T. \tag{4.116}$$

Example 2. We show that for ideal gases

$$\left(\frac{\partial E}{\partial V}\right)_T = 0, \quad \left(\frac{\partial E}{\partial p}\right)_T = 0.$$

Proof. (a)

$$\left(\frac{\partial E}{\partial V}\right)_T = \left[\frac{\partial}{\partial V}(A + TS)\right]_T$$

$$= -p + T\left(\frac{\partial S}{\partial V}\right)_T$$

$$= -p + T\left(\frac{\partial p}{\partial T}\right)_V.$$

Since for ideal gases

$$\left(\frac{\partial p}{\partial T}\right)_V = \frac{R}{V},$$

it follows

$$\left(\frac{\partial E}{\partial V}\right)_T = 0.$$

(b)

$$\left(\frac{\partial E}{\partial p}\right)_T = \left[\frac{\partial}{\partial p}(G - pV + TS)\right]_T$$

$$= -p\left(\frac{\partial V}{\partial p}\right)_T - T\left(\frac{\partial V}{\partial T}\right)_p.$$

For ideal gases the right hand side is equal to zero and consequently

$$\left(\frac{\partial E}{\partial p}\right)_T = 0.$$

Example 3. We show that for any substance

$$C_p = C_v + \frac{TV\alpha^2}{\kappa}.$$

To show it the following calculation is performed:

$$C_p - C_v = T\left[\left(\frac{\partial S}{\partial T}\right)_p - \left(\frac{\partial S}{\partial T}\right)_V\right] = T\left(\frac{\partial S}{\partial V}\right)_T\left(\frac{\partial V}{\partial T}\right)_p$$

$$= T\left(\frac{\partial p}{\partial T}\right)_V\left(\frac{\partial V}{\partial T}\right)_p = -\frac{T\left(\frac{\partial V}{\partial T}\right)_p^2}{\left(\frac{\partial V}{\partial p}\right)_T}$$

$$= \frac{TV\alpha^2}{\kappa}.$$

Problems

(1) Show that the efficiency of the Carnot cycle of a mole of a gas obeying the equation of state

$$p(V - b) = RT$$

is given by

$$\eta = 1 - \frac{T_2}{T_1}.$$

Assume that the specific heats of the gas are constants.

(2) Show that the efficiency of the Carnot cycle of a mole of a van der Waals gas obeying the equation of state

$$\left(p + \frac{a}{V^2}\right)(V - b) = RT$$

is given by

$$\eta = 1 - \frac{T_2}{T_1}.$$

Assume that the specific heats of the gas are constants.

(3) By using Eq. (4.87) show that Eq. (4.86) still holds true.

(4) Prove the following relations:

(a) $\left(\dfrac{\partial T}{\partial V}\right)_S = -\dfrac{T}{C_v}\left(\dfrac{\partial p}{\partial T}\right)_V,$

(b) $\left(\dfrac{\partial T}{\partial p}\right)_S = \dfrac{T}{C_p}\left(\dfrac{\partial V}{\partial T}\right)_p,$

(c) $\left(\dfrac{\partial V}{\partial p}\right)_S = \dfrac{C_v}{C_p}\left(\dfrac{\partial V}{\partial p}\right)_T,$

(d) $\left(\dfrac{\partial T}{\partial p}\right)_H = \dfrac{-V + T\left(\frac{\partial V}{\partial T}\right)_p}{C_p},$

(e) $\left(\dfrac{\partial T}{\partial V}\right)_E = \dfrac{p - T\left(\frac{\partial p}{\partial T}\right)_V}{C_v},$

(f) $\left(\dfrac{\partial H}{\partial G}\right)_E = \dfrac{C_p(\kappa p - \alpha T) - (1 - \alpha T)(C_p - \alpha p V)}{-S(\kappa p - \alpha T) - (C_p - \alpha p V)}.$

(5) Show the following:

(a) $\left(\dfrac{\partial E}{\partial S}\right)_p = T - p\left(\dfrac{\partial T}{\partial p}\right)_S = T\left[1 - \dfrac{p}{C_p}\left(\dfrac{\partial V}{\partial T}\right)_p\right],$

(b) $\left(\dfrac{\partial G}{\partial T}\right)_V = -S + V\left(\dfrac{\partial p}{\partial T}\right)_V = -S + \dfrac{V\alpha}{\kappa}$,

where α and κ are defined in Example 1.

(6) The following change is maintained isobaric and isothermal:

$$C_6H_6(l) + \tfrac{7}{2}O_2(g) \to 6CO_2(g) + 3H_2O(l).$$

It is found that $\Delta G = -3174.9 \text{ kJ mol}^{-1}$. Find ΔA at the same condition at $p = 1$ atm and $T = 298.2\,\text{K}$.

(7) Show that dQ/T is an exact differential by using the fundamental relation for E. Also show that for a Carnot cycle working with an ideal gas its cyclic integral vanishes.

(8) An extended strip of rubber has a length l when subjected to a tensile force f. If the volume change on extension may be neglected, show that

$$\left(\frac{\partial E}{\partial l}\right)_T = f - T\left(\frac{\partial f}{\partial T}\right)_l.$$

Show that the small temperature rise, ΔT, which takes place in a slow adiabatic stretching, is given by

$$\frac{\Delta T}{T} = \int_{l_0}^{l} \frac{1}{C_l}\left(\frac{\partial f}{\partial T}\right)_l dl,$$

where l_0 and l refer to the initial and final lengths, respectively, and C_l is the heat capacity of the rubber at constant extension. [Note that the work is equal to $f\,dl$ so that $dE = T\,dS + f\,dl$. For the second part of the problem calculate $(\partial T/\partial l)_S$ in terms of C_l and $(\partial f/\partial T)_l$. New "Maxwell's relations" are necessary for the present situation.]

(9) Between 0 and 60 K the molar heat capacity of Ag(s) is given approximately by the following expression:

$$c_p = \left(0.23\,T + 2.5 \times 10^{-3}\,T^2 - 1.9 \times 10^{-5}\,T^3\right) \text{J mol}^{-1}.$$

Calculate the entropy of Ag(s) at 60 K. Note $S(T = 0) = 0$.

(10) Find ΔA for O_2 $(g,\ 300.15\,\text{K},\ 2\ \text{atm}) \to O_2$ $(g,\ 300.15\,\text{K},\ 1\ \text{atm})$, assuming that the gas is ideal.

(11) In the temperature range of 298 to 2000 K the molar heat capacity of NH_3 (g) obeys the equation,

$$c_p = \left(29.75 + 2.51 \times 10^{-2}\,T - 1.55 \times 10^5\,T^{-2}\right) \text{J K}^{-1}\,\text{mol}^{-1}.$$

Find the entropy change at constant pressure arising from the temperature change from $T = 300$ to $1000\,\mathrm{K}$.

(12) At $0^\circ\mathrm{C}$, aluminum has the following properties: atomic weight $= 2.70 \times 10^{-2}$ kg mol^{-1}, density $= 2.70$ kg M^{-3}, $c_p = 0.930 \times 10^3$ J kg^{-1} K^{-1}, $\alpha = 71.4 \times 10^{-6}$ K^{-1}, $\kappa = 1.34 \times 10^{-21}$ m^2 N^{-1}. Calculate at $0^\circ\mathrm{C}$ the molar heat capacity at constant volume and the ratio of heat capacities γ.

(13) Find the entropy change for the reversible process of taking n moles of an ideal gas with constant heat capacity from T_1 and p_1 to T_2 and p_2.

(14) A gas obeys the following molar heat capacity equation and equation of state:

$$c_p = a + bT + cT^2$$

and

$$pv = RT + B_2(T)p + B_3(T)p^2.$$

Calculate the enthalpy and entropy changes for the reversible process from (T_1, p_1) to (T, p).

(15) Show Eq. (4.72).

(16) Show that for the Carnot cycle using an ideal gas as the working substance

$$\oint dE = 0, \quad \oint dH = 0, \quad \oint dA = 0, \quad \oint dG = 0, \quad \oint dS = 0.$$

Assume constant specific heats.

(17) Calculate the efficiency of the Rankine cycle and show

$$\oint dS = 0$$

for the cycle.

(18) Obtain the efficiency of the Otto cycle. Compare the efficiency of this cycle with the efficiency of the Carnot cycle. Also show that for the cycle

$$\oint dS = 0.$$

(19) Calculate the efficiency of the Diesel cycle and compare it with the efficiency of the Carnot cycle. Show that the efficiency of the Diesel cycle is greater than the efficiency of an Otto cycle at the same peak pressure. Also show that for the cycle

$$\oint dS = 0.$$

(20) The van der Waals constants for nitrogen are as follows:

$$a = 1.39 \text{ L}^2 \text{ atm mol}^{-2}, \quad b = 3.92 \times 10^{-2} \text{ L mol}^{-1}.$$

Taking the virial form for the van der Waals equation of state

$$\frac{pv}{RT} = 1 + \left(b - \frac{a}{RT}\right)\frac{1}{V} + \cdots,$$

calculate the enthalpy change $\Delta h = h(T,p) - h(T,0)$ arising from compression of the gas from $p = 0$ to 100 atm at constant $T = 300$ K.

(21) By using the van der Waals equation of state for real gas, calculate the reversible work done on the system due to a volume change from V_1 to V_2 at temperature T.

(22) Mercury vapor at 630.15 K and 1 atm pressure is heated to 823.15 K and its pressure is increased to 5 atm. Calculate the entropy change in the entropy units, the vapor being treated as an ideal monatomic gas.

(23) Calculate the nonideality corrections for enthalpy, internal energy, and entropy by using the van der Waals equation of state.

(24) Calculate the molar heat capacity difference $c_p - c_v$ for nitrogen at 298.15 K and 200 atm pressure, to first order in p, using the van der Waals equation of state, the van der Waals constants a and b being 1.39 L^2 atm mol^{-2} and 3.92×10^{-2} L mol^{-1}, respectively.

(25) Calculate for nitrogen the change of molar heat capacity c_p when the pressure is increased to 100 atm at 298.15 K, using the van der Waals equation of state. The constants are given in Problem 20.

(26) The molar heat capacity of solid iodine at temperature between 298.15 K to the melting point (386.75 K) at 1 atm pressure is given by the relation

$$c_p = \left[54.64 + 1.34 \times 10^{-3}(T - 298.15)\right] \text{ J K}^{-1} \text{ mol}^{-1}.$$

Determine the increase of entropy accompanying the change of 1 mole of iodine from 298.15 to 386.75 K.

Chapter 5

Equilibrium Conditions and Thermodynamic Stability

We have seen in Chapter 4 that the second law of thermodynamics supplies various fundamental equations involving thermodynamic functions associated with the entropy for reversible processes or when the system is in equilibrium. These fundamental equations provide means for us to analyze thermophysical data and correlate them. In addition to this, the laws of thermodynamics also enable us to understand under what conditions the systems reach equilibrium and how they behave in the neighborhood of the equilibrium state. This aspect of thermodynamics is physically richer than the former. The former, in fact, forms the equilibrium conditions and the latter provides the conditions for stability of equilibrium.

5.1 Inequalities

Since the second law of thermodynamics gives rise to inequalities involving (compensated) heat and entropy (or calortropy) as we have seen, we make use of them to examine the behavior of the systems in the neighborhood of equilibrium. From the second law of thermodynamics we have been able to deduce the differential form for the calortropy

$$d\Psi = \frac{dQ}{T} + dN. \tag{5.1}$$

Owing to the fact that $dN \geq 0$ for an infinitesimal process we deduce the inequality

$$d\Psi \geq \frac{dQ}{T}. \tag{5.2}$$

Since the calortropy Ψ reaches a maximum as the system tends toward equilibrium, if we imagine a virtual variation of the state from equilibrium that arises from variation in heat δQ, then since $\max(\Psi) = S$ at equilibrium it follows

$$\delta \Psi \leq \frac{\delta Q}{T}. \tag{5.3}$$

This can be made the starting point of a thermodynamic stability theory. This, however, is in contrast to the equilibrium thermodynamics practiced to this day, since the conventional thermodynamic stability theory is formulated in terms of the entropy obeying the Clausius–Duhem inequality, Eq. (4.65); one, in fact, begins with the observation that the entropy is maximum at equilibrium, that is, the entropy always increases if the system is not in equilibrium, and therefore a variation of entropy from equilibrium

$$\delta S = S - S_{\text{eq}} \tag{5.4}$$

must be such that it satisfies the inequality

$$\delta S \leq \frac{\delta Q}{T} \tag{5.5}$$

for all arbitrary virtual variations if the system is in equilibrium. This reasoning is troublesome and illogical because the entropy S is not defined for irreversible processes, but only for reversible processes or for a system at equilibrium only. After all, if the system is varied from the state of equilibrium by any means, it must be in a nonequilibrium state characterized by the calortropy. The troublesome feature vanishes if the calortropy is employed instead of S. This difficulty arises when the calortropy is confused with the entropy as was discussed in Chapter 4, but here we would like to add a point not discussed there.

The change in calortropy accompanying an irreversible process from one state to another in a system should obey the inequality

$$\Delta \Psi \geq \int_{(irr)} \frac{\text{d}Q}{T}, \tag{5.6}$$

where the equality holds at equilibrium. If a cycle is constructed that returns the system from the final state to the initial state through a reversible process in interaction with the surroundings (see Fig. 4.6), Inequality (5.6) can be expressed in terms of the entropy change complementary to the irreversible process in question and, as was shown in Chapter 4, we obtain the Clausius–Duhem inequality

$$\Delta S \geq \int_{(irr)} \frac{\text{d}Q}{T}. \tag{5.7}$$

This deduction for ΔS, however, is for a finite process only, not for an infinitesimal process for which one might infer the inequality

$$dS \geq \dbar Q/T, \tag{5.8}$$

which in turn would imply Inequality (5.5). We now would like to show that while Inequality (5.2) is a natural consequence of the second law of thermodynamics expressed in terms of the calortropy differential

$$d\Psi = \frac{\dbar Q}{T} + dN, \tag{5.9}$$

Inequality (5.7) does not imply Inequality (5.5). This can be seen as shown below:

In the case of a cycle depicted in Fig. 4.6, it is possible to write

$$\Delta S = \int_{(rev)} dS = \int_{(irr)} \frac{\dbar Q}{T}. \tag{5.10}$$

However, since the line integral along the reversible process in Eq. (5.10) is not the same as the line integral along the irreversible process on the right hand side of Inequality (Eq. (5.7)), an inequality $dS \geq \dbar Q/T$ cannot be inferred from the integral inequality (Eq. (5.7)). Put in another way, Inequality (5.7) can be written as

$$\int_{(rev)} dS - \int_{(irr)} \frac{\dbar Q}{T} \geq 0, \tag{5.11}$$

but for this to be written as a differential form holding everywhere in the reversible or irreversible segment of the cycle, it is necessary to cast it either in the form

$$\int_{(rev)} \left(dS - \frac{\dbar Q}{T} \right) \geq 0 \tag{5.12}$$

or in the form

$$\int_{(irr)} \left(dS - \frac{\dbar Q}{T} \right) \geq 0, \tag{5.13}$$

so that, under the continuity assumption, there follows the conclusion that the integrand is positive everywhere along the path of integration, that is, the inequality

$$dS \geq \frac{\dbar Q}{T}.$$

From this one may infer the inequality for δS given in Inequality (5.5).

However, neither of Inequalities (5.12) and (5.13) is possible to obtain from Inequality (5.11) since

$$\int_{(irr)} \frac{dQ}{T} \neq \int_{(rev)} \frac{dQ}{T}. \tag{5.14}$$

Thus we see that Inequality (5.5) is not necessarily true for an infinitesimal process. It should be replaced by Inequality (5.3) if we wish to examine the question of thermodynamic stability of a system near equilibrium. In retrospect, this conclusion should come as natural because the calortropy is a nonequilibrium extension of entropy that is defined for equilibrium only.

5.2   Equilibrium Conditions

Thermodynamic equilibrium is a singular state in the manifold of nonequilibrium thermophysical states of a thermodynamic system—the thermodynamic manifold, and the thermodynamic second law shows that this singular state is marked by inequalities, namely, such as the Clausius inequality or Inequality (Eq. (5.3)). The thermodynamic inequality (Eq. (5.3)) is expected to give rise to equilibrium conditions.

Inequality (5.3) and the identity of $\Psi_e \equiv \Psi$(equil.) with the Clausius entropy S—namely, the entropy of a reversible process—at equilibrium suggests that as the parameters (x) characterizing the state of the system are varied toward those (x_e) corresponding to equilibrium, the calortropy increases to a maximum where it becomes identical with S. This behavior is schematically explained in Fig. 5.1. We take advantage of this fact to formulate a thermodynamic stability theory.

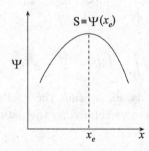

Fig. 5.1   The maximum value of the calortropy is S. Here x is a property characterizing the state of the system which is not in equilibrium. As x is varied and the system reaches equilibrium, the calortropy acquires a maximum value and the corresponding value of x is x_e, the equilibrium value of x, where $S = \Psi(x_e)$.

Let $\{x\}$ denote a set of macroscopic variables by which the thermodynamic state of the system is described. This set will be called the thermodynamic variables. For example, the set $\{x\}$ may be considered to be the variables $(E, v, c_i, \Phi_{qi} : r \leq i \leq 1; q \geq 1)$ where c_i denotes the mass fraction of species i defined by the ratio of the mass density ρ_i of species i to the total mass density ρ, and Φ_{qi} the variables characterizing the nonequilibrium state of the substance, such as the stress, heat flux, diffusion fluxes, and so on. The values of the set $\{x\}$ at equilibrium are denoted by $\{x_e\}$ at which Ψ acquires a maximum. In the case of the example taken for $\{x\}$ the set $\{x_e\}$ simply consists of $(E, v, c_i, : r \leq i \leq 1)$ because $\Phi_{qi} = 0$ at equilibrium and $\{x_e\}$ obey the equilibrium Gibbs relation. The calortropy Ψ is defined in the space of the full set $\{x\}$ of variables, whereas the (Clausius) entropy S lives in the space of the subset $\{x_e\}$ that corresponds to the equilibrium manifold of states, for example, $(E, v, c_i, : r \leq i \leq 1)$. With this understanding we may simply write $\Psi(x)$ and $S(x)$ to indicate their dependence on the thermodynamic variables. The results of the discussions in Chapter 4 and the previous paragraph suggest that there holds the inequality[1]

$$\Psi(x) \leq S(x). \tag{5.15}$$

The system is now varied from its equilibrium state. For an infinitesimal variation from equilibrium

$$\delta\Psi = \Psi(x) - \Psi(x_e) = \Psi(x) - S(x_e),$$
$$\delta S = S(x) - S(x_e),$$

and furthermore since $\Psi(x_e) = S(x_e)$ at equilibrium it follows that

$$\delta\Psi \leq \delta S. \tag{5.16}$$

Since $\delta S = \delta Q_{rev}/T$, Inequality (5.16) may be written as

$$\delta\Psi \leq \frac{\delta Q_{rev}}{T}, \tag{5.17}$$

if a virtual variation from equilibrium is performed on the system at equilibrium. Since the first law may be written in the case of a reversible process as

$$\delta E = \delta Q_{rev} + \delta W, \tag{5.18}$$

---

[1]The difference $S(x) - \Psi(x) \equiv S_{rel}$ is called the relative Boltzmann entropy in kinetic theory of fluids. See B. C. Eu, *J. Chem. Phys.* **106**, 2388 (1997) and B. C. Eu, *Nonequilibrium Statistical Mechanics* (Kluwer, Dordrecht, 1998).

where δW stands for the reversible work, we may write Eq. (5.17) in the form

$$\delta E + p\,\delta V - T\,\delta\Psi \geq \delta W_n, \qquad (5.19)$$

where

$$\delta W_n = \delta W + p\,\delta V, \qquad (5.20)$$

which denotes works other than the pressure–volume work. In the expression for the first law of thermodynamics we have replaced the derivative symbol d with the variation symbol δ. It must be noted that δQ_{rev} appears instead of δQ as in Inequality (5.2) since the entropy S is defined for a reversible process only. This also means that δW should be a reversible work that does not include the internal work associated with an irreversible process (see Appendix A for the meaning of dissipative internal work). Therefore, if a closed system at a constant energy and volume is to be in equilibrium, it must satisfy the equilibrium condition

$$(\delta\Psi)_{E,V} \leq -\delta W_n. \qquad (5.21)$$

If there is no internal work and the pressure–volume work is the only work then $\delta W_n = 0$, and the equilibrium condition (Eq. (5.19)) becomes

$$\delta E + p\,\delta V - T\,\delta\Psi \geq 0. \qquad (5.22)$$

It is possible to deduce that

$$(\delta\Psi)_{E,V} \leq 0. \qquad (5.23)$$

Henceforth we will confine our discussion to the case of $\delta W_n = 0$. For a closed system at constant energy and volume the equilibrium condition becomes

$$(\delta E)_{\Psi,V} \geq 0. \qquad (5.24)$$

This implies that the internal energy is at minimum at equilibrium, if Ψ and V are kept constant. This is schematically shown in Fig. 5.2. If T and V are kept constant, Condition (5.22) becomes

$$(\delta E)_{T,V} - T(\delta\Psi)_{T,V} \geq 0. \qquad (5.25)$$

It must be emphasized that the calortropy replaces the entropy in the aforementioned inequalities in contrast to the traditional formulation of thermodynamic stability theory, where the entropy S is used. We have earlier pointed out why the calortropy Ψ should replace the Clausius entropy S in the inequalities mentioned.

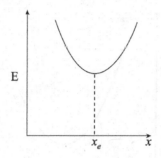

Fig. 5.2 The internal energy near equilibrium. The behavior of E near $x = x_e$ is deducible from Inequality (5.17) or (5.19).

Since at equilibrium $\Psi_e = S$, the inequalities mentioned become equalities. For example, we obtain from Inequality (5.22) the equation

$$\delta E = T \, \delta S - p \, \delta V. \tag{5.26}$$

This relation suggests that E may be regarded as a function of S and V at equilibrium and in general as a function of Ψ and V.

It is often useful to work with thermodynamic functions which are functions of variables other than Ψ. To find such a function we introduce a Legendre transformation

$$A = E - T\Psi. \tag{5.27}$$

The new thermodynamic function A is called the Helmholtz free energy or the work function.[2] Performing a variation of A and using Eq. (5.25) yields the inequality

$$\delta A + p \, \delta V + \Psi \, \delta T \geq 0, \tag{5.28}$$

and it yields the equilibrium condition at constant temperature and volume

$$(\delta A)_{T,V} \geq 0. \tag{5.29}$$

In other words, the Helmholtz free energy is minimum at equilibrium if T and V are kept constant; see Fig. 5.3. At equilibrium where nonequilibrium fluxes vanish the calortropy becomes the entropy and the inequality becomes the equality holding for δA

$$\delta A = - S \, \delta T - p \, \delta V, \tag{5.30}$$

---

[2]Although the same terminology is used as for the equilibrium case, it must be noted that the Helmholtz free energy defined in terms of Ψ is a generalization of the equilibrium Helmholtz free energy $A = E - TS$. It may be called nonequilibrium work function or nonequilibrium Helmholtz free energy. As the system approaches the equilibrium state, it tends to the equilibrium Helmholtz free energy or work function.

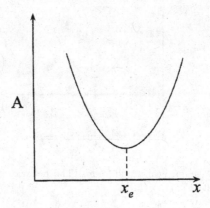

Fig. 5.3   The Helmholtz free energy near equilibrium.

which implies that A at equilibrium is a function of T and V. These are not the most convenient variables for laboratory experimental purposes, however. We may introduce a more suitable function by taking another Legendre transformation as follows:

$$G = A + pV = H - T\Psi. \tag{5.31}$$

By taking a variation of G and making use of Eq. (5.28) we find

$$\delta G \geq V\,\delta p - \Psi\,\delta T \tag{5.32}$$

as another form of the equilibrium condition. At constant T and p

$$(\delta G)_{T,p} \geq 0. \tag{5.33}$$

That is, G has a minimum at $x = x_e$ and at equilibrium there holds the form

$$\delta G = -S\,\delta T + V\,\delta p, \tag{5.34}$$

which means that G at equilibrium is a function of T and p. If p and T are kept constant $\delta G = 0$ at the minimum at equilibrium as shown in Fig. 5.4.

Still another equivalent form of the equilibrium condition can be derived in terms of enthalpy. With the definition (Eq. (3.12)) for H we may write the equilibrium condition (Eq. (5.22)) in the form

$$\delta H \geq T\,\delta\Psi + V\,\delta p. \tag{5.35}$$

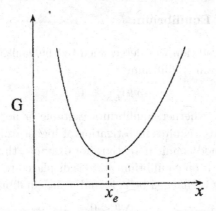

Fig. 5.4   The Gibbs free energy near equilibrium.

If Ψ and p are kept constant, it becomes

$$(\delta H)_{p,\Psi} \geq 0. \tag{5.36}$$

Since the equality holds at equilibrium and $\Psi_e = S$, there follows

$$\delta H = T\,\delta S + V\,\delta p, \tag{5.37}$$

which implies that H at equilibrium is a function of S and p; see Fig. 5.5. Equilibrium thermodynamics is based on equivalent fundamental relations (5.26), (5.30), (5.34), and (5.37) whose power and utility have been already demonstrated in the previous chapter.

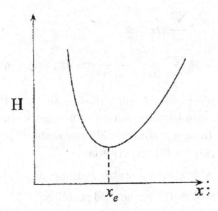

Fig. 5.5   The enthalpy near equilibrium.

5.3 Stability of Equilibrium

The equilibrium condition was determined by the vanishing first variation of the calortropy from equilibrium

$$(\delta\Psi)_{E,V} = 0,$$

but it does not say whether equilibrium is stable or not. The stability of the equilibrium state requires investigation of higher order variations of Ψ or equivalent thermodynamic potentials, for example, the second variation, $\delta^2\Psi$. If the system at an equilibrium state is displaced to an arbitrary state then the calortropy is changed by $\Delta\Psi$, and if the equilibrium is stable then

$$\Delta\Psi < 0. \tag{5.38}$$

If the equilibrium is of neutral stability then

$$\Delta\Psi = 0, \tag{5.39}$$

and if it is unstable then

$$\Delta\Psi > 0. \tag{5.40}$$

These criteria can be put in the context of virtual variation since

$$\Delta\Psi = \delta\Psi + \delta^2\Psi + \delta^3\Psi + \cdots. \tag{5.41}$$

Here $\delta\Psi$ is called the first variation, $\delta^2\Psi$ the second variation, $\delta^3\Psi$ the third variation, etc. If the complete set of thermodynamic variables is denoted by $\{x_i : i \geq 1\}$ then various variations are expressible in the forms

$$\delta\Psi = \sum_i \frac{\partial\Psi}{\partial x_i}\delta x_i, \tag{5.42}$$

$$\delta^2\Psi = \frac{1}{2}\sum_{i,j} \frac{\partial^2\Psi}{\partial x_i\partial x_j}\delta x_i\,\delta x_j, \tag{5.43}$$

$$\delta^3\Psi = \frac{1}{3!}\sum_{i,j,k} \frac{\partial^3\Psi}{\partial x_i\partial x_j\partial x_k}\,\delta x_i\,\delta x_j\,\delta x_k, \quad \text{etc.} \tag{5.44}$$

Here we are interested in continuous variation in $\{x_i\}$. If $\Delta\Psi < 0$ is to be satisfied, the first nonvanishing variation in Ψ must be of an even order for the same variation from a stable equilibrium state. Therefore, we obtain from Conditions (5.38)–(5.40) the criteria

$$\delta^2\Psi < 0 \quad \text{for stable equilibrium,} \tag{5.45}$$

$$\delta^2\Psi = 0 \quad \text{for neutral stability,} \tag{5.46}$$

$$\delta^2\Psi > 0 \quad \text{for unstable equilibrium.} \tag{5.47}$$

For further discussion on stability of thermodynamic systems on the basis of these criteria, the reader is referred to the literature.[3]

Since we are only interested in stability of the system in equilibrium submanifold, the nonequilibrium variables $\{\Phi_q : l < q \leq n\}$ in the manifold $\{x_i : 1 \leq i \leq n\}$ may be set equal to zero. We may then consider Ψ in the equilibrium submanifold $\{y_j : 1 \leq j \leq l\}$ of the manifold $\{x_i : 1 \leq i \leq n\}$, where $l < n$. In this case, the calortropy Ψ and its derivatives are evaluated in the submanifold $\{y_j\}$.

To make discussion as simple as possible let us confine the discussion to a single component fluid. When explicitly written out, $\delta^2\Psi$ is given by

$$\delta^2\Psi = \tfrac{1}{2}\left[\frac{\partial^2\Psi}{\partial E^2}(\delta E)^2 + \left(\frac{\partial^2\Psi}{\partial E\partial V} + \frac{\partial^2\Psi}{\partial V\partial E}\right)\delta E\,\delta V \right.$$
$$\left. + \frac{\partial^2\Psi}{\partial V^2}(\delta V)^2 + \cdots \right], \tag{5.48}$$

where the ellipsis represents the second derivatives of Ψ with respect to also the nonequilibrium fluxes. This expression is evaluated at the equilibrium submanifold

$$\{y_j\} = \{x_1, \ldots, x_l; x_i = 0, l < i \leq n\},$$

namely, the nonequilibrium fluxes are set equal to zero:

$$\delta^2\Psi|_{\{x_j\}=\{y_j\}} = \tfrac{1}{2}\left[\frac{\partial^2\Psi}{\partial E^2}(\delta E)^2 + \left(\frac{\partial^2\Psi}{\partial E\partial V} + \frac{\partial^2\Psi}{\partial V\partial E}\right)\delta E\,\delta V \right.$$
$$\left. + \frac{\partial^2\Psi}{\partial V^2}(\delta V)^2 + \cdots \right]_{\{x_j\}=\{y_j\}}.$$

It is convenient to remove the cross terms in this equation. First, by using the extended Gibbs relation, this equation may be cast in the form

$$\delta^2\Psi|_{\{x_j\}=\{y_j\}} = \tfrac{1}{2}\left[\delta E\,\delta\left(\frac{1}{T}\right) + \delta V\,\delta\left(\frac{p}{T}\right)\right]_{\{x_j\}=\{y_j\}}, \tag{5.49}$$

which then can be put in the form

$$\delta^2\Psi|_{\{x_j\}=\{y_j\}} = \tfrac{1}{2}\left[-\frac{C_v}{T^2}(\delta T)^2 - \frac{1}{T\kappa V}(\delta V)^2\right], \tag{5.50}$$

---

[3]B. C. Eu, *Generalized Thermodynamics: The Thermodynamics of Irreversible Processes and Generalized Hydrodynamics* (Kluwer, Dordrecht, 2002).

where

$$C_v = \left(\frac{\partial E}{\partial T}\right)_{V,\{\Phi_q\}}\Bigg|_{\{\Phi_q\}=0}, \tag{5.51}$$

$$\kappa = -\frac{1}{V}\left(\frac{\partial V}{\partial p}\right)_{T,\{\Phi_q\}}\Bigg|_{\{\Phi_q\}=0}. \tag{5.52}$$

Here we have used the extended Gibbs relation, which reduces to the equilibrium Gibbs relation at equilibrium. C_v is the isochoric heat capacity and κ is the isothermal compressibility at $\{\Phi_q\} = 0$, that is, at equilibrium. This quadratic form is negative if the following conditions are satisfied:

$$C_v > 0, \qquad \kappa > 0. \tag{5.53}$$

That is, if these conditions are satisfied then the equilibrium state is stable. These conditions also imply that

$$C_p > C_v \tag{5.54}$$

as a consequence of the stability condition of equilibrium. This consideration can be obviously generalized to the case of open systems and multiphase systems. For a brief discussion on the extended Gibbs relation and examples of its applicaiton, see Chapter 19 on nonequilibrium processes.

Chapter 6

The Third Law of Thermodynamics

Since we are from now on concerned with reversible processes only, the calortropy coincides with the Clausius entropy. Therefore, in this and subsequent chapters except for Chapter 19, we study the thermodynamics of reversible processes or systems at equilibrium by using only the concept of entropy. In this chapter, only pure (single-component) systems are considered in connection with the third law.

We have seen that the entropy of a system at temperature T can be calculated within a constant of integration. Thus, if the volume is constant, the entropy at temperature T is given by the formula

$$S(T) = S_0 + \int_0^T dT' \frac{C_v(T')}{T'}, \tag{6.1}$$

where S_0 is the value of the entropy at the absolute zero. If the value of S_0 is known, the absolute entropy can be obtained at any temperature. It is indeed possible to find the absolute entropy for perfect crystals according to the third law of thermodynamics.

In 1902, T. W. Richards experimentally observed that ΔS vanishes as the temperature tends to the absolute zero. This observation enabled W. Nernst to formulate his heat theorem. For an isothermal reversible process a Helmholtz free energy change ΔA may be written as

$$\Delta A = \Delta E - T \, \Delta S. \tag{6.2}$$

This may be cast in the form

$$\Delta A - \Delta E = T \left(\frac{\partial \, \Delta A}{\partial T} \right)_V. \tag{6.3}$$

Based on the experimental data available to him, Nernst observed that this difference always vanishes as T tends to zero and thus concluded that the difference not only vanishes, but also ΔA and ΔE have the same tangent asymptotically as T tends to zero. Therefore there should hold the limits

$$\lim_{T \to 0} \frac{\partial \, \Delta A}{\partial \, T} = \lim_{T \to 0} \frac{\partial \Delta E}{\partial T} = 0. \tag{6.4}$$

Consequently, by differentiating Eq. (6.3) with T, we find

$$\lim_{T \to 0} \frac{\partial \, \Delta A}{\partial T} = \lim_{T \to 0} \left(\frac{\partial \, \Delta E}{\partial T} - T \frac{\partial \, \Delta S}{\partial T} - \Delta S \right).$$

The limits in Eq. (6.4) suggest the following limit

$$\lim_{T \to 0} \left(T \frac{\partial \, \Delta S}{\partial T} + \Delta S \right) = 0$$

or equivalently

$$\lim_{T \to 0} \Delta S = 0. \tag{6.5}$$

This is called Nernst's heat theorem. Based on this experimental observation, Max Planck stated the third law of thermodynamics as follows:

The entropy of perfect crystals is equal to zero at the absolute zero of temperature independently of pressure and volume.

That is, independently of pressure and volume

$$\lim_{T \to 0} S\,(T) = 0. \tag{6.6}$$

Since the entropy in a reversible temperature change near the absolute zero at constant volume is given by the formula

$$S(T) = S_0 + \int_0^T dT' \, \frac{C_v(T')}{T'}, \tag{6.7}$$

the third law means $S_0 = 0$, for $S(T) \to 0$ independently of p and V as $T \to 0$. The undetermined constant S_0 is thereby removed. The third law not only allows to determine the entropy absolutely, but also has a number of thermodynamic implications. We discuss some of them in the following.

(1) *Specific heat tends to zero as T approaches the absolute zero.*

Since

$$S(T) = \int_0^T dT' \, \frac{C_v(T')}{T'} \tag{6.8}$$

and $S(T) \to 0$ as $T \to 0$ by the third law, the integral must be a decreasing function of T in the neighborhood of the absolute zero of temperature. That is, near $T = 0$

$$S(T) = cT^\alpha, \tag{6.9}$$

where $\alpha \geq 0$ and c is a constant. This implies that $C_v(T)$ cannot be a constant in the neighborhood of $T = 0$. Indeed, according to Debye's theory of simple solids the exponent α is equal to 3. That is,

$$C_v(T) = aT^3 \tag{6.10}$$

and consequently the entropy near absolute zero behaves like

$$S(T) = \frac{1}{3}aT^3. \tag{6.11}$$

Experiments show that the specific heats of amorphous glasses are approximately linear in T near the absolute zero of temperature: $C_v(T) = aT$. The entropy is then $S(T) = aT$ near $T = 0$.

(2) $\left(\frac{\partial V}{\partial T}\right)_p = 0$ and $\left(\frac{\partial p}{\partial T}\right)_V = 0$ as $T \to 0$.

To show it we recall one of Maxwell's relations and the third law:

$$\lim_{T \to 0} \left(\frac{\partial V}{\partial T}\right)_p = \lim_{T \to 0} - \left(\frac{\partial S}{\partial p}\right)_T = 0,$$

independently of p at $T = 0$ by the third law. The second identity may be shown similarly:

$$\lim_{T \to 0} \left(\frac{\partial p}{\partial T}\right)_V = \lim_{T \to 0} \left(\frac{\partial S}{\partial V}\right)_T = 0,$$

independently of V at $T = 0$ by the third law. The first identity also implies that the isobaric expansion coefficient tends to zero as $T \to 0$. Since we may write the second identity in terms of the isobaric expansion coefficient and the isothermal compressibility

$$\left(\frac{\partial p}{\partial T}\right)_V = -\frac{\left(\frac{\partial V}{\partial T}\right)_p}{\left(\frac{\partial V}{\partial p}\right)_T} = \frac{\alpha}{\kappa},$$

it also implies that α tends to zero faster than κ as $T \to 0$.

(3) *The absolute zero of temperature cannot be reached by any finite number of processes; it may be only reached asymptotically.*

In 1926, Giauque and Debye independently showed that low temperature can be achieved by adiabatic demagnetization (Fig. 6.1). In this technique a paramagnetic salt (gadolinium sulfate, for example) is repeatedly magnetized and then demagnetized. Demagnetization is accompanied by a decrease in temperature and it is found possible to reach 0.0014 K by the procedure. The thermodynamic basis of this method is described in Chapter 18.

As a magnetic field H is turned on a paramagnetic salt, the entropy decreases from its value at $H = 0$ because of the paramagnetic ordering effect of the magnetic field. Let us set

$$S_0 = \lim_{T \to 0} S(H = 0)$$

and assume that $S_0 \neq 0$ contrary to the third law. Then it would be possible to attain $T = 0$ in a single step of adiabatic demagnetization by starting the procedure at an appropriate temperature. This then would allow the system to reach $T = 0$ in a finite number of steps. But since $S_0 = 0$, it is practically impossible to attain $T = 0$ by a finite number of steps of adiabatic demagnetization since the temperature decrease achieved in each step gets smaller and smaller as the absolute zero is approached.

As a corollary to this theorem, we may state that *it is practically impossible to attain the unit efficiency for reversible Carnot cycles.* Since by the Carnot theorem and equivalently the second law of thermodynamics the efficiency of reversible cycles is always larger than those of irreversible cycles the present corollary may be regarded as a consequence of the second law of thermodynamics. Indeed, this should be so since the second law gives rise to entropy for reversible processes, and hence the corollary must be an attribute of the second law.

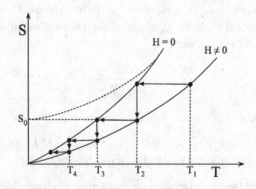

Fig. 6.1   Cooling by the Debye–Giauque adiabatic demagnetization method.

Problems

(1) Show that according to the third law of thermodynamics

$$\lim_{T \to 0} (C_p - C_v)/TV = 0.$$

(2) Assuming that a simple monatomic liquid forms a simple crystal obeying the Debye law of specific heat $C_v = cT^3$ when it crystallizes, calculate the entropy of the liquid at $T > T_m$ where T_m is the melting point of the crystal. The heat of melting of the crystal is ΔH_m. The pressure remains fixed at p throughout the process.

Chapter 7

Thermodynamics of Mixtures and Open Systems

The thermodynamics of mixtures requires a new concept unfamiliar to the thermodynamics of pure substances. It is embodied by chemical potentials first introduced by J. W. Gibbs. The thermodynamics of mixtures and of open systems is formulated by making use of the concept. It would be most appropriate from the viewpoint of the thermodynamic laws the concept is incorporated into the representations of the thermodynamic laws. Since Gibbs takes an axiomatic approach, in deference to his method we will follow his formulation first. Then the former approach is followed.

To discuss a theory of mixtures the composition variables must be specified. The mole numbers of components are used for the purpose. Thus the Gibbs space for a mixture requires an extension to account for the composition of the mixture; it is now spanned by the variable set $(S, E, V, n_1, n_2, \ldots, n_r)$, where n_i denote the mole numbers of the species in the mixture. The present chapter also prepares us for the studies of open systems in the subsequent chapters, where mixtures of various kinds are treated thermodynamically.

7.1   Chemical Potentials

7.1.1   *The Gibbs Theory*

Let us consider a mixture consisting of $n_1, n_2, \ldots, n_r$ moles of components, $1, 2, \ldots, r$. The internal energy of the mixture then may be regarded as a function of $S, V, n_1, n_2, \ldots, n_r$:

$$E = E(S, V, n_1, n_2, \ldots, n_r). \tag{7.1}$$

It must be noted here that this assumption of a function E in the manifold of $S, V, n_1, n_2, \ldots,$ and n_r presumes that there is a variable called entropy

S already established on the basis of the laws of thermodynamics. This variable S, however, is fundamentally different from the rest of the variables spanning the manifold because it is not a mechanical quantity as the others are. This should be clearly kept in mind. It would have been better if the fundamental relation for E were deduced for a mixture on the basis of the laws of thermodynamics, as will be done later, but in this subsection we first follow the conventional Gibbsian approach, which is axiomatic. For this reason the connection with the thermodynamic laws is unfortunately obscured.

The differential of E is easily obtained from the functional relation in Eq. (7.1):

$$dE = \left(\frac{\partial E}{\partial S}\right)_{V,n} dS + \left(\frac{\partial E}{\partial V}\right)_{S,n} dV + \sum_{i=1}^{r} \left(\frac{\partial E}{\partial n_i}\right)_{S,V,n_{j\neq i}} dn_i. \qquad (7.2)$$

If we identify the first two derivatives in Eq. (7.2), in analogy to the case of the thermodynamics of a closed system, with temperature and pressure

$$T = \left(\frac{\partial E}{\partial S}\right)_{V,n}, \qquad (7.3)$$

$$p = -\left(\frac{\partial E}{\partial V}\right)_{T,n}, \qquad (7.4)$$

and, in addition, denote by μ_i the derivative of E with respect to n_i

$$\mu_i = \left(\frac{\partial E}{\partial n_i}\right)_{S,V,n_{j\neq i}}, \qquad (7.5)$$

the differential is then written in the form

$$dE = T\,dS - p\,dV + \sum_{i=1}^{r} \mu_i\,dn_i. \qquad (7.6)$$

Here μ_i is called the chemical potential of component i of the mixture. It was introduced for the first time by J. W. Gibbs, and plays very important roles in thermodynamics. It is useful also to note that temperature and pressure are given by the same derivatives as those in Eq. (4.105) except that $n = \{n_i\}$ must also be kept constant while differentiating E. The definitions (Eqs. (7.3)–(7.5)) are mathematical and based on analogy to the mathematical structure of the thermodynamics of a closed single component system; they are not deduced by applying the laws of thermodynamics to mixtures as has been for a closed single-component system. To gain a more physically cogent formulation that closely makes use of the thermodynamic laws we consider an alternative approach in the following.

7.1.2 Alternative Consideration

From the strict standpoint of thermodynamics the differential form (Eq. (7.6)) for an open system should have been deduced from the proper physical interpretation of what $đQ$ should be in the expression

$$dS = \frac{đQ}{T}, \tag{7.7}$$

when the system is open with regard to matter. When the system exchanges matter with the surroundings, energy is also carried by matter, and the heat transfer between the system and the surroundings must be corrected for the amount of energy carried by the matter transferred. This correction may be written as

$$-\sum_{i=1}^{r} \mu_i \, dn_i, \tag{7.8}$$

where μ_i is the amount of energy (equivalently heat) carried by a mole of species i and the sign convention on dn_i is taken such that dn_i is taken negative when matter is given up by the system to the surroundings, and positive in the opposite case. Therefore the net heat taken up by the system is[1]

$$đQ = đQ' - \sum_{i=1}^{r} \mu_i \, dn_i, \tag{7.9}$$

where $đQ'$ is the heat change that would occur in the expression for the first law of thermodynamics, if the system were not open. It is important to take note of the difference in the meanings of heats in the two cases of open and closed systems. With this $đQ$ substituted into Eq. (7.7) and with the help of the first law of thermodynamics we find the differential form for dE presented for an open system in the same form as by the previous axiomatic

---

[1] Equivalently, one may regard the term (7.8) as a work performed when substances are transferred between the system and the surroundings which act as reservoirs of matter. In this manner of looking at the situation the work term in the first law of thermodynamics is modified as follows:

$$dW \Rightarrow dW + \sum_{i=1}^{r} \mu_i \, dn_i$$

under the same sign convention for dn_i as for Eq. (7.9).

method (Gibbs' method). In the present manner of approaching the subject, the derivatives (Eqs. (7.3)–(7.5)) are consequences of the differential form (Eq. (7.6)) derived from the second law of thermodynamics in the case of reversible processes in an open system. However, in the Gibbs method the existence of $S(E, V, n_1, \ldots, n_r)$ is implicitly assumed in analogy to the thermodynamics of closed systems in the original Gibbs theory. We have presented it first in this section out of historical deference to the axiomatic method of Gibbs, which is mathematically more elegant.

7.1.3   *Fundamental Relations for Open Systems*

Upon substitution of Eq. (7.7) and the first law of thermodynamics

$$dE = \text{d}Q' - p\,dV \tag{7.10}$$

in the case of pressure–volume work there follows the fundamental relation (Eq. (7.6)) for E. If there are other kinds of work present then they must be added to the right-hand side of Eq. (7.10).

By introducing the following Legendre transformations

$$
\begin{aligned}
H &= E + pV, \\
A &= E - TS, \\
G &= H - TS,
\end{aligned}
\tag{7.11}
$$

taking differential forms of H, A, and G, and making use of Eq. (7.6) we easily find the fundamental relations for thermodynamic potentials[2] H, A, and G:

$$dH = T\,dS + V\,dp + \sum_{i=1}^{r} \mu_i\,dn_i, \tag{7.12}$$

$$dA = -S\,dT - p\,dV + \sum_{i=1}^{r} \mu_i\,dn_i, \tag{7.13}$$

$$dG = -S\,dT + V\,dp + \sum_{i=1}^{r} \mu_i\,dn_i, \tag{7.14}$$

which are the generalizations of Eqs. (4.101), (4.102), (4.103), and (4.104) to an r-component mixture. These differential forms obviously imply that

---

[2]We call these functions thermodynamic potentials since their differential one-forms are monogenic, that is, the differential forms are integrable to scalar functions H, A, and G.

H, A, and G are a function of S, p, $n_1, n_2, \ldots, n_r$; T, V, $n_1, n_2, \ldots, n_r$; and T, p, $n_1, n_2, \ldots, n_r$, respectively:

$$H = H(S, p, n_1, n_2, \ldots, n_r),$$
$$A = A(T, V, n_1, n_2, \ldots, n_r),$$
$$G = G(T, p, n_1, n_2, \ldots, n_r).$$

The fundamental relations for H, A, and G also imply that the chemical potentials can be expressed in different ways depending on which one of E, H, A, and G is used:

$$\mu_i = \left(\frac{\partial E}{\partial n_i}\right)_{S,V,n'} = \left(\frac{\partial H}{\partial n_i}\right)_{S,p,n'} = \left(\frac{\partial A}{\partial n_i}\right)_{T,V,n'} = \left(\frac{\partial G}{\partial n_i}\right)_{T,p,n'}. \qquad (7.15)$$

Here n' denotes the set $\{n_j\}$ excluding n_i. Nevertheless, the last equality in Eq. (7.15) gives the most convenient interpretation as the partial molar Gibbs free energy of species i as will be seen shortly. In this regard, it should be noted that T and p are one of the most readily applicable parameters that can be kept constant among similar constraints such as S and p, T and V, etc.

7.2 Partial Molar Properties

Let M denote an extensive thermophysical property. Extensive thermodynamic properties may be generally considered functions of T, p, n_1, $n_2, \ldots, n_r$:

$$M = M(T, p, n_1, n_2, \ldots, n_r). \qquad (7.16)$$

By definition, extensive properties depend on the masses of species or the mole numbers, and they scale linearly, being a homogeneous function of degree 1. For example, take the internal energy E. If the mass of the system is increased then the internal energy will increase in direct proportion to the increase in mass. This implies, in mathematical terms, that extensive quantities are first degree homogeneous functions of n_1, n_2, $\ldots$, n_r of Euler. That is, if the mole numbers are increased by a factor λ for all components then M increases by the same factor λ. This may be expressed by the relation

$$M(T, p, \lambda n_1, \lambda n_2, \ldots, \lambda n_r) = \lambda M(T, p, n_1, n_2, \ldots, n_r). \qquad (7.17)$$

This is a special case of functions called Euler's homogeneous functions of degree q, and for the present case $q = 1$. We digress a little to discuss about Euler's homogeneous functions.

If a function $M(x_1, x_1, \ldots, x_r)$ has the property

$$M(\lambda x_1, \lambda x_2, \ldots, \lambda x_r) = \lambda^q M(x_1, x_2, \ldots, x_r),$$

then M is called a homogeneous function of Euler of degree q. A differentiation of the equation with respect to λ yields

$$\sum_{i=1}^{r} \left(\frac{\partial M}{\partial \lambda x_i} \right) x_i = q\lambda^{q-1} M(x_1, x_2, \ldots, x_r). \tag{7.18}$$

If λ is set equal to 1, there follows the relation

$$M(x_1, x_2, \ldots, x_r) = q^{-1} \sum_{i=1}^{r} x_i \left(\frac{\partial M}{\partial x_i} \right). \tag{7.19}$$

This mathematical relation is useful for discussing properties of mixtures in thermodynamics. In particular, if $q = 1$ and $x_i = n_i$, we obtain

$$M(T, p, n_1; n_2, \ldots, n_r) = \sum_{i=1}^{r} \left(\frac{\partial M}{\partial \lambda n_i} \right)_{T,p,n'} \frac{d\lambda n_i}{d\lambda}$$

$$= \sum_{i=1}^{r} n_i \left(\frac{\partial M}{\partial n_i} \right)_{T,p,n'} \quad (\lambda = 1). \tag{7.20}$$

Here the derivative is an intensive quantity and may be abbreviated by

$$\overline{M}_i = \left(\frac{\partial M}{\partial n_i} \right)_{T,p,n'}, \tag{7.21}$$

which allows to write M in the form

$$M(T, p, n_1, n_2, \ldots, n_r) = \sum_{i=1}^{r} n_i \overline{M}_i. \tag{7.22}$$

In this form, the extensive property M appears as an additive sum of contributions weighted by the mole number n_i of the components making up the mixture. It, however, should be kept in mind that $\overline{M}_i$ are generally functions of $n_1, n_2, \ldots, n_r$ or more precisely, $x_i = n_i/n$ $(i = 1, \ldots, r)$ where $n = n_1 + n_2 + \cdots + n_r$.

These mathematical properties are used in our discussion on the thermodynamics of mixtures. In particular, $\overline{M}_i$ is called the partial molar M of i.

Since the chemical potential of i may be given by

$$\mu_i = \left(\frac{\partial G}{\partial n_i}\right)_{T,p,n'}, \tag{7.23}$$

we see that it is the partial molar Gibbs free energy of i in a mixture consisting of r components. In this regard, note that μ_i is not a partial molar property of E, H, or A although it may be given by the derivative of E, H, and A as in Eq. (7.15). Therefore from Eq. (7.20) follows the relation

$$G = \sum_{i=1}^{r} n_i \mu_i. \tag{7.24}$$

In this form the extensive property G appears as an additive sum of $n_i \mu_i$ of the components making up the mixture. Here lies one of the utilities of the notion of partial molar properties in the thermodynamics of mixtures.

We now further develop consequences and thermodynamic relations of partial molar properties and, in particular, of chemical potentials. Since

$$\left(\frac{\partial G}{\partial T}\right)_{p,n} = -S, \tag{7.25}$$

$$\left(\frac{\partial G}{\partial n_i}\right)_{T,p,n'} = \mu_i, \tag{7.26}$$

Euler's condition on exact differentials yields the Maxwell relation

$$-\left(\frac{\partial \mu_i}{\partial T}\right)_{p,n} = \left(\frac{\partial S}{\partial n_i}\right)_{T,p,n'}. \tag{7.27}$$

By definition, the right-hand side of Eq. (7.27) is the partial molar entropy of i:

$$\bar{s}_i = \left(\frac{\partial S}{\partial n}\right)_{T,p,n'}. \tag{7.28}$$

Therefore we obtain

$$\bar{s}_i = -\left(\frac{\partial \mu_i}{\partial T}\right)_{p,n}. \tag{7.29}$$

Similarly, since

$$V = \left(\frac{\partial G}{\partial p}\right)_{T,n}, \tag{7.30}$$

by differentiating it with n_i, we obtain the relation

$$\bar{v}_i = \left(\frac{\partial \mu_i}{\partial p}\right)_{T,n}, \tag{7.31}$$

where $\bar{v}_i$ is the partial molar volume of i:

$$\bar{v}_i = \left(\frac{\partial V}{\partial n_i}\right)_{T,p,n'}. \tag{7.32}$$

Since G is a function of n_i, $i = 1, 2, \ldots, r$, the Maxwell relations for single component systems must be generalized. The additional Maxwell relations involve chemical potentials. The examples are

$$\left(\frac{\partial \mu_i}{\partial n_j}\right)_{T,p,n_k} = \left(\frac{\partial \mu_j}{\partial n_i}\right)_{T,p,n_k} \quad (k \neq i \text{ or } j). \tag{7.33}$$

As G is a function of T, p, n_1, $n_2, \ldots, n_r$, the chemical potentials must be functions of the same variables. If $n = \sum_{i=1}^{r} n_i$ is fixed, there are $r - 1$ independent mole numbers, and the chemical potentials may be regarded as functions of T, p, x_1, x_2, $\ldots$, x_{r-1} where x_i is the mole fraction of i:

$$x_i = \frac{n_i}{n}, \tag{7.34}$$

which normalizes to unity:

$$\sum_{i=1}^{r} x_i = 1. \tag{7.35}$$

Equations (7.29), (7.32), and (7.33) together imply the following differential form for μ_i:

$$d\mu_i = -\bar{s}_i\, dT + \bar{v}_i\, dp + \sum_{j=1}^{r-1} \mu_{ij}\, dx_j, \tag{7.36}$$

where

$$\mu_{ij} = \left(\frac{\partial \mu_i}{\partial x_j}\right)_{T,p,x_{k \neq j}}. \tag{7.37}$$

From Eq. (7.24) follows the differential form

$$dG = \sum_{i=1}^{r} \mu_i\, dn_i + \sum_{i=1}^{r} n_i\, d\mu_i.$$

Since we also have the fundamental relation

$$dG = -S\, dT + V\, dp + \sum_{i=1}^{r} \mu_i\, dn_i, \tag{7.38}$$

by comparing these two equations we arrive at the following important equation

$$-S \, dT + V \, dp = \sum_{i=1}^{r} n_i \, d\mu_i. \tag{7.39}$$

This is called the Gibbs–Duhem equation and plays important roles in studies of thermodynamics and equilibria in particular. We will have many occasions to use it in subsequent chapters. Incidentally, we remark that the Gibbs–Duhem equation is an integrability condition of the differential form (7.38), and Eq. (7.24) is an integral of differential form (7.39) in the space spanned by T, p, $n_1, \ldots, n_r$. The reason is that if Eqs. (7.38) and (7.39) are added side by side there follows the integral G of the differential form (7.38). It is easy to verify that the integrability condition for the fundamental relation

$$dS = T^{-1} \left(dE + p \, dV - \sum_{i=1}^{r} \mu_i \, dn_i \right) \tag{7.40}$$

is then given by

$$Ed\left(\frac{1}{T}\right) + Vd\left(\frac{p}{T}\right) - \sum_{i=1}^{r} n_i d\left(\frac{\mu_i}{T}\right) = 0, \tag{7.41}$$

which can be shown to be equivalent to the Gibbs–Duhem relation (Eq. (7.39)). Upon addition of Eqs. (7.40) and (7.41) there follows the integral of the differential form (Eq. (7.40)) within a constant of integration:

$$S = T^{-1} \left(E + pV - \sum_{i=1}^{r} \mu_i n_i \right). \tag{7.42}$$

Equation (7.24) and the Legendre transformations (Eq. (7.11)) imply that it is, of course, another way of writing the Gibbs free energy.

Lastly, multiplying n_i to Eq. (7.36) and summing over i, and then making use of the Gibbs–Duhem equation (Eq. (7.39)) on the resulting equation, we obtain the equation

$$\sum_{i=1}^{r} \sum_{j=1}^{r-1} n_i \mu_{ij} \, dn_j = 0, \tag{7.43}$$

which relates μ_{ij} to one another. This equation will be found useful for the theory of mixtures.

7.3   Measurement of Partial Molar Properties

Thermodynamic functions are easily evaluated for mixtures once the partial molar properties are tabulated as a function of temperature, pressure, and concentrations. It then is obviously important to examine how we might measure them in the laboratory. Here we discuss some experimental procedures for measuring partial molar properties. To prepare for the discussion we first examine another useful relation underlying such measurements.

Let M be an extensive property. Then

$$dM = dM(T, p, n_1, n_2, \ldots, n_r)$$

$$= \left(\frac{\partial M}{\partial T} \right)_{p,n} dT + \left(\frac{\partial M}{\partial p} \right)_{T,n} dp + \sum_{i=1}^{r} \left(\frac{\partial M}{\partial n_i} \right)_{T,p,n_{j \neq i}} dn_i. \quad (7.44)$$

But we know

$$M = \sum_{i=1}^{r} n_i \overline{M}_i, \quad (7.45)$$

where $\overline{M}_i$ is the partial molar property of i. On taking the differential form of M by using Eq. (7.45) we obtain

$$dM = \sum_{i=1}^{r} \overline{M}_i \, dn_i + \sum_{i=1}^{r} n_i \, d\overline{M}_i. \quad (7.46)$$

Comparing it with Eq. (7.44) yields a relation similar to the Gibbs–Duhem equation,

$$\sum_{i=1}^{r} n_i \, d\overline{M}_i = \left(\frac{\partial M}{\partial p} \right)_{p,n} dp + \left(\frac{\partial M}{\partial T} \right)_{T,n} dT. \quad (7.47)$$

If T and p are kept constant, it reduces to

$$\sum_{i=1}^{r} n_i \, d\overline{M}_i = 0. \quad (7.48)$$

Dividing it with dn_1, we obtain

$$\sum_{i=1}^{r} n_i \left(\frac{\partial \overline{M}_i}{\partial n_1} \right)_{T,p,n_2,\ldots,n_r} = 0. \quad (7.49)$$

In terms of mole fractions, this equation is expressible as

$$\sum_{i=1}^{r} x_i \left(\frac{\partial \overline{M}_i}{\partial x_1} \right)_{T,p,x_2,\ldots,x_{r-1}} = 0. \quad (7.50)$$

This relation will be useful for devising methods of measuring partial molar properties discussed below.

7.3.1 Method of Intercepts

We wish to devise methods of measuring partial molar property M for a two-component mixture composed of n_1 and n_2 moles of components 1 and 2, respectively. Setting $r = 2$ and dividing Eq. (7.45) with $n = n_1 + n_2$, we obtain the mean molar value of M:

$$M_m = x_1\overline{M}_1 + x_2\overline{M}_2. \tag{7.51}$$

A differentiation of Eq. (7.51) with respect to x_1 yields

$$\left(\frac{\partial M_m}{\partial x_1}\right)_{T,p} = \overline{M}_1 - \overline{M}_2, \tag{7.52}$$

for which we have made use of Eq. (7.50) for $r = 2$:

$$x_1\left(\frac{\partial \overline{M}_1}{\partial x_1}\right)_{T,p} + x_2\left(\frac{\partial \overline{M}_2}{\partial x_1}\right)_{T,p} = 0.$$

Since $x_1 + x_2 = 1$, we may eliminate x_1 or x_2 from Eq. (7.51) and, by making use of Eq. (7.52), obtain the equation in terms of x_2

$$\begin{aligned} M_m(x_2) &= \overline{M}_1(x_2) + \left[\overline{M}_2(x_2) - \overline{M}_1(x_2)\right]x_2 \\ &= \overline{M}_1(x_2) - \left(\frac{\partial M_m}{\partial x_2}\right)_{T,p} x_2, \end{aligned} \tag{7.53}$$

or the equation in terms of x_1

$$\begin{aligned} M_m(x_1) &= \overline{M}_2(x_1) + \left[\overline{M}_1(x_1) - \overline{M}_2(x_1)\right]x_1 \\ &= \overline{M}_2(x_1) + \left(\frac{\partial M_m}{\partial x_1}\right)_{T,p} x_1. \end{aligned} \tag{7.54}$$

If M_m is measured and plotted against x_2, say, a curve in Fig. 7.1 is obtained. The tangent line to $M_m(x_2)$ at $x_2 = x$ is

$$\begin{aligned} y(x_2) &= A - \left(\frac{\partial M_m}{\partial x_2}\right)_{T,p} |_{x_2=x} x_2 \\ &= A + \left[\overline{M}_2(x) - \overline{M}_1(x)\right]x_2, \end{aligned} \tag{7.55}$$

where A is the intercept at $x_2 = 0$. Since at $x_2 = x$

$$M_m(x) = \overline{M}_1(x) + \left[\overline{M}_2(x) - \overline{M}_1(x)\right]x$$

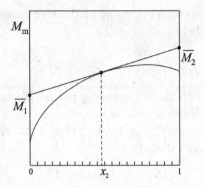

Fig. 7.1   Method of intercepts. The partial molar properties $\overline{M}_1$ and $\overline{M}_2$ are the intercepts of the tangent to the curve at x_2.

and

$$y(x) = M_m(x),$$

it follows

$$A = \overline{M}_1(x).$$

Therefore the tangent line to the curve at $x_2 = x$ is given by

$$y(x_2) = \overline{M}_1(x) + \left[\overline{M}_2(x) - \overline{M}_1(x) \right] x_2. \qquad (7.56)$$

Hence $\overline{M}_1(x)$ at $x_2 = x$ is given by the intercept of this tangent line, and varying the position of x continuously between $0 \leq x_2 \leq 1$ the intercept $\overline{M}_1(x)$, the partial molar value of M, can be measured as a function of x for component 1. Similarly, by plotting M_m against x_1 and using the tangent equation

$$y(x_1) = \overline{M}_2(x) + \left[\overline{M}_1(x) - \overline{M}_2(x) \right] x_1, \qquad (7.57)$$

$\overline{M}_2(x)$ at $x_1 = x$ is obtained from the intercept of the tangent line.

7.3.2   *Direct Method*

Since by definition

$$\overline{M}_i = \left(\frac{\partial M}{\partial n_i} \right)_{T,p,n_{j \neq i}},$$

we see that partial molar properties can be measured by simply following the change in M as the mole number of a component is changed while the mole numbers of other components are kept constant. The slope of the curve for M vs the mole number can then be obtained graphically.

7.3.3 Method of Apparent Molar Property

We illustrate this method with the example of two components. In this method, we measure apparent molar property Ω_M defined by

$$\Omega_M = \frac{(M - n_1 M_1)}{n_2}, \tag{7.58}$$

where M_1 is the molar property of pure component 1. By rearranging the terms in the form

$$M - n_1 M_1 = n_2 \Omega_M$$

and differentiating it with n_2 at constant p and T we obtain

$$\Omega_M = \overline{M}_2 - n_2 \left(\frac{\partial \Omega_M}{\partial n_2} \right)_{T,p,n_1}. \tag{7.59}$$

This equation suggests that $\overline{M}_2(a)$ at $n_2 = a$ can be determined by measuring the position of the intercept of the tangent line to Ω_M at $n_2 = a$ as schematically shown in Fig. 7.2.

7.3.4 Density Dependence of $\overline{M}_i$

Partial molar properties thus measured can be summarized with a power series in mole numbers. For example, in the case of two-component mixtures we may express them in the series of mole fractions

$$\overline{M}_1 = M_{10} + M_{11} x_2 + M_{12} x_2^2 + \cdots, \tag{7.60}$$

$$\overline{M}_2 = M_{20} + M_{21} x_1 + M_{22} x_1^2 + \cdots. \tag{7.61}$$

In this mode of representation, M_{10} and M_{20} are, respectively, the molar property of the pure substance 1 and that of pure substance 2, and

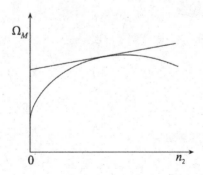

Fig. 7.2 The method of apparent molar property. The intercept of the tangent is $\overline{M}_2(a)$ at $n_2 = a$.

M_{11}, M_{21}, and so on are parameters determined experimentally. These parameters as well as M_{10} and M_{20} are, of course, dependent on temperature and pressure.

Problems

(1) The volume of aqueous solution of NaCl at 298.15 K and 1 atm depends on the molality m of NaCl as follows: in the units of L

$$V = 1.00294 + 0.01640\, m + 0.00214\, m^{3/2} + 2.7 \times 10^{-6}\, m^{5/2}.$$

Find molal volumes $\bar{V}_{NaCl}$ and $\bar{V}_{H_2O}$.

(2) When 1.158 moles of water are dissolved in 0.842 moles of ethanol the volume of the solution is 0.06816 L at 298.15 K. If $\bar{V}_{H_2O} = 0.01698\,$L mol^{-1} in this solution, how large is $\bar{V}_{ethanol}$?

(3) Compare the partial molar volumes of the components with their molar volumes, if $H_2O(l)$ and $C_2H_5OH(l)$ have molecular weights of 1.802×10^{-2} and 4.602×10^{-2} kg mol^{-1}, and densities of 0.9970 and 0.7852 kg L^{-1}, respectively, at the same temperature.

(4) Show the equivalence of Eqs. (7.39) and (7.41).

(5) Show that the specific heat per volume c_v may be given by

$$c_v = T\left[\left(\frac{\partial^2 p}{\partial T^2}\right)_V - n\left(\frac{\partial^2 \mu}{\partial T^2}\right)_V\right],$$

where $n = N/V$. This can be useful for studying the T dependence of c_v, especially, near the critical point. [Hint: Use the Gibbs–Duhem equation for a single component.]

Chapter 8

Heterogeneous Equilibria

We have defined the term *heterogeneous* as the state of a system in which the intensive properties vary with position in space. For example, if the density changes discontinuously across the boundary of two homogeneous regions then the overall system is heterogeneous. Such homogeneous regions of a heterogeneous system are called the *phases* of the system. The thermodynamic theory developed up to this point in the previous chapters assumes the system is homogeneous. However, since thermodynamic systems are often heterogeneous it is necessary to generalize the theory so as to treat them thermodynamically. The desired generalization was originally achieved in a complete form by J. W. Gibbs in his monumental work,[1] and we follow his theory.

8.1 Equilibrium Conditions for a Multiphase System

Let us imagine a system of ν phases composed of r chemically inert components. The entire system is isolated from the surroundings. Suppose that these phases of different thermodynamic states, each of which is in internal equilibrium by itself, are artificially separated by a device of walls, which are thermal insulators, rigid, and impermeable to matter (molecules), so that the whole system is kept from reaching equilibrium. The thermal insulators keep the phases from reaching thermal equilibrium and the rigid walls prevent them from reaching mechanical equilibrium, while impermeable

---

[1] J. W. Gibbs, *Trans. Connecticut Acad. II*, 382–404 (1873) reprinted in *The Scientific Papers of J. Willard Gibbs* (Dover, New York, 1961).

membranes forbid them from attaining material equilibrium. Let us assume that there are ν phases and the states of the ν phases are specified, respectively, by T_α, p_α, $n_1^{(\alpha)}$, $n_2^{(\alpha)}$, ..., $n_r^{(\alpha)}$, $\alpha = 1, 2, \ldots, \nu$, where T_α means the temperature, p_α the pressure, and $n_i^{(\alpha)}$ the mole number of species i in phase α, which adds up to the total mole number n_i of species i in the system:

$$n^{(\alpha)} = \sum_{i=1}^{r} n_i^{(\alpha)}. \tag{8.1}$$

Therefore there are $(r-1)$ independent mole fractions, say, $x_1^{(\alpha)}$, $x_2^{(\alpha)}$, ..., $x_{r-1}^{(\alpha)}$ ($x_i^{(\alpha)} = n_i^{(\alpha)}/n^{(\alpha)}$), and the thermodynamic state of each phase may be specified by $(r+1)$ variables, T_α, p_α, $x_1^{(\alpha)}$, $x_2^{(\alpha)}$, ..., $x_{r-1}^{(\alpha)}$ ($\alpha = 1, 2, \ldots, r$). As the constraints of thermal insulators, rigid walls, and impermeable membranes are removed in part or all, the phases will eventually reach equilibrium to the extent the constraints are removed. We now wish to find the necessary and sufficient conditions for equilibria between different phases.

The equilibrium condition previously established for a single phase system can be easily generalized to a multiphase system. In order for the system to be in equilibrium the internal energy variation must be

$$(\delta E)_{\Psi, V, n_1, \ldots, n_r} \geq 0. \tag{8.2}$$

We further assume that the phases are sufficiently large in mass and size, so that the surface (interfacial) contributions to thermodynamic extensive quantities are negligible compared with their bulk values. Since the phases are internally in equilibrium, $\Psi^{(\alpha)} = S^{(\alpha)}$ for each phase. This does not necessarily mean that the total calortropy Ψ of the entire system is equal to the total entropy of the system since it may not be in equilibrium. Nevertheless, there does not arise the need to know Ψ for the discussion of equilibrium conditions under the assumption of internal equilibrium of phases since here we are not interested in how the whole system reaches equilibrium in time and space, but in the equilibrium conditions only. The extensive variables for the whole system under the assumption of no interfacial contribution then may be given simply as sums of the contributions from the ν separate phases:

$$E = \sum_{\alpha=1}^{\nu} E^{(\alpha)}, \quad S = \sum_{\alpha=1}^{\nu} S^{(\alpha)}, \quad V = \sum_{\alpha=1}^{\nu} V^{(\alpha)}, \quad n_i = \sum_{\alpha=1}^{\nu} n_i^{(\alpha)}. \tag{8.3}$$

Since the individual phases are assumed to be internally in equilibrium, the variation in $E^{(\alpha)}$ is given by

$$\delta E^{(\alpha)} = T_\alpha \; \delta S^{(\alpha)} - p_\alpha \; \delta V^{(\alpha)} + \sum_{i=1}^{r} \mu_i^{(\alpha)} \delta n_i^{(\alpha)} \quad (\alpha = 1, 2, \ldots, \nu), \quad (8.4)$$

where $\mu_i^{(\alpha)}$ are chemical potentials of species i in phase α and $\Psi^{(\alpha)}$ is replaced by $S^{(\alpha)}$ because the phases are internally in equilibrium.

Substitution of this formula into Eq. (8.2) yields the inequality

$$\sum_{\alpha=1}^{\nu} \left(T_\alpha \; \delta S^{(\alpha)} - p_\alpha \; \delta V^{(\alpha)} + \sum_{i=1}^{r} \mu_i^{(\alpha)} \; \delta n_i^{(\alpha)} \right) \geq 0. \quad (8.5)$$

Since S, V, and n_i are kept constant, this inequality is subject to the conditions

$$\delta S = \sum_{\alpha=1}^{\nu} \delta S^{(\alpha)} = 0, \quad (8.6)$$

$$\delta V = \sum_{\alpha=1}^{\nu} \delta V^{(\alpha)} = 0, \quad (8.7)$$

$$\delta n_i = \sum_{\alpha=1}^{\nu} \delta n_i^{(\alpha)} = 0 \; (i = 1, 2, \ldots, r). \quad (8.8)$$

Inequality (8.5) is the general equilibrium condition for the heterogeneous system we are considering here. To better understand its significance in more detail we consider special cases by removing the constraints initially imposed on the nature of phase boundaries; namely, by making them movable, diathermal, or permeable.

8.1.1 *Mechanical Equilibrium*

Let us assume that the boundary between phases 1 and 2 is made movable, so that phases 1 and 2 are allowed to attain mechanical equilibrium, while all other constraints are still kept imposed. This means that

$$\delta S^{(\alpha)} = 0, \quad \delta n_i^{(\alpha)} = 0$$

for all α and

$$\delta V^{(\alpha)} = 0$$

for all α except for $\alpha = 1, 2$. The physical meanings of these conditions are that there are no heat and material exchanges between subsystems (i.e., phases) and the subsystems for $\alpha \neq 1, 2$ do not change their volume. Such conditions can be achieved if the subsystems are not allowed to reach thermal, material, and mechanical equilibrium, or the walls between subsystems are kept adiabatic, impermeable to matter, and mechanically rigid. Then, since by Eq. (8.7)

$$\delta V^{(1)} = -\delta V^{(2)},$$

the equilibrium condition (Eq. (8.5)) takes the inequality

$$(p_2 - p_1)\, \delta V^{(1)} \geq 0. \tag{8.9}$$

If $\delta V^{(1)} \geq 0$ then $p_2 \geq p_1$. If $\delta V^{(1)} \leq 0$ then $p_2 \leq p_1$. But $\delta V^{(1)}$ is arbitrary; see Fig. 8.1. Therefore Condition (8.9) can be satisfied only if

$$p_1 = p_2. \tag{8.10}$$

This is the condition for mechanical equilibrium between phases 1 and 2. Obviously, if the wall is movable, the pressure on both phases must be equalized for mechanical equilibrium to be established.

However, if $\delta V^{(1)} \geq 0$ is the only possible variation then the equilibrium condition is

$$p_2 \geq p_1.$$

This situation is schematically illustrated in Fig. 8.2, where there are catches on one side of the wall, so that it can move to the right, but not to the left. In the opposite case, that is, if $\delta V^{(1)} \leq 0$ is the only possible variation then the equilibrium condition is

$$p_2 \leq p_1.$$

This situation is illustrated in Fig. 8.3, where the wall can move only to the left. If the wall is rigid, then

$$\delta V^{(1)} = \delta V^{(2)} = 0$$

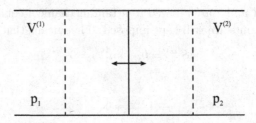

Fig. 8.1 The wall is movable in either direction in such a way that pressures are equalized on both sides.

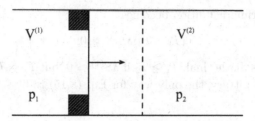

Fig. 8.2 The wall can move only toward right from the catches.

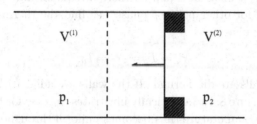

Fig. 8.3 The wall can move only toward left from the catches.

and there can be no condition imposed on p_1 and p_2; there is therefore no question of mechanical equilibrium.

One can repeat the same argument with different pairs of phases and finally obtain the conditions

$$p_1 = p_2 = \cdots = p_\nu \tag{8.11}$$

for mechanical equilibrium, if all the phase boundaries are deformable. In this case, the phase boundaries will deform or move until the pressure becomes uniform over the whole system of phases.

8.1.2 *Thermal Equilibrium*

Suppose the boundary of phases 1 and 2 is made diathermal, so that phases 1 and 2 are allowed to exchange heat so that they can reach thermal equilibrium, but all other constraints are still maintained intact. In this case,

$$\delta V^{(\alpha)} = 0, \quad \delta n_i^{(\alpha)} = 0 \tag{8.12}$$

for all α, and

$$\delta S^{(\alpha)} = 0 \tag{8.13}$$

for all α except for $\alpha = 1, 2$. But from Eq. (8.6) we obtain

$$\delta S^{(1)} = -\delta S^{(2)} \tag{8.14}$$

and the equilibrium condition becomes

$$(T_1 - T_2)\delta S^{(1)} \geq 0. \tag{8.15}$$

Since the wall is diathermal, $T_1 \geq T_2$ if $\delta S^{(1)} > 0$, but $T_1 \leq T_2$ if $\delta S^{(1)} < 0$. Since $\delta S^{(1)}$ is arbitrary, the only way for Eq. (8.15) to be satisfied is

$$T_1 = T_2. \tag{8.16}$$

This is the condition for thermal equilibrium. That is, the temperatures of phases 1 and 2 must be equal at thermal equilibrium. Repeating the same argument for other pairs of phases we find the thermal equilibrium conditions for ν phases:

$$T_1 = T_2 = \cdots = T_\nu, \tag{8.17}$$

when all the walls are diathermal. If the walls are adiabatic, Eq. (8.17) is not required. Figure 8.4 schematically illustrates the case for Eq. (8.16). At equilibrium the temperature thus becomes uniform throughout the phases, if the walls are all diathermal.

8.1.3 *Material Equilibrium*

Suppose that the boundary between phases 1 and 2 is made permeable to species i so that phases 1 and 2 are allowed to reach material equilibrium with respect to species i while other constraints imposed are unchanged. Consequently there hold

$$\delta V^{(\alpha)} = 0, \quad \delta S^{(\alpha)} = 0 \tag{8.18}$$

for all α, but

$$\delta n_j^{(\alpha)} = 0 \tag{8.19}$$

for all j and α except for $j = i$ and $\alpha = 1, 2$. In that case, by Eq. (8.8)

$$\delta n_i^{(1)} = -\delta n_i^{(2)} \tag{8.20}$$

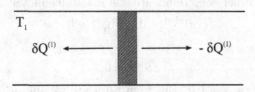

Fig. 8.4   The wall is diathermal, allowing thermal equilibrium to be reached.

and the material equilibrium condition (Eq. (8.5)) takes the form
$$(\mu_i^{(1)} - \mu_i^{(2)}) \, \delta n_i^{(1)} \geq 0. \tag{8.21}$$
Since the membrane is permeable to i, $\delta n_i^{(1)}$ is arbitrary. Therefore, if $\delta n_i^{(1)} > 0$ then
$$\mu_i^{(1)} \geq \mu_i^{(2)},$$
and if $\delta n_i^{(1)} < 0$ then
$$\mu_i^{(1)} \leq \mu_i^{(2)}.$$
Hence the only way for Eq. (8.21) to be satisfied is for the chemical potentials of i in the two phases to be equal:
$$\mu_i^{(1)} = \mu_i^{(2)}. \tag{8.22}$$
This is the condition for material equilibrium when the membrane is permeable to i. By repeating the same argument for other pairs of phases and also for other species we obtain
$$\mu_i^{(1)} = \mu_i^{(2)} = \cdots = \mu_i^{(\nu)} \tag{8.23}$$
for all $i = 1, 2, \ldots, r$ for the conditions for material equilibrium when the membranes are permeable to all species. These conditions mean that chemical potentials become uniform throughout the phases at material equilibrium, if the membranes are permeable to all species.

By collecting the results (Eqs. (8.11), (8.17), and (8.23)) obtained earlier, we reach the conclusion that the necessary conditions for equilibrium of a heterogeneous systems are
$$p_1 = p_2 = \cdots = p_\nu,$$
$$T_1 = T_2 = \cdots = T_\nu,$$
$$\mu_1^{(1)} = \mu_1^{(2)} = \cdots = \mu_1^{(\nu)}, \tag{8.24}$$
$$\vdots$$
$$\mu_r^{(1)} = \mu_r^{(2)} = \cdots = \mu_r^{(\nu)}.$$
These are also the sufficient conditions. It can be shown as follows: If $T_\alpha = T$, $p_\alpha = p$, and $\mu_i^{(\alpha)} = \mu_i$ for all α and i, that is, if Eq. (8.24) holds, then
$$(\delta E)_{\Psi, V, n_1, \ldots, n_r} = \sum_{\alpha=1}^{\nu} \left(T_\alpha \, \delta S^{(\alpha)} - p_\alpha \, \delta V^{(\alpha)} + \sum_{i=1}^{r} \mu_i^{(\alpha)} \, \delta n_i^{(\alpha)} \right)$$
$$= T \sum_{\alpha=1}^{\nu} \delta S^{(\alpha)} - p \sum_{\alpha=1}^{\nu} \delta V^{(\alpha)} + \sum_{i=1}^{r} \mu_i \sum_{\alpha=1}^{\nu} \delta n_i^{(\alpha)},$$

but this must be identically equal to zero

$$(\delta E)_{\Psi, V, n_1, \ldots, n_r} = 0, \qquad (8.25)$$

since Eqs. (8.6)–(8.8) must hold. This is the condition for equilibrium when the calortropy (entropy if in equilibrium), volume, and mole numbers are fixed. Therefore Eq. (8.24) is the necessary and sufficient conditions for ν heterogeneous phases to be at equilibrium when the phase boundaries are free from constraints mentioned earlier. By this the sufficiency of Conditions (8.24) is proved.

The same conclusion as for the equilibrium conditions in Eq. (8.24) can be arrived at, if the uncompensated heat is calculated by means of Eqs. (4.38) and (8.4) subject to the constraints (Eqs. (8.6)–(8.8)) and it is set equal to zero at equilibrium. We elaborate on this statement. For the ν phase system there holds the general relation

$$\sum_{\alpha=1}^{\nu} \left(T_\alpha \, \delta \Psi^{(\alpha)} - \delta E^{(\alpha)} - p_\alpha \, \delta V^{(\alpha)} + \sum_{i=1}^{r} \mu_i^{(\alpha)} \delta n_i^{(\alpha)} \right) = \delta N, \qquad (8.26)$$

for which we have used Eqs. (7.9) and (7.10). Here N is the uncompensated heat. Since $\Psi^{(\alpha)} = S^{(\alpha)}$ for all α on account of the assumption of internal equilibrium of phases and, furthermore, since

$$E = \sum_{\alpha=1}^{\nu} E^{(\alpha)} = \text{constant}$$

on account of the assumption that the entire system is isolated and hence

$$\sum_{\alpha=1}^{\nu} \delta E^{(\alpha)} = 0,$$

Eq. (8.26) can be written as

$$\sum_{\alpha=1}^{\nu} \left(T_\alpha \, \delta S^{(\alpha)} - p_\alpha \, \delta V^{(\alpha)} + \sum_{i=1}^{r} \mu_i^{(\alpha)} \delta n_i^{(\alpha)} \right) = \delta N. \qquad (8.27)$$

This variation is subject to the constraints of Eqs. (8.6)–(8.8). On applying the Lagrange multiplier method to this variational problem we obtain

$$\sum_{\alpha=1}^{\nu} \left[(T_\alpha - T) \, \delta S^{(\alpha)} - (p_\alpha - p) \, \delta V^{(\alpha)} + \sum_{i=1}^{r} \left(\mu_i^{(\alpha)} - \mu_i \right) \delta n_i^{(\alpha)} \right] = \delta N$$

$$\geq 0, \quad (8.28)$$

where T, p, and μ_i are the Lagrange multipliers. Since at equilibrium or for a reversible process

$$\delta N = 0$$

and variations $\delta S^{(\alpha)}$, $\delta V^{(\alpha)}$, and $\delta n_i^{(\alpha)}$ are arbitrary, there follow the equilibrium conditions

$$T = T_\alpha, \quad p = p_\alpha, \quad \mu_i = \mu_i^{(\alpha)},$$

which are the same as conditions in Eq. (8.24).

8.2 Gibbs Phase Rule

For phases composed of r non-reacting components there are $(r+1)$ intensive variables, T_α, p_α, $x_1^{(\alpha)}$, $x_2^{(\alpha)}$, ..., $x_{r-1}^{(\alpha)}$, where $x_i^{(\alpha)}$ is the mole fraction of species i in phase α. Consequently, there are $\nu(r+1)$ intensive variables for a heterogeneous system with ν phases. Since there are $(2+r)(\nu-1)$ conditions to be satisfied by these variables according to the equilibrium conditions just established, the number of free variables is

$$f = \nu(r+1) - (2+r)(\nu-1)$$
$$= r + 2 - \nu. \tag{8.29}$$

This is the number of thermodynamic degrees of freedom and is called the Gibbs phase rule. This rule holds subject to the validity of the assumptions and approximations made to derive Equilibrium Conditions (8.24). Let us recall that Equilibrium Conditions (8.24) are derived under the assumptions of deformable, diathermal, and permeable boundaries (or interfaces). Other assumptions are that there are negligible contributions to E and S from the interfaces, so that *E and S are sums of bulk phase contributions only; that there are uniform normal pressures on phases; and there should be no chemical reactions in the system.*

Consider, for example, a single-component liquid in equilibrium with its own vapor. Since $r = 1$ and $\nu = 2$ in this case, the Gibbs phase rule gives $f = 1$. Therefore the temperature or the pressure can be the variable in this case, and when either one of them is given, the state of the system is completely determined. For instance, if the temperature is given, the vapor pressure is uniquely determined. If three phases of a substance, for example, water, ice, and its vapor, are in equilibrium then $f = 0$. This means that the triple point has no thermodynamic degree of freedom and therefore it is uniquely given and invariant for the substance. This invariance was one of the reasons for choosing the triple point of water as the reference point of temperature scale discussed in Chapter 2. We shall consider some simple cases to illustrate applications of the Gibbs theory of heterogeneous phase equilibria in subsequent chapters.

8.3   One-Component, Two-Phase Systems

Let us assume that the system consists of two phases of a single species, for example, vapor and liquid, or liquid and solid, or vapor and solid of a substance. As is clear from the Gibbs phase rule, $f = 1$ in this case, and it implies that p may be regarded as a function of T, and vice versa. Let us then find out the relationship between the two variables.

Specializing Equilibrium Conditions (8.24) to the case of two phases, we obtain for the equilibrium conditions

$$\mu^{(\alpha)} = \mu^{(\beta)},$$
$$T_\alpha = T_\beta = T, \tag{8.30}$$
$$p_\alpha = p_\beta = p,$$

where α and β denote the two phases; see Fig. 8.5. Since there now is no need for a subscript to distinguish species, we have dropped the species subscript from chemical potentials. Differentiating the first equation in Eq. (8.30) yields

$$d\mu^{(\alpha)} = d\mu^{(\beta)}. \tag{8.31}$$

Since T and p are the variables we may write this equation in the form

$$\left(\frac{\partial \mu^{(\alpha)}}{\partial T}\right)_p dT + \left(\frac{\partial \mu^{(\alpha)}}{\partial p}\right)_T dp = \left(\frac{\partial \mu^{(\beta)}}{\partial T}\right)_p dT + \left(\frac{\partial \mu^{(\beta)}}{\partial p}\right)_T dp. \tag{8.32}$$

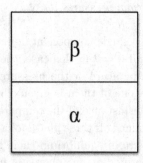

Fig. 8.5   Two-phase equilibrium: phases α and β are in equilibrium across the boundary.

Recalling that

$$\left(\frac{\partial \mu^{(i)}}{\partial T}\right)_p = -s^{(i)}, \quad \left(\frac{\partial \mu^{(i)}}{\partial p}\right)_T = v^{(i)} \ (i = \alpha, \beta), \tag{8.33}$$

where $s^{(i)}$ and $v^{(i)}$ are, respectively, the molar entropy and volume of the substance in phase i, and by rearranging the terms in Eq. (8.32) we obtain the equation

$$\left(v^{(\beta)} - v^{(\alpha)}\right) dp = \left(s^{(\beta)} - s^{(\alpha)}\right) dT. \tag{8.34}$$

This may be written in the form of a differential equation

$$\frac{dp}{dT} = \frac{\Delta s}{\Delta v}, \tag{8.35}$$

where Δs and Δv are, respectively, the molar entropy and the molar volume change accompanying the phase change from α to β:

$$\Delta s = s^{(\beta)} - s^{(\alpha)}, \quad \Delta v = v^{(\beta)} - v^{(\alpha)}. \tag{8.36}$$

If we denote the molar enthalpies of phases α and β by $h^{(\alpha)}$ and $h^{(\beta)}$, the first equation in Eq. (8.36) may be written

$$h^{(\alpha)} - Ts^{(\alpha)} = h^{(\beta)} - Ts^{(\beta)},$$

where the temperature is the same on both sides since the two phases are in equilibrium. By using this relation we may cast Eq. (8.35) in the form

$$\frac{dp}{dT} = \frac{\Delta h}{T \, \Delta v}, \tag{8.37}$$

where Δh is the molar heat of phase transformation, namely, the molar heat of melting, vaporization, and so on:

$$\Delta h = h^{(\beta)} - h^{(\alpha)}. \tag{8.38}$$

Equation (8.37) is called the Clapeyron equation. It describes the temperature dependence of p, and vice versa.

If one of two phases is in a condensed state, such as solid or liquid, and the other is vapor, there is a large difference in molar volumes of phases α and β. In that case, since the volume of the condensed phase may be neglected, compared with that of the vapor phase, we may approximate

$$\Delta v \approx v^{(\beta)}, \tag{8.39}$$

where the phase β is designated the vapor phase. There then follows from Eq. (8.37) the approximate equation

$$\frac{dp}{dT} = \frac{\Delta h}{Tv^{(\beta)}}. \tag{8.40}$$

Furthermore, if the vapor pressure of phase β is low the vapor may be regarded as an ideal gas, and hence the ideal gas equation of state may be used to a good approximation. Eliminating $v^{(\beta)}$ by using the ideal gas equation of state we then obtain from Eq. (8.41) the equation

$$\frac{d\ln p}{dT} = \frac{\Delta h}{RT^2}. \tag{8.41}$$

This is called the Clapeyron–Clausius equation. It is often employed for discussing liquid–vapor or solid–vapor equilibria.

In the case of solid–liquid equilibria it is not generally possible to approximate Δv with $v^{(\beta)}$, since the volumes of the two phases are comparable. In fact, the sign of Δv is crucial for determining the slope of the $p - T$ curve. Thus, in the convention taking the heat of fusion Δh positive, the slope has two different signs (Fig. 8.6)

$$\frac{dp}{dT} > 0 \quad \text{if } \Delta v > 0, \qquad \frac{dp}{dT} < 0 \quad \text{if } \Delta v > 0. \tag{8.42}$$

In the case of $\Delta v > 0$ the p–T curve has a positive slope, whereas in the case of $\Delta v < 0$ it has a negative slope. The former corresponds to an expansion of volume on melting whereas the latter corresponds to a contraction of volume. A typical example for a volume contraction on melting is the ice–water phase transition.

It has a significance in the recreational sport of ice skating.[2] Since ice melts under pressure according to the prediction of Eq. (8.37), the water

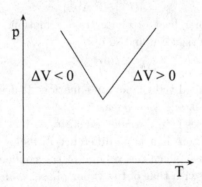

Fig. 8.6 Two different cases of the slope of the p vs T curves at the triple point. The solid–vapor and liquid–vapor coexistence curves are omitted from the figure for brevity (see Fig. 8.9 below).

---

[2]For a recent account of physics underlying this phenomenon, see R. Rosenberg, *Physics Today*, **58**(12), 50 (2005).

generated serves the role of a lubricant and thus making skating facilitated. It is possible to estimate the temperature change arising from a pressure difference Δp in the case of ice–water equilibrium. Since Δh and Δv do not change much if the temperature is not widely different from the melting point, they may be regarded as independent of temperature. The Clapeyron equation then may be easily integrated to yield

$$T \exp(-p\, \Delta v/\Delta h) = C, \tag{8.43}$$

where C is the integration constant. Take T_0 for the melting point at pressure p_0, for example, 1 atm. Eliminating the integration constant C with this set of variables (T_0, p_0) we obtain from Eq. (8.43)

$$T \exp(-p\, \Delta v/\Delta h) = T_0 \exp(-p_0\, \Delta v/\Delta h). \tag{8.44}$$

Setting $T = T_0 + \Delta T$ we finally obtain from Eq. (8.44) the formula for the temperature variation:

$$\Delta T = T_0[\exp(\Delta p\, \Delta v/\Delta h) - 1]. \tag{8.45}$$

Since $\Delta v \leq 0$ for the ice–water transition, the right-hand side is negative and as the pressure is increased from p_0, the melting point decreases below the melting point of ice at p_0. This results in the melting of ice under pressure p, if the ambient temperature is maintained at T_0.

8.3.1 *Vapor Pressure and Measurement of Δh*

The Clapeyron–Clausius equation can be used for measuring the heat of vaporization or sublimation. If Δh does not depend on T too strongly over the temperature range of interest we may regard it as a constant and integrate Eq. (8.41) over temperature. The result of integration is

$$\ln p = -\frac{\Delta h}{RT} + C, \tag{8.46}$$

where C is the integration constant. Let us denote by p_0 the pressure at temperature T_0. It is then possible to eliminate C in favor of T_0 and p_0 and obtain

$$\ln\left(\frac{p}{p_0}\right) = \frac{\Delta h}{R}\left(\frac{1}{T_0} - \frac{1}{T}\right). \tag{8.47}$$

This equation suggests that it is possible to obtain Δh by plotting $\ln(p/p_0)$ against T^{-1} to measure the slope of the line. This procedure is illustrated

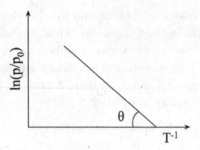

Fig. 8.7 Schematic drawing of the logarithm of vapor pressure against $1/T$. The slope gives Δh.

in Fig. 8.7. If the line intersects with the T^{-1} axis at an angle θ, the heat of vaporization is given by the formula

$$\Delta h = R \tan \theta.$$

The assumption that Δh is constant with respect to T can be removed and the procedure described earlier can be refined with T-dependent Δh. For the purpose of integration it would require fitting Δh to a function of T and experimentally determining the parameters in the fitting function for Δh. Such functions are often available in the literature on thermophysical data; for example, the National Institute of Science and Technology (NIST) reference data base (NIST webbook).

When the Clapeyron–Clausius equation or the Clapeyron equation is integrated for various two-phase equilibria and the results are assembled together, we obtain a surface in the three-dimensional space of p, V, and T as schematically shown in Fig. 8.8. It is often convenient to cut this surface with a plane in order to display the cross section as shown in Fig. 8.9, which shows schematically the essential features of phase equilibria of a single component system. Such a diagram is called a phase diagram.

8.3.2 Ramsay–Young Rule

If the heat of vaporization is assumed to be constant, the Clapeyron–Clausius equation is easily integrated to the form

$$\ln p = \frac{A}{T} + C, \tag{8.48}$$

where

$$A = -\frac{\Delta h_v}{R},$$

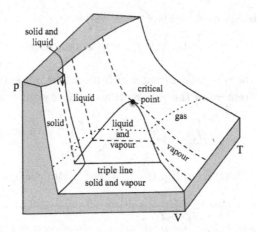

Fig. 8.8   Three-dimensional phase surface.

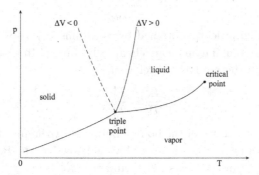

Fig. 8.9   Phase diagram for a one-component system in the $p - T$ plane. This is a cut of the surface shown in Fig. 8.8 such that it includes the critical and triple points. Two alternative cases are shown for two different signs of ΔV.

Δh_v being the heat of vaporization, and C is the integration constant. The parameter depends on the nature of the substance. Let us consider two substances a and b at temperature T_a and T_b, respectively. Assuming that the approximation implied by Eq. (8.48) holds for substances a and b, we obtain

$$\ln p_a = \frac{A_a}{T_a} + C_a, \quad \ln p_b = \frac{A_b}{T_b} + C_b. \qquad (8.49)$$

If the two substances have the same vapor pressure at T_a and T_b, respectively, there holds the relation

$$\frac{A_a}{T_a} + C_a = \frac{A_b}{T_b} + C_b. \qquad (8.50)$$

Rearranging this equation yields

$$\frac{T_a}{T_b} = \frac{T_a}{A_b}(C_a - C_b) + \frac{A_a}{A_b}. \tag{8.51}$$

Let (T_a', T_b') be the temperatures at which the vapor pressures are equal to p', which is different from the vapor pressure $p_a = p_b$ at which Eq. (8.51) holds. Then we have

$$\frac{T_a'}{T_b'} = \frac{T_a'}{A_b}(C_a - C_b) + \frac{A_a}{A_b}. \tag{8.52}$$

Subtracting Eq. (8.52) from Eq. (8.51), there follows the equation

$$\frac{T_a}{T_b} - \frac{T_a'}{T_b'} = C_1(T_a - T_a'), \tag{8.53}$$

where

$$C_1 = \frac{(C_a - C_b)}{A_b}. \tag{8.54}$$

This result was empirically obtained by W. Ramsay and S. Young. They found that C_1 is small if a and b are chemically similar. In that case, $C_1 \simeq 0$ and consequently there holds the relation

$$\frac{T_a}{T_b} = \frac{T_a'}{T_b'}. \tag{8.55}$$

This means that the ratio of the boiling points of two similar liquids should have the same value at all pressures. This is known as the Ramsay–Young rule. For example, in the case of H_2O and ethanol,

$$p_{H_2O} = 0.0122 \text{ m Hg at } T = 287.5 \text{ K},$$

$$p_{\text{ethanol}} = 0.122 \text{ m Hg at } T = 273.2 \text{ K}.$$

Thus the temperature ratio is

$$\frac{287.5}{273.2} = 1.052.$$

On the other hand, the normal boiling points are

$$T = 373.2 \text{ K for } H_2O,$$

$$T = 351.5 \text{ K for ethanol},$$

in which case the temperature ratio is

$$\frac{373.2}{351.5} = 1.062.$$

This value has about 1% deviation from 1.052 and indicates the rule holds within the error. This rule allows a quick approximate determination of the boiling point of a substance from the boiling point of another similar substance if the boiling point of the latter is known at a pressure and the boiling point of the former is desired at the same pressure.

8.4 Two-Component Systems

For a two-component system the Gibbs phase rule is

$$f = 4 - \nu,$$

which, depending on the number of phases, will be;

$$
\begin{aligned}
f &= 0 \quad \text{for } \nu = 4 \\
&= 1 \quad \text{for } \nu = 3 \\
&= 2 \quad \text{for } \nu = 2 \\
&= 3 \quad \text{for } \nu = 1.
\end{aligned}
$$

The last case is a homogeneous two-component system which does not interest us here. The case of $f = 0$ corresponds to a point, which is invariant, in a three-dimensional space of intensive thermodynamic variables, for example, T, p, and x_1. The case of $f = 1$ can be described by the Clapeyron–Clausius equation whereas the case of $f = 2$ is described by a differential form of two independent variables. Since we have already examined the Clapeyron–Clausius equation, we shall consider the latter case ($f = 2$) in this section.

Since there are two phases and two components in this case, the equilibrium conditions are

$$
\begin{aligned}
T_1 &= T_2 = T, \\
p_1 &= p_2 = p, \\
\mu_i^{(1)} &= \mu_i^{(2)} = \mu_i \quad (i = 1, 2).
\end{aligned}
\tag{8.56}
$$

Taking differentials, we obtain the last condition in the form

$$d\mu_i^{(1)} = d\mu_i^{(2)} \quad (i = 1, 2). \tag{8.57}$$

Let us take T, p, $x_2^{(\alpha)}$ ($\alpha = 1, 2$) for independent variables. The chemical potentials then are functions thereof.

Applying Eq. (7.36) to the present case, we obtain differential forms for chemical potentials

$$d\mu_i^{(\alpha)} = -\bar{s}_i^{(\alpha)} dT + \bar{v}_i^{(\alpha)} dp + \phi_i^{(\alpha)} dx_2^{(\alpha)} \quad (i = 1, 2; \ \alpha = 1, 2), \qquad (8.58)$$

where $\bar{s}_i^{(\alpha)}$ and $\bar{v}_i^{(\alpha)}$ are, respectively, the partial molar entropy and partial molar volume of component i in phase α, and

$$\phi_i^{(\alpha)} = \left(\frac{\partial \mu_i^{(\alpha)}}{\partial x_2^{(\alpha)}} \right)_{T,p}. \qquad (8.59)$$

Since in general

$$\frac{S^{(\alpha)}}{n} = \sum_{i=1}^{r} x_i^{(\alpha)} \bar{s}_i^{(\alpha)},$$

and

$$\frac{V^{(\alpha)}}{n} = \sum_{i=1}^{r} x_i^{(\alpha)} \bar{v}_i^{(\alpha)} \quad \left(n = \sum_{i=1}^{r} n_i \right),$$

the Gibbs–Duhem equation for the present case may be written in the form

$$\sum_{i=1}^{2} x_i^{(\alpha)} d\mu_i^{(\alpha)} + \sum_{i=1}^{2} x_i^{(\alpha)} (\bar{s}_i^{(\alpha)} dT - \bar{v}_i^{(\alpha)} dp) = 0 \quad (\alpha = 1, 2). \qquad (8.60)$$

This equation leads us to a pair of equations which are comparable to the Clapeyron equation for two-phase equilibria of a single component system. In order to obtain one of them let us take $\alpha = 1$ in Eq. (8.60):

$$\sum_{i=1}^{2} x_i^{(1)} d\mu_i^{(1)} + \sum_{i=1}^{2} x_i^{(1)} (\bar{s}_i^{(1)} dT - \bar{v}_i^{(1)} dp) = 0. \qquad (8.61)$$

On elimination of $d\mu_i^{(1)}$ from this equation by using Eqs. (8.57) and (8.58) there follows the differential form

$$\sum_{i=1}^{2} x_i^{(1)} \phi_i^{(2)} dx_2^{(2)} + \Delta\bar{s}^{(1)} dT - \Delta\bar{v}^{(1)} dp = 0, \qquad (8.62)$$

where

$$\Delta\bar{v}^{(1)} = \sum_{i=1}^{2} x_i^{(1)} \Delta\bar{v}_i, \qquad \Delta\bar{s}^{(1)} = \sum_{i=1}^{2} x_i^{(1)} \Delta\bar{s}_i, \qquad (8.63)$$

$$\Delta\bar{v}_i = \bar{v}_i^{(1)} - \bar{v}_i^{(2)}, \qquad \Delta\bar{s}_i = \bar{s}_i^{(1)} - \bar{s}_i^{(2)}. \qquad (8.64)$$

Since at equilibrium

$$\Delta \bar{s}_i = \frac{\Delta \bar{h}_i}{T}, \tag{8.65}$$

where $\Delta \bar{h}_i$ is the enthalpy change for component i owing to the phase change

$$\Delta \bar{h}_i = \bar{h}_i^{(1)} - \bar{h}_i^{(2)}, \tag{8.66}$$

we may recast Eq. (8.62) in a more suitable form

$$-\frac{\Delta \bar{h}^{(1)}}{T} dT = -\Delta \bar{v}_1^{(1)} \, dp + \left(\sum_{i=1}^{2} x_i^{(1)} \phi_i^{(2)} \right) dx_2^{(2)}, \tag{8.67}$$

with $\Delta \bar{h}^{(1)}$ defined by

$$\Delta \bar{h}^{(1)} = \sum_{i=1}^{2} x_i^{(1)} \Delta \bar{h}_i. \tag{8.68}$$

A similar analysis can be made by taking $\alpha = 2$ in Eq. (8.60). By defining

$$\Delta \bar{v}^{(2)} = \sum_{i=1}^{2} x_i^{(2)} \Delta \bar{v}_i, \quad \Delta \bar{h}^{(2)} = -\sum_{i=1}^{2} x_i^{(2)} \Delta \bar{h}_i, \tag{8.69}$$

we also obtain a differential form equivalent to Eq. (8.67):

$$-\frac{\Delta \bar{h}^{(2)}}{T} dT = -\Delta \bar{v}^{(2)} dp + \left(\sum_{i=1}^{2} x_i^{(2)} \phi_i^{(1)} \right) dx_2^{(1)}. \tag{8.70}$$

Note that Eq. (8.67) is in reference to $x_2^{(2)}$ in phase 2 whereas Eq. (8.70) is in reference to $x_2^{(1)}$ in phase 1. The solutions of these two equations will give relations for T and p in reference to the concentration of component 2 either in phase 1 or in phase 2.

Since it is not easy to integrate the differential forms in the three-dimensional space of T, p, $x_2^{(1)}$ or T, p, $x_2^{(2)}$, we shall consider a couple of special cases. These special cases mathematically correspond to dividing the space with a plane and studying the relations between two variables in the plane.

If the pressure is kept constant, we obtain first-order differential equations from Eqs. (8.67) and (8.70):

$$\begin{aligned} \left(\frac{d \ln T}{dx_2^{(2)}} \right)_p &= -\frac{1}{\Delta \bar{h}^{(1)}} \sum_{i=1}^{2} \phi_i^{(2)} x_i^{(1)}, \\ \left(\frac{d \ln T}{dx_2^{(1)}} \right)_p &= -\frac{1}{\Delta \bar{h}^{(1)}} \sum_{i=1}^{2} \phi_i^{(2)} x_i^{(2)}. \end{aligned} \tag{8.71}$$

For the purpose of integrating these equations it is necessary to know the $x_2^{(2)}$ dependence of $\Delta \bar{h}^{(1)}$, $\phi_i^{(2)}$, and $x_2^{(1)}$ or the $x_2^{(1)}$ dependence of $\Delta \bar{h}^{(1)}$, $\phi_i^{(2)}$, and $x_2^{(2)}$. Assuming they are known, when these two differential equations are integrated, there arise relations between T and $x_2^{(1)}$ or T and $x_2^{(2)}$. Schematically, these relations may be represented by the curves in Figs. 8.10 and 8.11. Notice that the two curves are not necessarily of the same shape.

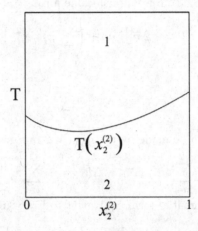

Fig. 8.10 Phase diagram for a two-component system in $T - x_2^{(2)}$ plane. The superscript refers to the phase whereas the subscript refers to the component. Numerals 1 and 2 refer to the phases.

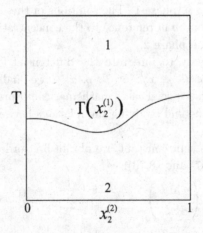

Fig. 8.11 Phase diagram for a two-component system in $T - x_2^{(1)}$ plane. The meanings of the symbols are similar to the Fig. 8.10.

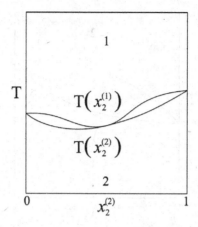

Fig 8.12 Phase diagram for a two-component system in T–x_2 plane. This is a combined figure of Figs. 8.10 and 8.11.

This is generally the case. By combining the two figures into one we usually represent the phase diagrams in a single figure, and in this particular case obtain the $T - x_2$ curves for a two-phase system shown in Fig. 8.12:

On the other hand, if the temperature is kept constant, we obtain from Eqs. (8.67) and (8.70) the following pair of equations:

$$\left(\frac{dp}{dx_2^{(2)}}\right)_T = \frac{1}{\Delta\overline{v}^{(1)}} \sum_{i=1}^{2} \phi_i^{(2)} x_i^{(1)},$$

$$\left(\frac{dp}{dx_2^{(1)}}\right)_T = \frac{1}{\Delta\overline{v}_i^{(2)}} \sum_{i=1}^{2} \phi_i^{(1)} x_i^{(2)}.$$

$$(8.72)$$

By integrating these equations, it is possible to find the pressure–concentration relations for the system at a given T. For example, they may look as shown in Figs. 8.13 and 8.14. In Figs. 8.12 and 8.14 two curves coincide at a value of x. At that temperature and pressure two phases acquire the same composition. Such a mixture is called *azeotropic*. When a mixture becomes azeotropic, fractional distillation is no longer possible, unless the variable kept fixed, namely, T or p, is appropriately changed. Water–alcohol mixtures or benzene–ethanol mixtures can be azeotropic at certain conditions. We will study phase equilibria of two-component solutions in more detail in later chapters, where thermodynamics of solutions is dealt with.

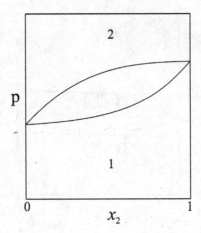

Fig. 8.13   Phase diagram for a two-component system in $p-x_2$ plane.

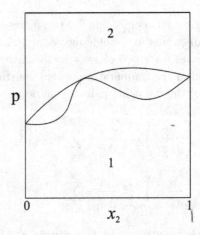

Fig. 8.14   Phase diagram for a two-component system in $p-x_2$ plane.

If x_2 is kept constant then there results from Eqs. (8.67) and (8.70) Clapeyron equations, which we have already studied. Therefore we do not consider this case here.

It is, of course, possible to study multicomponent mixtures, but the principles required are the same and the general theory developed can be followed, except that the phase diagrams become more complex and the equations involved are more complicated. These topics are referred to the monographs specializing on the subjects.

Problems

(1) The vapor pressure of a liquid is represented by the expression

$$\ln p = -\frac{A}{T} + B - CT + DT^2.$$

Find the temperature dependence of the heat of vaporization under the assumption that the vapor is ideal.

(2) Assume that the chemical potentials for a binary mixture are given by

$$\mu_i^{(\alpha)} = \mu_i^{0(\alpha)} + RT \ln x_i^{(\alpha)} \quad (1, 2; \ i = 1, 2),$$

when two phases (liquid and vapor) are in equilibrium. Obtain from Eq. (8.71) the differential equations for T with regard to variables $x_i^{(\alpha)}$. Integrate these differential equations to obtain the relations to $x_i^{(\alpha)}$ ($\alpha = 1, 2$).

Chapter 9

Thermodynamics of Real Fluids

While discussing the basic thermodynamic concepts and principles in the previous chapters, we have indicated some procedures to compute thermodynamic quantities of ideal gases. In this and the following chapters we apply those concepts and principles to study the thermodynamic properties of real gases and liquids. Since the results for ideal gases can be obtained at zero pressure or low density limits from those which we will obtain, we will not separately consider the ideal gas thermodynamics, but only indicate the ideal gas results whenever necessary and possible. The thermodynamics of real fluids requires a suitable generalization of an equation of state beyond the ideal gas equation of state as well as the caloric equation of state and the constitutive relations for chemical potentials—generally called constitutive equations.

9.1   Constitutive Equations

The thermodynamic state of a system at equilibrium is described by the equilibrium Gibbs relation (fundamental relation), which is a differential form for the entropy for an r component mixture

$$T \, dS = dE + p \, dV - \sum_{i=1}^{r} \mu_i \, dn_i \tag{9.1}$$

or in the energy representation

$$dE = T \, dS - p \, dV + \sum_{i=1}^{r} \mu_i \, dn_i. \tag{9.2}$$

This differential form in principle gives the internal energy E in terms of independent variables S, V, and n_i, $i = 1, 2, \ldots, r$, or in the entropy mode,

159

the entropy S in terms of E, V, and n_i, $i = 1, 2, \ldots, r$, if the differential form is integrated. For a closed system $dn_i = 0$, and we have

$$dE = T\,dS - p\,dV. \tag{9.3}$$

These equilibrium Gibbs relations are not complete for the description of thermodynamic states of a system since the derivatives are not specified in terms of the independent variables chosen. That is, for example, in the case of the energy mode the derivatives of the differential form are

$$T = \left(\frac{\partial E}{\partial S}\right)_{V,n}, \tag{9.4}$$

$$p = \left(\frac{\partial E}{\partial V}\right)_{S,n}, \tag{9.5}$$

$$\mu_i = \left(\frac{\partial E}{\partial n_i}\right)_{S,V,n_{j\neq i}}, \tag{9.6}$$

which evidently must be functions of $s = S/n$, $v = V/n$, $n_i(i = 1, 2, \ldots, r)$, but their precise functional forms in terms of the independent variables are not revealed by the differential form (Eqs. (9.1) or (9.2)) itself. In thermodynamics, the functional forms of such derivatives must be supplied on the empirical grounds. They are called the constitutive relations. They are characteristics of the substance of interest, indicating their thermophysical properties in terms of material parameters.

It is common to take the caloric equation of state — i.e., c_v or c_p as a function of T and v or p — in place of $T = T(s, v, n_1, \ldots, n_r)$. Thus we choose the following set of constitutive equations:

$$c_v = c_v(T, v, x_1, \ldots, x_{r-1}) \tag{9.7}$$

or

$$c_p = c_p(T, p, x_1, \ldots, x_{r-1}), \tag{9.8}$$

and

$$p = f(T, v, x_1, \ldots, x_{r-1}), \tag{9.9}$$

$$\mu_i = \mu_i(T, p, x_1, \ldots, x_{r-1}) \quad (i = 1, 2, \ldots, r), \tag{9.10}$$

where it is assumed that there are r components in the system. These relations reflect the properties of the substance in question, and hence the name constitutive equations (or relations). Especially, Eq. (9.9) is called the equation of state, and it must be empirically determined through experiment or theoretically derived by means of statistical mechanics. There are numerous empirical equations of state known in thermodynamics. Different substances also have different equations of state, although the same class of substances may have the same mathematical form for the equation of state with different parameters which reflect the characteristics of the substances.

9.1.1 *Ideal Gas Equation of State*

For ideal gases, as shown earlier, the equation of state takes the universal form

$$pV = nRT, \tag{9.11}$$

where n is the mole number of the gas in volume V. This equation of state holds only when the pressure is sufficiently low. Thus, more rigorously, we may write it in the form

$$\lim_{p \to 0} \frac{pV}{nRT} = 1.$$

Since by definition

$$\left(\frac{\partial E}{\partial V} \right)_T = 0$$

or

$$\left(\frac{\partial E}{\partial p} \right)_T = 0$$

for ideal gases, the internal energy of ideal gases depends on the temperature only.

9.1.2 *Caloric Equation of State*

Since specific heats of monatomic ideal gases are known to be independent of T, we will specifically denote them by $C_p^* = nc_p^*$ and $C_v^* = nc_v^*$:

$$c_p^* = \lim_{p \to 0} c_p, \quad c_v^* = \lim_{p \to 0} c_v. \tag{9.12}$$

If the temperature is not so high as to excite electrons in atoms into the excited states, the specific heats of monatomic gases are those of the translational degrees of freedom and have the following values: $c_p^* = \frac{5}{2}R$ and

$c_v^* = \frac{3}{2}R$ per mole. Specific heats of ideal diatomic and polyatomic gases depend on temperature because the excitation of internal motions depends on temperature, and they are sometimes a rather strong function of T. It is generally possible to split specific heats into the translational and internal contributions:

$$c_p(T) = c_p^* + c_p^{\text{int}}(T), \qquad c_v(T) = c_v^* + c_v^{\text{int}}(T). \qquad (9.13)$$

Note that for ideal gases the translational part is the same for all molecules regardless of whether they are monatomic, diatomic, or polyatomic. It is possible to derive explicitly the contributions of the internal degrees of freedom to the specific heats by using the method of statistical mechanics, but it is sufficient for our purpose here to express them in a power series of T as indicated in Eq. (3.39), Chapter 3. In this work, we are consciously avoiding blending thermodynamics with statistical mechanics — the molecular theory — as much as possible.

9.1.3   *Ratio of Specific Heats and Compressibility*

The ratio of specific heats C_p and C_v can be measured from compressibility data. To see this we calculate the polytropic ratio

$$\gamma = \frac{C_p}{C_v} = \frac{c_p}{c_v}. \qquad (9.14)$$

This can be easily expressed as

$$\gamma = \frac{\left(\frac{\partial V}{\partial p}\right)_T}{\left(\frac{\partial V}{\partial p}\right)_S}, \qquad (9.15)$$

which can be shown to be true as follows:

$$\gamma = \frac{\left(\frac{\partial S}{\partial T}\right)_p}{\left(\frac{\partial S}{\partial T}\right)_V} = \frac{\partial(S,p)}{\partial(T,p)}\frac{\partial(T,V)}{\partial(S,V)}$$

$$= \frac{\partial(T,V)}{\partial(T,p)}\frac{\partial(S,p)}{\partial(T,V)}\frac{\partial(T,V)}{\partial(S,V)} = \frac{\partial(T,V)}{\partial(T,p)}\frac{\partial(S,p)}{\partial(S,V)}$$

$$= \frac{\left(\frac{\partial V}{\partial p}\right)_T}{\left(\frac{\partial V}{\partial p}\right)_S}.$$

This implies that it is simply the ratio of isothermal compressibility κ_T to adiabatic compressibility κ_S defined by

$$\kappa_S = -v^{-1}\left(\frac{\partial v}{\partial p}\right)_S, \tag{9.16}$$

that is,

$$\gamma = \frac{\kappa_T}{\kappa_S}. \tag{9.17}$$

This shows that γ can be measured if isothermal and adiabatic compressibilities are determined. We will discuss a way of determining the polytropic ratio γ in the following.

9.1.4 *Sound Wave Velocity and Polytropic Ratio*

A small amplitude, long wavelength compressional oscillatory motion propagating in a compressible fluid is called a sound wave. At each point of the fluid, compression and rarefaction are alternately caused by the sound wave. The velocity of such a sound wave is related to the polytropic ratio γ as will be shown. Therefore the former can be used for measuring the latter.

As preparation for the discussion, we examine the thermodynamics of compression and rarefaction related to such a wave in an ideal gas. We assume that the gas obeys the ideal constitutive relations and also is nondissipative, that is, there is no viscosity and thermal conductivity of the fluid. This latter assumption is not generally true, but simplifies the discussion in making the point.

A sound wave in an ideal monatomic gas is an adiabatic process and hence isentropic. For we find that the equilibrium Gibbs relation can be written as

$$dS = T^{-1}(dE + p\, dV)$$
$$= nRd\left[\ln\left(T^{3/2}V\right)\right]. \tag{9.18}$$

For an adiabatic process the argument of the logarithmic function is indeed a constant as shown in a previous chapter, and hence

$$dS = 0. \tag{9.19}$$

Therefore a sound wave is an isentropic process.

If we denote by p_0 and ρ_0 the equilibrium pressure and mass density, and by p' and ρ' their variations caused by the sound wave then

$$p = p_0 + p', \quad \rho = \rho_0 + \rho', \tag{9.20}$$

where

$$p' \ll p_0, \quad \rho' \ll \rho_0.$$

In view of these disparities in magnitude of the variables the equation of continuity for ρ and the momentum balance equation may be linearized as follows:

$$\frac{\partial \rho'}{\partial t} = -\rho_0 \nabla \cdot \mathbf{u}, \tag{9.21}$$

$$\rho_0 \frac{\partial \mathbf{u}}{\partial t} = -\nabla p', \tag{9.22}$$

where $\mathbf{u}$ is the velocity of the fluid. Since the flow process in the sound wave is adiabatic, or isentropic, and the pressure varies together with the density, p may be expanded in ρ', and to the first order in ρ' we obtain

$$p' = \left(\frac{\partial p}{\partial \rho_0}\right)_S \rho'. \tag{9.23}$$

Furthermore, we have

$$\rho_0 \left(\frac{\partial p}{\partial \rho_0}\right)_S = -v \left(\frac{\partial p}{\partial v}\right)_S = \kappa_S^{-1}; \tag{9.24}$$

where $v = 1/\rho_0$, specific volume. From Eqs. (9.21) and (9.23) follows the equation

$$\frac{\partial p'}{\partial t} + \rho_0 \left(\frac{\partial p}{\partial \rho_0}\right)_S \nabla \cdot \mathbf{u} = 0. \tag{9.25}$$

It is convenient to introduce a scalar potential $\phi(\mathbf{r}, t)$ such that

$$\mathbf{u} = \nabla \phi. \tag{9.26}$$

Then we find, by combining Eqs. (9.22) and (9.26), the relation between p' and ϕ:

$$p' = -\rho_0 \frac{\partial \phi}{\partial t}. \tag{9.27}$$

On substitution of Eqs. (9.26) and (9.27) into Eq. (9.25) we obtain the wave equation for ϕ:

$$\frac{\partial^2 \phi}{\partial t^2} = c^2 \nabla^2 \phi, \tag{9.28}$$

where c is the speed of sound wave defined by

$$c = \sqrt{\left(\frac{\partial p}{\partial \rho_0}\right)_S}. \tag{9.29}$$

Since for an ideal gas

$$\left(\frac{\partial p}{\partial \rho_0}\right)_S = \frac{1}{\kappa_S \rho_0} = \frac{\gamma}{\kappa_T \rho_0} = \frac{\gamma RT}{m}, \tag{9.30}$$

where m is the molecular mass of the gas, we obtain the sound wave speed in the form

$$c = \sqrt{\frac{\gamma RT}{m}}. \tag{9.31}$$

This is a desired relation we are looking for. The general solution of the wave equation, Eq. (9.28), is given by

$$\phi(\mathbf{r}, t) = f_1(\mathbf{r} - ct) + f_2(\mathbf{r} + ct). \tag{9.32}$$

Here we will confine the discussion to an one-dimensional sound wave in the direction of x axis. Since the x-component of the velocity u_x is given by

$$u_x = \frac{\partial}{\partial x} f_1(x - ct), \tag{9.33}$$

but

$$p' = \rho_0 c \frac{\partial}{\partial x} f_1(x - ct), \tag{9.34}$$

we find

$$u_x = \frac{p'}{\rho_0 c} = \frac{\rho'}{\rho_0} c, \tag{9.35}$$

for which we have used Eq. (9.23), $p' = c^2 \rho'$. The discussion given here and, especially, Formula (9.31) show that the polytropic ratio γ can be determined by measuring the sound wave speed u_x. The discussion given in this subsection is not of equilibrium thermodynamics — in fact, of hydrodynamics in essence, but it shows the role played by equilibrium thermodynamics in the derivation of the wave equation. It underscores the importance of the concepts in equilibrium thermodynamics even for investigation of dynamic problems, although such concepts require some subtle assumptions on time and spatial scales of the processes involved.

It is useful to mention an empirical relation relating the sound velocity, density, and surface tension of a liquid, known as the Auerbach relation: $c = \left(\sigma/6.33 \times 10^{-10}\rho\right)^{2/3}$ where σ is the surface tension and ρ is the density of the liquid.[1] According to S. Blairs,[2] this relation may be

---

[1] See N. Auerbach, *Experientia* **4**, 473 (1948).

[2] *J. Colloid Interface Sci.* **302**, 312 (2006).

written in a form $c = Av^{1/6}\left(\gamma\sigma/\rho\right)^{1/2}$, where A is a constant, which remains empirical.

9.2 Virial Equation of State

One of the most widely used equations of state is the virial equation of state, which is often called the virial expansion[3] for pressure. This equation of state has a firm statistical mechanical theory to support it, although it does not describe the liquid regime as well as the van der Waals equation of state does. In fact, the latter historically preceded the former. It is common to express the virial expansion, in the case of a single-component system, in the form

$$\frac{pv}{RT} = 1 + B(T)v^{-1} + C(T)v^{-2} + \cdots, \tag{9.36}$$

where the coefficients B, C, and so on, are called the second, third virial coefficient, and so on. They are functions of T alone. Alternatively, the virial expansion may be written as a power series of pressure

$$pv = RT + B'(T)p + C'(T)p^2 + \cdots. \tag{9.37}$$

Since the two forms of the virial expansion must be equivalent, the virial coefficients in the two modes of expansion are related to each other. By substituting Eq. (9.36) into Eq. (9.37) and comparing the resulting series in inverse power of v with Eq. (9.36) we find their relationships:

$$B' = B,$$

$$C' = \frac{1}{RT}(C - B^2), \text{ etc.}$$

Second virial coefficients vary with temperature as shown in Fig. 9.1 and this is evident from an approximate van der Waals form, Eq. (9.45), given below. The temperature at which $B(T)$ vanishes is called the Boyle temperature: $B(T_B) = 0$. The van der Waals equation predicts the Boyle temperature at $T_B = a/Rb$. Compared with the experimental value, the van der Waals equation predicts the Boyle temperature too high. If the pressure is moderate, the virial expansion for pressure may be truncated at the v^{-1} term inclusive:

$$\frac{pv}{RT} = 1 + B(T)v^{-1}. \tag{9.38}$$

---

[3]M. Kammerlingh-Onnes for the first time proposed the virial expansion as an alternative to the van der Waals equation of state.

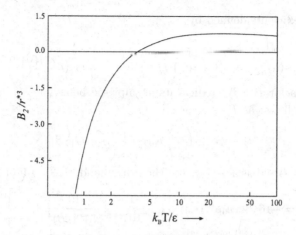

Fig. 9.1 The temperature dependence of the second virial coefficient. It has a zero at $T = T_B$, the Boyle point. $r^* \equiv \sigma$, and ε is the well depth of the potential.

Then at the Boyle temperature, at which the second virial coefficient vanishes, the gas represented by this equation of state behaves as if it is ideal, since there $B(T_B) = 0$.

Before closing this section, it is interesting to note that by using the statistical mechanical formula for $B(T)$ it is possible to derive[4] the temperature dependence of the second virial coefficient of the Lennard–Jones fluid obeying the potential

$$U(r) = 4\varepsilon\left[\left(\frac{r}{\sigma}\right)^{12} - \left(\frac{r}{\sigma}\right)^{6}\right].$$

It is given by the formula

$$B_2 = -4\sqrt{2}\Gamma\left(\frac{3}{4}\right)v_0 t^{1/4}\times$$

$$\left[6tM\left(\frac{7}{4},\frac{3}{2},t\right) - (1+4t)M\left(\frac{3}{4},\frac{1}{2},t\right)\right.$$

$$\left. + \frac{5\Gamma(\frac{1}{4})}{3\Gamma(\frac{3}{4})}t^{3/2}M\left(\frac{9}{4},\frac{5}{2},t\right) + \frac{\Gamma(\frac{1}{4})}{2\Gamma(\frac{3}{4})}\sqrt{t}\,(1-4t)\,M\left(\frac{5}{4},\frac{3}{2},t\right)\right]. \quad (9.39)$$

Here $t = \varepsilon\beta$ $(\beta = 1/k_B T)$, $v_0 = \pi\sigma^3/6$, $\Gamma(x)$ is a Gamma function, and $M(a, b, t)$ is Kummer's confluent hypergeometric function

$$M(a, b, t) = \sum_{n=0}^{\infty}\frac{(a)_n}{(b)_n}\frac{t^n}{n!} \quad (9.40)$$

---

[4]See B. C. Eu, *Exact Analytic Second Virial Coefficient for the Lennard–Jones Fluid*, e-print arXiv: physics/0909.3326v1 (2009).

with the coefficients defined by

$$(a)_0 = 1,$$
$$(a)_n = a(a+1)(a+2)\cdots(a+n-1) \ (n \geq 1).$$

This formula for $B_2(\beta)$ predicts its asymptotic behaviors — the limiting laws — as follows: as $T \to \infty$ or $\varepsilon\beta \to 0$

$$B_2 = 4\sqrt{2}\Gamma\left(\frac{3}{4}\right)v_0(\varepsilon\beta)^{1/4}[1 + O(\varepsilon\beta)] \qquad (9.41)$$

and thus $B_2 \to +0$ as $T \to \infty$. On the other hand, as $T \to 0$ or $\varepsilon\beta \to \infty$

$$B_2(T) = -16\sqrt{2\pi}v_0 e^{\varepsilon\beta}(\varepsilon\beta)^{\frac{3}{2}}\left[1 + \frac{19}{16\varepsilon\beta} + \frac{105}{512(\varepsilon\beta)^2} + \cdots\right] \qquad (9.42)$$

and thus $B_2 \to -\infty$ as $T \to \infty$. These asymptotic formulas are consistent with the experimental behavior of B shown in Fig. 9.1. As a matter of fact, with an appropriate choice of potential parameters Formula (9.38) predicts experimental data of simple fluids in excellent accuracy. The exact value of the Boyle point is also computable from Formula (9.39).

9.3   van der Waals Equation of State

As the density of gas increases, the ideal gas equation of state begins to show its inadequacy. To remedy it van der Waals, based on an insightful argument on dense fluids, proposed an equation of state

$$\left(p + \frac{a}{v^2}\right)(v - b) = RT, \qquad (9.43)$$

where a and b are parameters related to the attractive dispersion force between molecules and the size of the molecule, respectively. Note that $v = V/n$ is the specific volume.

The van der Waals equation of state may be cast into a virial expansion if it is expanded in a power series of b/v:

$$\frac{pv}{RT} = 1 + \left(b - \frac{a}{RT}\right)v^{-1} + b^2 v^{-2} + \cdots. \qquad (9.44)$$

By comparing it with the virial expansion, Eq. (9.36), we therefore obtain the second virial coefficient in terms of the van der Waals parameters a and b:

$$B(T) = b - \frac{a}{RT}. \qquad (9.45)$$

Since b and a are known to be related to the repulsive and the attractive force, respectively, the aforementioned relation shows that the second virial coefficient is related to the intermolecular forces. In fact, the second virial coefficient can be a source for information on intermolecular forces. We note in passing that intermolecular forces can be deduced also from data on transport coefficients such as viscosities and from molecular beam experiments. In the van der Waals theory the Boyle temperature is given by

$$T_B = \frac{a}{Rb}. \tag{9.46}$$

The van der Waals equation of state describes continuous pressure–volume isotherms even in the subcritical regime, although it is known that they are discontinuous in the liquid–vapor coexistence regime. Despite this difficulty, if the Maxwell construction is employed as shown in the following, it works impressively well, covering both gas and liquid regimes. It is capable of describing the phase transition unlike the virial equation of state earlier. In Fig. 9.2 pressure is plotted against volume for various

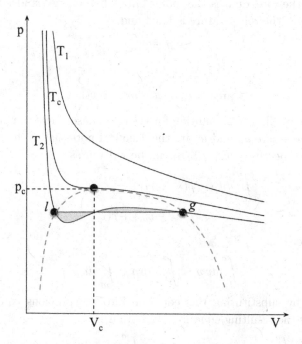

Fig. 9.2 Isotherms of the van der Waals equation of state. l and g stand for the coexisting liquid and vapor at equilibrium, respectively, determined by the Maxwell construction. The broken curve represents the liquid–vapor coexistence curve.

temperatures by using Eq. (9.43). The van der Waals equation is continuous and gives rise to loops for temperatures below the critical temperature T_c. The states corresponding to the loops between (v_{min}, p_{min}) and (v_{max}, p_{max}) are not thermodynamically stable since in the interval

$$\left(\frac{\partial p}{\partial v}\right)_T > 0, \tag{9.47}$$

and therefore they are not realizable. The actual transition may be regarded to occur along a horizontal line between points l and g. J. C. Maxwell suggested to construct the horizontal lines such that the areas of the shaded regions are equal. This procedure is called the Maxwell construction. It can be rationalized as follows.

We calculate the area under the curve between points l and g. Since the area under the curve is an integral of p over an interval of v, we have

$$\int_l^g dv\, p = -\int_l^g [dg - d(pv) + s\, dT], \tag{9.48}$$

where g in the integrand is the molar Gibbs free energy, and s the molar entropy. Since the temperature is fixed and

$$g = \mu,$$

we find

$$\text{area} = g_g - g_l + p\,(v_g - v_l).$$

Since by the equilibrium condition for the two-phase single component system $\mu_g = \mu_l$, where μ_g and μ_l are the chemical potential of the gas and of the liquid, respectively, Eq. (9.48) can be written as

$$\int_l^g dv\, p = p(v_g - v_l)$$
$$= p(v_g - v_m) + p(v_m - v_l). \tag{9.49}$$

On the other hand,

$$\int_l^g dv\, p = \int_l^m dv\, p + \int_m^g dv\, p. \tag{9.50}$$

Therefore, by substituting this equation into the previous equation and rearranging the resulting equation, we obtain

$$\int_m^g dv\, p - p\,(v_g - v_m) = -\int_l^m dv\, p + p\,(v_m - v_l), \tag{9.51}$$

which implies that the shaded areas in Fig. 9.2 are equal.

Since the p–v curve at $T = T_c$ has an inflection point, the critical temperature T_c is determined by putting the first two derivatives equal to zero:

$$\left(\frac{\partial p}{\partial v}\right)_{T_c} = 0, \tag{9.52}$$

$$\left(\frac{\partial^2 p}{\partial v^2}\right)_{T_c} = 0. \tag{9.53}$$

Twice differentiating the van der Waals equation of state and setting the derivatives equal to zero yields two equations corresponding to those in Eqs. (9.52) and (9.53):

$$p_c + \frac{a}{v_c^2} - \frac{2a}{v_c^3}(v_c - b) = 0, \tag{9.54}$$

$$-\frac{4a}{v_c^3} + \frac{6a}{v_c^4}(v_c - b) = 0. \tag{9.55}$$

At the critical point the van der Waals equation is

$$\left(p_c + \frac{a}{v_c^2}\right)(v_c - b) = RT_c. \tag{9.56}$$

When these three equations are solved for p_c, v_c, and T_c, there follow three expressions for v_c, p_c, and T_c given in terms of parameters a and b only:

$$v_c = 3b,$$
$$p_c = \frac{a}{27b^2}, \tag{9.57}$$
$$T_c = \frac{8a}{27Rb}.$$

It is amusing to note that elimination of a and b from three equations in Eq. (9.57) yields an equation

$$p_c v_c = \frac{3}{8} RT_c,$$

which looks like an ideal gas equation of state at the critical point except for the factor 3/8. Equation (9.57) may be used to obtain the parameters a and b from the critical data. Some values of a and b so determined are listed in Table 9.1. It is possible to calculate various thermodynamic properties of the system by using the van der Waals equation of state once the van der Waals constants are determined. The results thus obtained are not always completely satisfactory, but give useful approximations for thermodynamic

Table 9.1   van der Waals constants for various fluids.

Fluids	$a\,(\mathrm{L}^2\,\mathrm{atm}\,\mathrm{mol}^{-2})$	$b\,(\mathrm{L}\,\mathrm{mol}^{-1}) \times 10^2$
Ar	1.35	3.23
He	0.034	2.38
H_2	0.245	2.67
O_2	1.36	3.19
N_2	1.39	3.92
CO_2	3.60	4.28
NH_3	4.17	3.72
H_2O	5.46	3.30

quantities when no other information is available at high pressures where a nonideality correction must be made in order to properly account for thermodynamic data.

9.4   Law of Corresponding States

It is possible to define reduced temperature τ, reduced pressure π, and reduced volume ν by scaling T, p, and v with their critical values:

$$T = \tau T_c, \quad v = \omega v_c, \quad p = \pi p_c. \tag{9.58}$$

In the case of the van der Waals equation of state, since the critical data are given by Eq. (9.57), on elimination of the van der Waals constants a and b from the equation of state there follows a reduced equation of state

$$\left(\pi + \frac{3}{\omega^2}\right)(3\omega - 1) = 8\tau. \tag{9.59}$$

An interesting feature of this equation is that it does not contain parameters indicative of the substance. It is a universal relationship for the reduced variables involved and therefore is expected to hold for a class of substances. If equimolar amounts of two gases, whose $p-v-T$ relation may be represented by a van der Waals equation of state, are at the same reduced pressure and reduced volume then they are expected to be at the same reduced temperature. They are then said to be in the corresponding states. Equation (9.59) is an example of the law of corresponding states. In fact, the van der Waals equation is not the only equation of state for which the law of corresponding states holds. For example, the virial equation of state can be put into a reduced form if the reduced variables are suitably defined. Thus we may write the equation of state in a reduced form valid

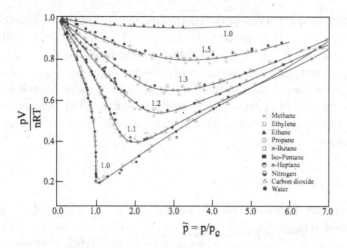

Fig. 9.3 Law of corresponding states. The compressibility factor pv/RT for several gases are plotted as a function of the reduced pressure p/p_c at various values of the reduced temperature T/T_c. Reproduced with permission from G. J. Su, *Ind. Eng. Chem.* **38**, 803 (1946) © 1946 American Chemical Society.

for a wide class of substances, independent of material parameters

$$\pi = f(\tau, \omega) \tag{9.60}$$

and compactly summarize the equation of state data in a single universal equation. As Fig. 9.3 shows, there are many fluids which obey the law of corresponding states to a good approximation. Notice that data for various gases follow the same reduced curve at a given reduced temperature.

9.5 Thermodynamic Functions

In the following subsections, we shall calculate various thermodynamic quantities by using the equations of state for real gases given in the previous section. Although there are many more equations of state known in the literature, we first will use either the van der Waals or the virial equation of state, since they have statistical mechanical foundations[5] and sufficiently

---

[5]It must be noted that the van der Waals equation of state can be derived in statistical mechanics, only if some intuitive approximations are made on the basis of physical arguments, whereas the virial equation of state can be justified rigorously by expanding the partition function in a density series by means of the cluster expansion. However, the cluster expansion has a limited radius of convergence.

well illustrate applications of the thermodynamic principles and methods so far developed. Then in the next chapter we will discuss the canonical equation of state, which can be given a full statistical mechanical support unlike the original van der Waals equation of state for which only an approximate derivation can be given. The thermodynamic equations developed are valid for any aggregate of matter up to a certain extent. It is only when we choose a particular equation of state that the calculations must be specialized to gases, liquids, or solids.

9.5.1   *Reversible Work*

Let us consider a pressure–volume work performed reversibly on the system at constant temperature. Assume that the volume is changed from v_1 to v_2. For a van der Waals gas the work done on the system is easily found:

$$W = -\int_{v_1}^{v_2} p\, dv$$

$$= -RT \ln\left(\frac{v_2 - b}{v_1 - b}\right) - a\left(\frac{1}{v_2} - \frac{1}{v_1}\right). \tag{9.61}$$

In order to identify the contribution arising from the nonideality of the gas we rearrange the terms and write this equation in the form

$$W = -RT \ln\left(\frac{v_2}{v_1}\right) + W_{\text{vdw}}, \tag{9.62}$$

where

$$W_{\text{vdw}} = -RT \ln\left(\frac{1 - bv_2^{-1}}{1 - bv_1^{-1}}\right) - a\left(\frac{1}{v_2} - \frac{1}{v_1}\right)$$

$$= (bRT - a)\left(\frac{1}{v_2} - \frac{1}{v_1}\right) + O(v^{-2}). \tag{9.63}$$

This is the correction to the work done on the ideal gas as a result of molecular interactions in the gas. The sign of the first-order correction depends on the temperature and the relative magnitudes of a and b. If the temperature is such that $b \geq a/RT$, then W_{vdw} is positive, otherwise it is negative for the process considered. If the virial equation of state (Eq. (9.36)) is used, the work is given by the formula

$$W = -RT \ln\left(\frac{v_2}{v_1}\right) + W_{\text{vir}}, \tag{9.64}$$

where

$$W_{\text{vir}} = -BRT\left(\frac{1}{v_1} - \frac{1}{v_2}\right) - \tfrac{1}{2}CRT\left(\frac{1}{v_1^2} - \frac{1}{v_2^2}\right) + \cdots. \tag{9.65}$$

The first-order term agrees with the first-order term in Eq. (9.63) by the van der Waals equation of state in view of Eq. (9.45). Therefore, to this order of approximation, the van der Waals and the virial equation of state have the same value for work.

9.5.2 Heat Change in Isothermal Expansion

We now calculate the heat change accompanying a reversible isothermal expansion for a van der Waals gas. Similar calculations can be, of course, performed for real gases obeying other kinds of equation of state. Since in the case of an isothermal expansion of a mole of a substance

$$dQ = T \, ds = T\left(\frac{\partial s}{\partial V}\right)_T dv = T\left(\frac{\partial p}{\partial T}\right)_v dv,$$

the reversible heat change accompanying the volume change from v_1 to v is given by the formula

$$Q = \int_{v_1}^{v} dv' \, T\left(\frac{\partial p}{\partial T}\right)_{v'}. \tag{9.66}$$

The van der Waals equation of state

$$\left(p + \frac{a}{v^2}\right)(v - b) = RT$$

may be arranged to the form

$$p = \frac{RT}{v - b} - \frac{a}{v^2},$$

which on differentiation with T yields

$$\left(\frac{\partial p}{\partial T}\right)_v = \frac{R}{v - b} - \frac{1}{v^2}\frac{da}{dT}.$$

It must be noted that the parameter a is generally dependent on T and the van der Waals attraction force constant. On substitution of this derivative into Eq. (9.66) in the case of a mole of a van der Waals gas and performing the integration we easily find

$$Q = \int_{v_1}^{v} dv' \left[\frac{RT}{v' - b} - \frac{T}{v'^2}\left(\frac{da}{dT}\right)\right]$$

$$= RT \ln\left(\frac{v - b}{v_1 - b}\right) + \left(\frac{1}{v} - \frac{1}{v_1}\right) T \frac{da}{dT}. \tag{9.67}$$

This may be recast in a form better exhibiting the nonideal part by using a procedure similar to that leading to Eq. (9.62):

$$Q = RT \ln\left(\frac{v}{v_1}\right) + Q_{\text{vdw}}, \tag{9.68}$$

where

$$Q_{\text{vdw}} = RT \ln\left(\frac{1 - bv^{-1}}{1 - bv_1^{-1}}\right) + \left(\frac{1}{v} - \frac{1}{v_1}\right) T \frac{da}{dT}. \tag{9.69}$$

The first term on the right in Eq. (9.68) is obviously the heat change due to the volume change for the ideal gas and Q_{vdw} is the correction to it which originates from the molecular interaction.

9.5.3 *Standard States*

Since in thermodynamics it is possible to determine only the changes in thermodynamic functions accompanying a reversible process from one state to another, we are not able to determine the absolute values of thermodynamic state functions. In order to tabulate them it is therefore desirable to choose a reference state and fix thermodynamic state functions relative to that reference state. It was indicated in Chapter 3 that specific heats are usually tabulated for the ideal gas in the case of gases, and at 1 atm ($1 \text{ atm} \equiv 1.01325 \times 10^5$ Pa) in the case of liquids and solids. The ideal gas state is chosen for gases, because specific heats are independent of pressure and volume for ideal gases.

In the case of enthalpy the reference state is the ideal gas state or the limit for the real gas as the pressure approaches the zero pressure, and the standard state is then the hypothetical ideal gas at $T = 298.15$ K and $p = 1$ atm in the case of real gases. For liquids or solids of elements the standard state is their most stable state at the pressure of 1 atm. At the standard state at $T = 298.15$ K the enthalpy is set equal to zero for liquids and solids of elements:

$$H(p = 1 \text{ atm}; \ T = 298.15 \text{ K}) = 0 \tag{9.70}$$

and the entropy is also taken equal to zero at $T = 0$:

$$S(p = 1 \text{ atm}; T = 0) = 0. \tag{9.71}$$

Since the entropy is a logarithmic function of pressure, the standard state entropy for a gas cannot be determined simply at $p = 0$ as in the case of enthalpy. The standard state entropy for a gas is defined as that of the hypothetical ideal gas at 1 atm pressure:

$$S^0 = \lim_{p \to 0} [(S(T, p) + R \ln p].$$
(9.72)

With this preparation, we are now able to calculate various thermodynamic functions at arbitrary temperature and pressure.

9.5.4 Enthalpy

Let us denote by h the molar enthalpy of a pure substance. Since it is convenient to take the experimentally most suitable variables as independent variables, we shall assume that h is a function of T and $p : h = h(T, p)$. Then the differential of h is given by

$$dh(T, p) = \left(\frac{\partial h}{\partial T} \right)_p dT + \left(\frac{\partial h}{\partial p} \right)_T dp.$$
(9.73)

We integrate this differential form from state (p_1, T_1) to state (p, T). Since dh is an exact differential in the space of T and p, the process in question can be arbitrarily and conveniently distorted into two steps: $(p_1, T_1) \to (p_1, T)$ (isobaric path) and $(p_1, T) \to (p, T)$ (isothermal path), as shown in Fig. 9.4. This distortion of the path is possible owing to the fact that dh is a total (exact) differential. First, we consider the isobaric process along the vertical path from (p_1, T_1) to (p_1, T).

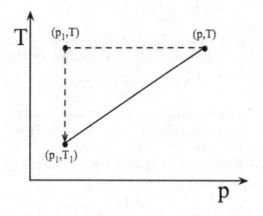

Fig. 9.4   Path of integration for h in $p-T$ plane.

The enthalpy change accompanying the isobaric process is

$$h(T, p_1) - h(T_1, p_1) = \int_{T_1}^{T} dT' \left(\frac{\partial h}{\partial T'}\right)_p (p_1) = \int_{T_1}^{T} dT' \, c_p(T', p_1),$$

(9.74)

where c_p is the molar specific heat at constant pressure. Since

$$\left(\frac{\partial h}{\partial p}\right)_T = v - T \left(\frac{\partial v}{\partial T}\right)_p,$$

(9.75)

the enthalpy change along the isothermal (horizontal) path is

$$h(T, p) - h(T, p_1) = \int_{p_1}^{p} dp' \left[v - T\left(\frac{\partial v}{\partial T}\right)_{p'}\right](T, p').$$

(9.76)

By combining Eqs. (9.74) and (9.76) we find the enthalpy change for the process from (p_1, T_1) to (p, T)

$$h(T, p) = h(T_1, p_1) + \int_{T_1}^{T} dT' \, c_p(T', p_1) + \int_{p_1}^{p} dp' \left[v - T\left(\frac{\partial v}{\partial T}\right)_{p'}\right](T, p').$$

(9.77)

If the initial pressure is taken equal to zero, namely, if $p_1 = 0$,

$$h(T, p) = h^*(T) + \int_{0}^{p} dp' \left[v - T\left(\frac{\partial v}{\partial T}\right)_{p'}\right](T, p'),$$

(9.78)

where

$$h^*(T) = h(T_1, 0) + \int_{T_1}^{T} dT' \, c_p^*(T')$$

(9.79)

with $c_p^* = \lim_{p \to 0} c_p$, then $h(T_1, 0)$ is the enthalpy of the substance at $T = T_1$, where we may take $T_1 = 298.15$ K. We thus see that

$$h^*(T) = \lim_{p \to 0} h(T, p),$$

that is, $h^*(T)$ is the molar enthalpy of a hypothetical ideal gas, and therefore the second term on the right-hand side of Eq. (9.78) is the nonideality correction for the enthalpy.

In the case of liquids and solids of elements, if we take $p_1 = 1$ atm and $T_1 = 298.15$ K, we obtain

$$h(T, p) = h^0(T) + \int_{1}^{p} dp' \left[v - T\left(\frac{\partial v}{\partial T}\right)_{p'}\right](T, p'),$$

(9.80)

where

$$h^0(T) = \int_{T_1}^{T} dT \; c_p(T, p = 1) \tag{9.81}$$

for which we have prescribed the standard enthalpy by the convention introduced in the previous section:

$$h(T_1 = 298.15 \text{ K}, \, p = 1 \text{ atm}) = 0.$$

Let us further examine Eq. (9.78). Since for ideal gases

$$v - T\left(\frac{\partial v}{\partial T}\right)_p = 0,$$

the integral in Eq. (9.78) vanishes and the enthalpy becomes that of an ideal gas:

$$h(T, p) = h^*(T).$$

The nonideality correction

$$h_{\text{real}} = \int_0^p dp' \left[v - T\left(\frac{\partial v}{\partial T}\right)_{p'} \right] (T, p') \tag{9.82}$$

may be calculated with the equation of state for the system. If we assume the virial equation of state (Eq. (9.37)), the integrand is calculated to be

$$v - T\left(\frac{\partial v}{\partial T}\right)_p = \left(B - T\frac{dB}{dT}\right) + \left(C' - T\frac{dC'}{dT}\right)p + \cdots.$$

Upon putting this result into the integral we obtain

$$h_{\text{real}} = \left(B - T\frac{dB}{dT}\right)p + \frac{1}{2}\left(C' - T\frac{dC'}{dT}\right)p^2 + \cdots. \tag{9.83}$$

If the second virial coefficient is estimated with the van der Waals theory form

$$B = b - \frac{a}{RT},$$

we obtain to first order in p

$$h_{\text{real}} = \left(b - \frac{2a}{RT}\right)p + O(p^2). \tag{9.84}$$

Observe that the real gas correction h_{real}, to first order in p, is positive if

$$b - \frac{2a}{RT} \geq 0,$$

but negative otherwise. Since parameter a is representative of the attractive part of the intermolecular force whereas b is a measure of the repulsive part, the enthalpy becomes smaller than $h^*(T)$ if the temperature is sufficiently low so that the attractive part of the intermolecular force becomes dominant over the repulsive part.

Since specific heats can be calculated with $h(T, p)$, we can obtain the pressure dependence of the nonideality correction for $c_p(T, p)$ as follows:

$$c_p(T, p) = c_p^*(T) + c_{p\,\text{real}}(T, p), \tag{9.85}$$

where

$$c_p^* = \left(\frac{\partial h^*}{\partial T}\right)_p,$$

$$\tag{9.86}$$

$$c_{p\,\text{real}} = \left(\frac{\partial h_{\text{real}}}{\partial T}\right)_p = -T\frac{d^2 B}{dT^2}p + O(p^2).$$

In the van der Waals theory

$$c_{p\,\text{real}} = \frac{2a}{RT^2}p + O(p^2). \tag{9.87}$$

This gives a rough idea of how c_p depends on T and p for real gases. We thus see that data for c_p^* and the equation of state would be sufficient for determining c_p at $p \neq 0$.

If the process $(p_1, T_1) \to (p, T)$ involves a phase transition at $T = T_m$, it is necessary to take the latent heat into account. For the purpose of calculating the enthalpy change for the process, we return to Eq. (9.77) and modify the integral. The range of integral must be split into two parts: (T_1, T_m) and (T_m, T); see Fig. 9.6 below for the integration path superimposed on the phase diagram. Thus we find

$$h(T, p) - h(T_1, p_1) = \Delta h_m(T_m, p_1) + \int_{T_1}^{T_m} dT\, c_p^{(1)}(T, p_1)$$

$$+ \int_{T_m}^{T} dT'\, c_p^{(2)}(T', p_1), \tag{9.88}$$

where Δh_m is the latent heat at $T = T_m$ and $p = p_1$, and $c_p^{(1)}$ and $c_p^{(2)}$ are, respectively, the specific heat of the low temperature phase 1 and of the high temperature phase 2. The value of $h(T_1, p_1)$ may be fixed by choosing an appropriate standard state.

9.5.5 *Internal Energy*

It is possible to calculate the volume dependence of the internal energy by using a method similar to that for the enthalpy since it is permissible to distort the path of integration into isochoric and isothermal paths because $d\mathcal{E}$ is an exact differential. Let us denote the molar internal energy by $\mathcal{E} = E/n$, where n is the mole number. It is convenient to regard $\mathcal{E}$ as a function of v and T. To investigate the T and v dependence of $\mathcal{E}$ we derive the following differential form for $\mathcal{E}$

$$d\mathcal{E}(T,v) = \left(\frac{\partial \mathcal{E}}{\partial T}\right)_v dT + \left(\frac{\partial \mathcal{E}}{\partial v}\right)_T dv$$

$$= c_v \, dT + \left[T\left(\frac{\partial p}{\partial T}\right)_v - p\right] dv, \qquad (9.89)$$

where c_v is the molar specific heat at constant volume. For this formula we have made use of $\mathcal{E} = \mathcal{A} + Ts$, where $\mathcal{A}$ is the molar work function and s the molar entropy, and one of Maxwell's relations. By integrating this differential form along the path $(T_1, v_1) \to (T, v_1)$ as shown in Fig. 9.5 we obtain

$$\mathcal{E}(T,v_1) - \mathcal{E}(T_1,v_1) = \int_{T_1}^{T} dT' \, c_v(T',v_1) \qquad (9.90)$$

and, integrating along the path $(T, v_1) \to (T, v)$,

$$\mathcal{E}(T,v) = \mathcal{E}(T,v_1) + \int_{v_1}^{v} dv' \left[T\left(\frac{\partial p}{\partial T}\right)_v - p\right](v'). \qquad (9.91)$$

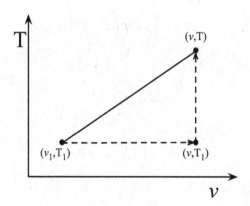

Fig. 9.5   Path of integration for $\mathcal{E}$ in T–v plane.

Combining the two equations (Eqs. (9.90) and (9.91)) yields the internal energy change for the process $(T_1, v_1) \to (T, v)$

$$\mathcal{E}(T, v) = \mathcal{E}(T_1, v_1) + \int_{T_1}^{T} dT' \, c_v(T', v_1) - \int_{v_1}^{v} dv' \left[p - T \left(\frac{\partial p}{\partial T} \right)_{v'} \right] (T, v').$$

$$(9.92)$$

Now let us take the limit $v_1 \to \infty$. This limit is equivalent to the limit $p \to 0$ at which the gas behaves as an ideal gas and therefore

$$\lim_{v_1 \to \infty} c_v(T, v_1) = c_v^*(T). \tag{9.93}$$

Furthermore, we may choose the same standard state for the internal energy as for the enthalpy and set

$$\mathcal{E}(T_1 = 298.15 \text{ K}, \, p_1 = 1 \text{ atm}, \, v_1 = \infty) = 0 \tag{9.94}$$

to remain consistent with the standard state for the enthalpy of the gas. Therefore, if we take the initial volume to be infinite so that the gas becomes ideal, the internal energy may be given in the form

$$\mathcal{E}(T, v) = \mathcal{E}^*(T) + \mathcal{E}_{\text{real}}(T, v), \tag{9.95}$$

where

$$\mathcal{E}^*(T) = \int_{T_1}^{T} dT' \, c_v^*(T') + \mathcal{E}(T_1, \infty)$$

$$= \int_{T_1}^{T} dT' \, c_v^*(T') \qquad \text{if } T_1 = 298.15 \text{ K}, \tag{9.96}$$

$$\mathcal{E}_{\text{real}} = \int_{v}^{v_1} dv' \left[p - T \left(\frac{\partial p}{\partial T} \right)_{v'} \right] (T, v'). \tag{9.97}$$

Notice that as $v \to \infty$,

$$\lim_{v \to \infty} \frac{pv}{RT} = 1$$

and therefore

$$p - T \left(\frac{\partial p}{\partial T} \right)_v \to 0,$$

which implies that

$$\lim_{v \to \infty} \mathcal{E}_{\text{real}}(T, v) = 0,$$

and consequently

$$\lim_{v \to \infty} \mathcal{E}(T, v) = \mathcal{E}^*(T). \tag{9.98}$$

We thus see that $\mathcal{E}^*(T)$ is the internal energy of a mole of the gas and $\mathcal{E}_{\text{real}}(T, v)$ is the nonideality correction. Note that this result is consistent with the definition of ideal gases

$$\left(\frac{\partial \mathcal{E}}{\partial v}\right)_T = 0,$$

which means that the internal energy of ideal gases is independent of v.

We may calculate the nonideality correction by using the virial equation of state (Eq. (9.36)). Since the virial equation of state gives

$$T\left(\frac{\partial p}{\partial T}\right)_v - p = \frac{RT^2}{v^2}\frac{dB}{dT} + \frac{RT^2}{v^3}\frac{dC}{dT} + \cdots,$$

substitution of this equation into the integral for $\mathcal{E}_{\text{real}}$ yields

$$\mathcal{E}_{\text{real}}(T, v) = -\frac{RT^2}{v}\frac{dB}{dT} - \frac{1}{2}\frac{RT^2}{v^2}\frac{dC}{dT} - \cdots \tag{9.99}$$

and therefore

$$\mathcal{E}(T, v) = \mathcal{E}^*(T) - \frac{RT^2}{v}\frac{dB}{dT} - \frac{1}{2}\frac{RT^2}{v^2}\frac{dC}{dT} - \cdots. \tag{9.100}$$

If the second virial coefficient is estimated from the van der Waals equation of state we find

$$\mathcal{E}_{\text{real}} = -\frac{a}{v} + O(v^{-2}), \tag{9.101}$$

which implies that the internal energy is smaller than the ideal gas value because of the intermolecular attraction, at least, to first order in v^{-2} in the van der Waals theory.

It is also possible to calculate the nonideality correction for c_v by using Eq. (9.101):

$$c_v(T, v) = c_v^*(T) + c_{v\,\text{real}}(T, v), \tag{9.102}$$

where

$$c_{v\,\text{real}} = -\frac{R}{v}\frac{d}{dT}\left(T^2\frac{dB}{dT}\right) - \frac{R}{2v^2}\frac{d}{dT}\left(T^2\frac{dC}{dT}\right) - \cdots \tag{9.103}$$

If Eq. (9.101) is used for calculating the specific heat, there follows

$$c_v(T, v) = c_v^*(T) + O(v^{-2}).$$

The specific heat in this case is equal to the ideal gas value c_v^* to second order in v^{-2}.

9.5.6 *Entropy*

Entropy may be calculated in basically the same way as for other thermodynamic functions. Let us denote the molar entropy by $s = S/n$ where n is the mole number of the substance. It may be considered a function of T and p. To find its dependence on T and p we integrate its differential form

$$ds(T, p) = \left(\frac{\partial s}{\partial T}\right)_p dT + \left(\frac{\partial s}{\partial p}\right)_T dp$$

$$= \frac{c_p}{T} dT - \left(\frac{\partial v}{\partial T}\right)_p dp. \qquad (9.104)$$

The second line is due to the definition of heat capacity and the use of a Maxwell relation. This differential form is now integrated from (T_1, p_1) to (T, p). Since s is a state function, the path may be distorted to the dotted line in Fig. 9.4. By integrating Eq. (9.104) along the isobaric path we obtain

$$s(T, p_1) = s(T_1, p_1) + \int_{T_1}^{T} dT \frac{c_p(T)}{T}. \qquad (9.105)$$

The integration along the isothermal path requires a little more careful consideration. Since along the isothermal path

$$ds(T, p) = -\left(\frac{\partial v}{\partial T}\right)_p dp$$

and the p-dependence of $(\partial v/\partial T)_p$ is roughly

$$\left(\frac{\partial v}{\partial T}\right)_p \sim \frac{R}{p}$$

and therefore is singular with respect to p, the differential form must be cast into a more suitable form as follows:

$$s(T, p) = \left[\frac{R}{p} - \left(\frac{\partial v}{\partial T}\right)_p\right] dp - \left(\frac{R}{p}\right) dp. \qquad (9.106)$$

On integration of this equation along the isothermal path, there follows

$$s(T,p) = s(T,p_1) - R\ln\left(\frac{p}{p_1}\right) + \int_{p_1}^{p} dp' \left[\frac{R}{p'} - \left(\frac{\partial v}{\partial T}\right)_{p'}\right](T,p').$$

(9.107)

When Eqs. (9.105) and (9.107) are combined, there results the entropy change for the process under consideration

$$s(T,p) = s(T_1,p_1) + \int_{T_1}^{T} dT' \frac{c_p(T',p_1)}{T'} - R\ln\left(\frac{p}{p_1}\right)$$

$$+ \int_{p_1}^{p} dp' \left[\frac{R}{p'} - \left(\frac{\partial v}{\partial T}\right)_{p'}\right](T,p').$$

(9.108)

Note that the specific heat c_p in Eq. (9.108) depends on the pressure p_1 at which it is measured. We consider this expression for systems in different states of aggregation.

9.5.6.1 *Real Gases*

If the substance is a real gas, the thermodynamic functions, such as the enthalpy, the free energy, and so on, are calculated in reference to the standard state which is taken as the state of a hypothetical ideal gas at $p = 1$ atm. For this purpose it is convenient to cast the entropy formula in Eq. (9.108), in reference to such standard state. This aim is easily achieved if we rearrange the terms in Eq. (9.108) and take the limit $p_1 \to 0$ as follows:

$$\lim_{p_1 \to 0} s(T,p) = \lim_{p_1 \to 0} \left\{ s(T_1,p_1) + R\ln p_1 - \int_{0}^{p_1} dp' \left[\frac{R}{p'} - \left(\frac{\partial v}{\partial T}\right)_{p'}\right](p') \right.$$

$$\left. + \int_{T_1}^{T} dT' \frac{c_p(T',p_1)}{T'} \right\}$$

$$- R\ln p + \int_{0}^{p} dp' \left[\frac{R}{p'} - \left(\frac{\partial v}{\partial T}\right)_{p'}\right](p').$$

(9.109)

Note that the following limits hold:

$$\lim_{p_1 \to 0} c_p(T,p_1) = c_p^*(T)$$

and

$$\lim_{p_1 \to 0} \int_{0}^{p_1} dp' \left[\frac{R}{p'} - \left(\frac{\partial v}{\partial T}\right)_{p'}\right](p') = 0$$

because the equation of state becomes that of the ideal gas as $p_1 \to 0$
where the gas obeys the ideal gas equation of state. Therefore Eq. (9.109)
becomes

$$\lim_{p_1 \to 0} s(T,p) = \lim_{p_1 \to 0} [s(T_1, p_1) + R \ln p_1] + \int_{T_1}^{T} dT' \frac{c_p^*(T')}{T'}$$

$$- R \ln p + \int_0^p dp' \left[\frac{R}{p'} - \left(\frac{\partial v}{\partial T} \right)_{p'} \right] (p'). \qquad (9.110)$$

According to the standard entropy of gas introduced in Eq. (9.72) we have

$$s^0 = \lim_{p_1 \to 0} [s(T_1, p_1) + R \ln p_1]. \qquad (9.111)$$

Note that s^0 depends on T_1. We thus finally obtain the entropy in the form

$$s(T,p) = s^*(T) - R \ln p + s_{\text{real}}, \qquad (9.112)$$

where

$$s^*(T) = s^0 + \int_{T_1}^{T} dT \frac{c_p^*(T)}{T} \qquad (9.113)$$

and

$$s_{\text{real}}(T,p) = \int_0^p dp' \left[\frac{R}{p'} - \left(\frac{\partial v}{\partial T} \right)_{p'} \right] (T, p').$$

Since in the case of an ideal gas

$$s_{\text{real}} = 0,$$

$s^*(T)$ may be interpreted as the molar entropy of the hypothetical ideal
gas at 1 atm pressure. Then s_{real} is the nonideality correction.

It is easy to calculate the nonideality correction by using the virial
equation of state (Eq. (9.37)). The result is

$$s_{\text{real}} = -\frac{dB}{dT} p - \frac{1}{2} \frac{dC'}{dT} p^2 - \cdots. \qquad (9.114)$$

If the van der Waals estimate (Eq. (9.45)) is used for the second virial
coefficient, s_{real} to first order in p is

$$s_{\text{real}} = -\frac{a}{RT^2} p + O(p^2). \qquad (9.115)$$

The entropy of a substance in a condensed phase may be calculated with
Eq. (9.108) in reference to the standard state prescribed in the previous
section. Thus one may choose $p_1 = 1$ atm. In that case, $s(T_1, p_1 = 1 \text{ atm})$

is the entropy of the substance at T_1 in reference to the standard state at which the entropy is prescribed equal to zero.

9.5.6.2 *Substances in Condensed Phase*

Since if the substance of interest is in a condensed phase the reference state is taken differently from that of a gas, there is no need for taking the limit of $p_1 \to 0$. To be general we consider the case of the pressure range involving two condensed phases. See Fig. 9.6. The path of integration $(p_1, T_0) \to (p, T)$ in the case may be distorted into two components $(p_1, T_0) \to (p_1, T)$ and $(p_1, T) \to (p, T)$ as indicated in Fig. 9.6, the first of which traverses the solid–liquid phase equilibrium curve at $T = T_m$ and the second of which is an isothermal process within the liquid phase. On application of Eq. (9.104) along the isobaric path $(p_1, T_0) \to (p_1, T)$ we find

$$s(T, p_1) = \int_0^{T_m} dT \frac{c_p^{(s)}(T, p_1)}{T} + \Delta s_m(T_m, p_1) + \int_{T_m}^{T} dT' \frac{c_p^{(l)}(T', p_1)}{T'},$$

where $c_p^{(s)}$ and $c_p^{(l)}$ are the molar specific heat of the solid and of the liquid at $p = p_1$, respectively, and T_m is the melting point at $p = p_1$, and

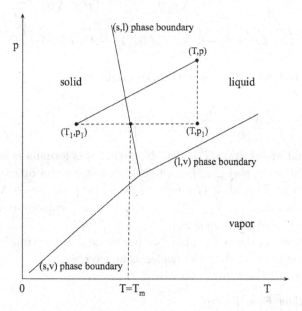

Fig. 9.6 A path crossing the solid–liquid phase boundary in p–T plane. The integration path $(p_0, T_0) \to (p, T)$ can be distorted into path $(p_0, T_0) \to (p_0, T) \to (p_0, T) \to (p, T)$. The melting point is $T = T_m$.

$\Delta s_m(T_m, p_1)$ is the discontinuity in entropy at $T = T_m$ and $p = p_1$ owing to the solid–liquid phase transition. The third law of thermodynamics is made use of for the first integral on the right. Along the isothermal path $(p_1, T) \rightarrow (p, T)$ we find

$$s(T,p) = s(T,p_1) - R\ln\left(\frac{p}{p_1}\right) + \int_{p_1}^{p} dp' \left[\frac{R}{p'} - \left(\frac{\partial v}{\partial T}\right)_{p'}\right](T,p').$$

Combining these two equations yields

$$s(T,p) = \int_0^{T_m} dT \frac{c_p^{(s)}(T,p_1)}{T} + \Delta s_m(T_m, p_1) + \int_{T_m}^{T} dT' \frac{c_p^{(l)}(T',p_1)}{T'}$$

$$- R\ln\left(\frac{p}{p_1}\right) + \int_{p_1}^{p} dp' \left[\frac{R}{p'} - \left(\frac{\partial v}{\partial T}\right)_{p'}\right](T,p'). \qquad (9.116)$$

The last term can be computed from the knowledge of the equation of state of the liquid. Defining the reference entropy at $p_1 = 1\,\mathrm{atm}$

$$s_m^0(T) = \lim_{p_1 \to 1}\left[s(T,p) - R\ln\left(\frac{p}{p_1}\right) + \int_{p_1}^{1} dp' \left[\frac{R}{p'} - \left(\frac{\partial v}{\partial T}\right)_{p'}\right](T,p')\right]$$

$$= \int_0^{T_m} dT \frac{c_p^{(s)}(T,1)}{T} + \Delta s_m(T_m, 1) + \int_{T_m}^{T} dT' \frac{c_p^{(l)}(T',1)}{T'}, \quad (9.117)$$

we express the entropy at $p_1 = 1\,\mathrm{atm}$ in the form

$$s(T,p) = s_m^0(T) - R\ln p + \int_1^p dp' \left[\frac{R}{p'} - \left(\frac{\partial v}{\partial T}\right)_{p'}\right](T,p') \qquad (9.118)$$

for the liquid at arbitrary T and p. Note that this formula is in the same form as that for a real gas, Eq. (9.112), except for the difference in the meanings of $s^*(T)$ and $s_m^0(T)$, arising from the presence of $\Delta s_m(T_m, 1)$ and the split integrals of $c_p^{(\alpha)}/T$ $(\alpha = s,l)$ for the entropy contributions from the solid and liquid phases.

The entropy in the case of liquid–vapor or solid–vapor transition can be found similarly. It is left to the reader as an exercise.

9.5.7   *Gibbs Free Energy*

Since the molar enthalpy and entropy of a substance are already calculated, the molar Gibbs free energy is easily obtained from them by using the formula

$$g = h - Ts.$$

In the gas phase excluding a phase transition, substitution of Eqs. (9.77) and (9.108) yields the molar free energy of a gas

$$g(T,p) = g^*(T) + RT \ln p + \int_0^p dp \left(v - \frac{RT}{p} \right), \qquad (9.119)$$

where

$$g^*(T) = h^*(T) - Ts^*(T). \qquad (9.120)$$

The meaning of $g^*(T)$ is clear from the meanings of h^* and s^*: it is the molar Gibbs free energy of the hypothetical ideal gas at 1 atm pressure. Therefore the last term in Eq. (9.119) is the nonideality correction. It vanishes for ideal gases:

$$\lim_{p \to 0} \int_0^p dp' \left(v - \frac{RT}{p'} \right) = 0.$$

This is easily verified. In this case, we have for the molar Gibbs free energy

$$g(T,p) = g^*(T) + RT \ln p, \qquad (9.121)$$

which is the molar Gibbs free energy that we have calculated previously for ideal gas.

Now the question arises as to what would be the free energy for the case of a process in which a phase transition is involved, for example, the process described in Fig. 9.6. Since the free energy is continuous across two phases in equilibrium for the isobaric process in Fig. 9.6, if we denote the free energies of the solid and liquid phases by $g^{(s)}$ and $g^{(l)}$, then

$$g^{(s)}(T_m, p) = g^{(l)}(T_m, p) \qquad (9.122)$$

the free energy can be still written as Formula (9.119). This is left to the reader as an exercise.

9.6 Fugacity

In analogy to the ideal gas formula for $g(T,p)$ we write the molar Gibbs free energy of a real gas in the form

$$g(T,p) = g^*(T) + RT \ln p_f. \qquad (9.123)$$

Comparing it with the formula in Eq. (9.119) yields

$$p_f = p \exp \int_0^p dp' \left(\frac{v}{RT} - \frac{1}{p'} \right).$$ (9.124)

This was introduced by G. N. Lewis who called it the fugacity of the real gas at T and p. Obviously, as the pressure tends to zero, the equation of state becomes that of the ideal gas and consequently the fugacity p_f approaches the pressure p. Therefore the fugacity is a measure of nonideality of the substance. The nonideality correction is sometimes expressed as excess free energy:

$$g_{ex} = RT \ln \left[\exp \int_0^p dp' \left(\frac{v}{RT} - \frac{1}{p'} \right) \right].$$ (9.125)

The fugacity then may be written as

$$p_f = p \exp \left(\frac{g_{ex}}{RT} \right) \equiv fp,$$ (9.126)

where f is called the fugacity coefficient:

$$f = \exp \int_0^p dp' \left(\frac{v}{RT} - \frac{1}{p'} \right).$$

The fugacity f tends to 1 as p decreases to zero, namely, as the gas becomes ideal.

Performing the integration in Eq. (9.124) with the virial equation of state (Eq. (9.36)) we obtain

$$p_f = p \exp \left(\frac{B}{RT} p + \frac{C'}{2RT} p^2 + \cdots \right).$$ (9.127)

If the pressure is not too high, the exponential factor may be expanded to yield a series in p:

$$p_f = p + \frac{B}{RT} p^2 + O(p^3).$$ (9.128)

In order to gain insight into the concept of fugacity let us use the van der Waals estimate for B:

$$B = b - \frac{a}{RT}.$$

To the first order in p, Eq. (9.128) becomes

$$p_f = p \left[1 + \frac{1}{RT} \left(b - \frac{a}{RT} \right) p \right].$$ (9.129)

In the domain of T where the repulsive force is dominant over the attractive force the fugacity is larger than the pressure predicted by the ideal gas equation of state, whereas in the opposite case the fugacity is smaller than the pressure value of the ideal gas value because of the attractive pull by the molecules. This is precisely the picture provided by the van der Waals theory, and the fugacity reflects this intermolecular interaction effect on thermodynamic functions and the Gibbs free energy in particular.

Since for pure substances the molar Gibbs free energy is simply the chemical potential

$$g = \mu(T, p),$$

the chemical potential of the substance may be written as

$$\mu(T, p) = \mu^*(T) + RT \ln p + \int_0^p dp' \left(v - \frac{RT}{p'} \right). \tag{9.130}$$

This can be put in a form analogous to the formula for an ideal gas, if the fugacity is used:

$$\mu(T, p) = \mu^*(T) + RT \ln p_f. \tag{9.131}$$

It is instructive to note that we could have obtained it if integration was performed on the differential form

$$d\mu(T, p) = -s \, dT + v \, dp \tag{9.132}$$

along the path indicated by the dotted line in Fig. 9.4. The meaning of $\mu^*(T)$ can be inferred from that of $g^*(T)$: that is, *it is the chemical potential of the hypothetical ideal gas at 1 atm pressure, and is called the standard chemical potential of the substance*. It may be evaluated from the standard enthalpy and the standard entropy. By definition

$$\mu^*(T) = h^*(T) - Ts^*(T), \tag{9.133}$$

where

$$h^*(T) = h^0 + \int^T dT' \, c_p^*(T'),$$

$$\tag{9.134}$$

$$s^*(T) = s^0 + \int^T dT' \, \frac{c_p^*(T')}{T'},$$

with h^0 and s^0 denoting the integration constants that may be determined from the standard enthalpy and the standard entropy, respectively.

As often is the case for gases, if the specific heat c_p can be expanded in a temperature series

$$c_p^*(T) = c_{p0}^* + c_{p1}^* T + c_{p2}^* T^2 + \cdots, \tag{9.135}$$

the integrals in Eq. (9.134) can be easily calculated to obtain

$$h^*(T) = h^0 + c_{p0}^* T + \tfrac{1}{2} c_{p1}^* T^2 + \cdots,$$

$$\tag{9.136}$$

$$s^*(T) = s^0 + c_{p0}^* \ln T + c_{p1}^* T + \cdots.$$

When these results are combined into Eq. (9.133), there arises a series for $\mu^*(T)$:

$$\mu^*(T) = \left(h^0 - s^0 T \right) + c_{p0}^* T (1 - \ln T) - \tfrac{1}{2} c_{p1}^* T^2 - \cdots. \tag{9.137}$$

An alternative way of calculating $\mu^*(T)$ is as follows. We first observe the identity

$$\left(\frac{\partial}{\partial T} \frac{\mu}{T} \right)_p = -\frac{h}{T^2}. \tag{9.138}$$

In the limit of zero pressure this equation takes the ideal gas form

$$\left[\frac{\partial}{\partial T} \frac{\mu}{T} \right]_{p=0} = -\frac{h^*(T)}{T^2}. \tag{9.139}$$

Integrating it over T at $p = 0$ yields

$$\mu^* = -T \int^T dT \, \frac{h^*(T)}{T^2} - IT, \tag{9.140}$$

where I is the integration constant. Substituting Eq. (9.136) and performing the integration, we obtain

$$\mu^*(T) = h^0 - c_{p0}^* T \ln T - \tfrac{1}{2} c_{p1}^* T^2 - \cdots - IT. \tag{9.141}$$

The integration constant I in Eq. (9.140) is identified by comparing Eq. (9.140) with Eq. (9.137):

$$I = s^0 - c_{p0}^*. \tag{9.142}$$

It is called Nernst's chemical constant. This constant may be experimentally determined from calorimetry data with the help of the third law of thermodynamics. These chemical constants are useful for a number of thermodynamic calculations and, especially, for chemical equilibrium constants.

9.7 Joule–Thomson Experiment

When the volume of a gas is expanded reversibly and adiabatically, it is accompanied by a change in temperature. The change can be studied by the Joule–Thomson experiment.

Let us consider a gas contained in a cylinder AB whose wall is thermally insulated; see Fig. 9.7. The gas in the cylinder is infinitesimally slowly compressed at pressure p by the piston on the left. The gas then effuses through the porous plug (the shaded part within the cylinder) into the right hand side of the cylinder. Pressure p' is applied to the piston on the right. At the end of the process, the piston on the left is brought to position A' while the piston at B is pushed to position B'. Let the area of the piston be A. Then the force pA is acted on the gas from the left. The opposing force is $p'A$ on the gas on the right-hand side of the porous plug. If the volume of the gas on the left is V, the path traversed is V/A on the left of the plug and V'/A on the right. The total work performed on the gas is then

$$\int_{in}^{fin} dW = pV - p'V'.$$

Since the experiment is done adiabatically and very slowly so that the irreversibility is minimized, we have

$$Q = \int_{in}^{fin} dQ = 0$$

and consequently by the first law of thermodynamics

$$E' - E = pV - p'V'.$$

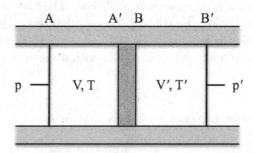

Fig. 9.7 Joule–Thompson experiment. A gas is adiabatically effused through the porous plug between A' and B under a difference in pressure p and p'. The (shaded) wall of the cylinder is thermally insulated.

By rearranging it we obtain

$$E + pV = E' + p'V'.$$

This relation means that the process is isoenthalpic. Therefore the molar enthalpy of the gas remains unchanged during the process:

$$h = h'. \qquad (9.143)$$

This condition must hold because the heat content of the gas remains unchanged owing to the particular setup of the experiment adiabatically performed.

Let us now examine the accompanying temperature change. This temperature change arising from the pressure change at an adiabatic condition is given by the Joule–Thomson coefficient

$$\mu_{JT} = \left(\frac{\partial T}{\partial p}\right)_h. \qquad (9.144)$$

This coefficient may be expressed in terms of more easily measurable quantities in the following manner:

$$\mu_{JT} = -\frac{\left(\frac{\partial h}{\partial p}\right)_T}{\left(\frac{\partial h}{\partial T}\right)_p}$$

$$= -c_p^{-1}\left[\left(\frac{\partial g}{\partial p}\right)_T + T\left(\frac{\partial s}{\partial p}\right)_T\right]$$

$$= c_p^{-1}\left[T\left(\frac{\partial v}{\partial T}\right)_p - v\right]$$

$$= v c_p^{-1}(T\alpha - 1), \qquad (9.145)$$

where α is the isothermal expansion coefficient. The first line follows upon use of the chain relation (Eq. (3.68)). Since $v \geq 0$ and $c_p \geq 0$, the Joule–Thomson coefficient is positive if $T\alpha \geq 1$ and it is negative if $T\alpha \leq 1$. Put in other words, the temperature decreases with decreasing pressure if

$$-\left(\frac{\partial h}{\partial p}\right)_T > 0,$$

and the temperature increases with decreasing pressure if

$$-\left(\frac{\partial h}{\partial p}\right)_T < 0.$$

As a matter of fact, if temperature is plotted against pressure at constant values of h, the isoenthalpic curves are obtained and they go through a

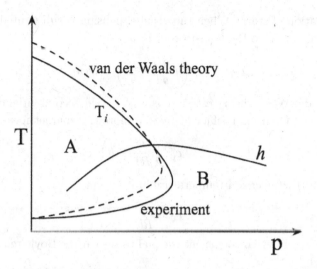

Fig. 9.8 Inversion curve, which is the locus of $\mu_{JT} = 0$. The solid curve denotes the experimental curve and the broken, the van der Waals theory prediction. In the inner region of the inversion curve, $\mu_{JT} > 0$ whereas $\mu_{JT} < 0$ in the outer region of the curve.

maximum. When the locus of the maximum is plotted in the T–p plane we obtain the inversion curve as shown in Fig. 9.8. The temperature at which

$$\mu_{JT} = 0 \tag{9.146}$$

is called the inversion temperature T_i which is determined from the equation

$$T_i \alpha(T_i) = 1, \tag{9.147}$$

the zero of μ_{JT}; see Eq. (9.145). Here

$$\alpha(T_i) = \left[v^{-1} \left(\frac{\partial v}{\partial T} \right)_p \right]_{T=T_i}. \tag{9.148}$$

In the domain A in Fig. 9.8, the Joule–Thomson coefficient is positive since the temperature along an isoenthalpic curve increases with the pressure, whereas in the domain B it is negative since the temperature decreases with increasing pressure. Thus in the former domain the gas cools as it expands, whereas in the latter it heats up on expansion. This cooling effect has an important technological application since it forms the thermodynamic foundation of the procedure for liquefying gases.

In order to better understand the inversion temperature and the inversion curve let us calculate the Joule–Thomson coefficient by using the virial equation of state. When the virial expansion is substituted into Eq. (9.145), we obtain to the lowest order in p

$$\mu_{JT} = c_p^{-1}\left(T\frac{dB}{dT} - B\right) + O(p).\tag{9.149}$$

If the van der Waals theory estimate is used for the second virial coefficient B, the Joule–Thomson coefficient takes a more transparent form

$$\mu_{JT} = c_p^{-1}\left(\frac{2a}{RT} - b\right),\tag{9.150}$$

which gives the inversion temperature in the form

$$T_i = \frac{2a}{Rb}.$$

Since the second virial coefficient is equal to zero at the Boyle temperature, we also see that

$$T_i = 2T_B$$

in the case of the van der Waals approximation, Eq. (9.150), for μ_{JT}; see Eq. (9.45). It is possible to obtain a more precise formula for the Joule–Thomson coefficient than Eq. (9.150), if we make use of the van der Waals equation of state without an approximation. Since from the van der Waals equation state follows

$$\left(\frac{\partial v}{\partial T}\right)_p = R[p + av^{-2} - 2av^{-3}(v-b)]^{-1},\tag{9.151}$$

the Joule–Thomson coefficient in the van der Waals theory is given by the formula

$$\mu_{JT} = c_p^{-1}\frac{2av^{-2}(v-b) - b(p + av^{-2})}{p + av^{-2} - 2av^{-3}(v-b)}.\tag{9.152}$$

Therefore the inversion temperature is determined by the solution of the equation

$$2av^{-2}(v-b) - b(p + av^{-2}) = 0.\tag{9.153}$$

Note that v must be regarded as a function of T and p in this equation. This equation can be cast into a relation for T_i and p.

On use of the van der Waals equation of state, it can be put into the form

$$2av^{-2}(v-b)^2 - bRT_i = 0,\tag{9.154}$$

which is easily solved for b/v:

$$\frac{b}{v} = 1 \pm \sqrt{\frac{bRT_i}{2a}}. \tag{9.155}$$

Using this in Eq. (9.153), we find a quadratic equation for T_i, which is easily solved for T_i:

$$T_i = \frac{10a}{9bR}\left[1 - \frac{3b^2p}{5a} \pm \frac{4}{5}\sqrt{1 - \frac{3b^2p}{a}}\right]. \tag{9.156}$$

The solutions are real for all p satisfying the condition

$$p \leq \frac{a}{3b^2}. \tag{9.157}$$

This result for the inversion temperature is correct only qualitatively, because of the approximate nature of the van der Waals equation of state, but still gives a fairly good idea of the pressure–temperature domain where the Joule–Thomson coefficient is negative. The inversion curve calculated with Eq. (9.155) is schematically plotted in Fig. 9.8 (the broken curve).

Equation (9.150) gives us useful insight into what happens in the gas as the pressure is changed. As the gas is compressed, the intermolecular distances will decrease and the molecules will feel stronger intermolecular attractions. In such a range of distance the interaction energy is dominated by the attractive potential. This energy is converted into the kinetic energy through collisions and the gas will be heated up as it is compressed. This is easily explained by Formula (9.150), which shows that the Joule–Thomson coefficient will be positive as long as the strength of the attractive forces is sufficiently large as to make $2a/RT$ larger than b which is a measure of the repulsive force. As the pressure is further increased, the intermolecular distances decrease further and the repulsive potential becomes predominant. The effect of this exchange in predominance of the attractive and repulsive potentials can be understood with Eq. (9.150), if we note that to the lowest order of approximation it is possible to write

$$\frac{2a}{RT} - b = \frac{2a}{pv} - b.$$

Therefore at a constant volume there will be a value of pressure at which $2a/pv - b$ changes its sign, resulting in a negative Joule–Thomson coefficient. This interpretation is in accord with the condition on pressure (Eq. (9.157)) above.

9.8   Liquefaction of Gases

Examination of the Joule–Thomson coefficient of a gas provides the information on the state of the gas from which to start its liquefaction successfully. If the temperature and pressure are in the domain bounded by the inversion curve (see the domain A in Fig. 9.8) at a given value of enthalpy, a simple adiabatic expansion of the gas will lower the temperature and, by repeating the process, it will be possible to lower the temperature below the critical temperature. Then mere compression of the cooled gas will be sufficient to put it into the liquid form.

Figure 9.9 shows a schematic diagram of the setup. A compressed gas at p and T is adiabatically expanded through the nozzle to attain pressure p_b and eventually temperature T_b, the boiling point of the liquid. The portion of the gas not liquefied attains thermal equilibrium with the gas on the verge of expansion at T. This gas is compressed to p and the cycle is repeated. Since the whole system is thermally insulated, the process is isoenthalpic.

Let us calculate the fraction ϵ of the gas liquefied. Since the process is isoenthalpic, the enthalpy of the substance before and after throttling must be equal to the sum of the enthalpies of the fraction liquefied and the remainder of the gas unliquefied. Therefore, for a mole of the gas

$$h_g(T,p) = (1 - \epsilon)h_g(T,p_b) + \epsilon h_l(T_b,p_b), \qquad (9.158)$$

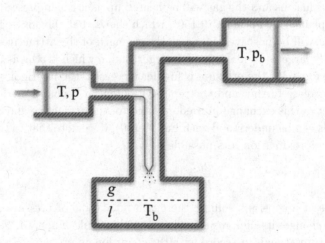

Fig. 9.9   Liquefaction of gas: A schematic illustration.

where h_g and h_l denote the molar enthalpy of the gas and of the liquid, respectively, at the state indicated. Solving this equation for ϵ, we obtain

$$\epsilon = \frac{h_g(T, p_b) - h_g(T, p)}{h_g(T, p_b) - h_l(T_b, p_b)}. \tag{9.159}$$

The enthalpy changes in Eq. (9.159) may be easily calculated from Eq. (9.77):

$$h_g(T, p_b) - h_g(T, p) = \int_p^{p_b} dp \left[v_g - T \left(\frac{\partial v_g}{\partial T} \right)_p \right],$$

$$h_g(T, p_b) - h_l(T_b, p_b) = \Delta h_v(T_b, p_b) + \int_{T_b}^T dT \, c_p^{(g)}(T, p_b),$$

where Δh_v is the heat of vaporization at T_b and p_b,

$$\Delta h_v(T_b, p_b) = h_g(T_b, p_b) - h_l(T_b, p_b),$$

$c_p^{(g)}$ is the heat capacity of the gas, and v_g denotes the molar volume of the gas. Substituting these results into Eq. (9.159) yields

$$\epsilon = \frac{-\int_p^{p_b} dp' \, \mu_{JT}(p') c_p^{(g)}(T, p')}{\Delta h_v(T_b, p_b) + \int_{T_b}^T dT' \, c_p^{(g)}(T', p_b)}. \tag{9.160}$$

Since $\Delta h_v \geq 0$ and $c_p^{(g)} \geq 0$, the denominator is always positive and hence it follows that

$$\epsilon > 0,$$

that is, the gas will liquefy only if

$$\int_{p_b}^p dp \, \mu_{JT}(p) c_p^{(g)}(T, p) > 0,$$

which implies that the condition for liquefaction is

$$\mu_{JT} > 0. \tag{9.161}$$

Therefore it can be concluded that the gas will liquefy only when T and p are in the domain bounded by the inversion curve, namely, the domain A in Fig. 9.8. In Table 9.2 are listed some maximum zero pressure inversion temperatures T_i^* for various gases. The maximum efficiency of liquefaction at a given operating temperature is achieved at the pressure determined by the equation

$$\left(\frac{\partial \epsilon}{\partial p} \right)_T = c_p^{(g)}(T, p) \mu_{JT}(T, p) \left(\Delta h_v + \int_{T_b}^T dT' \, c_p^{(g)} \right)^{-1} = 0. \tag{9.162}$$

Table 9.2 T_i^*.

CO$_2$	1500 K
Ar	723
N$_2$	621
H$_2$	195
He	35
Air	603

Since the specific heat is not equal to zero, the condition is satisfied only if p is such that

$$\mu_{JT}(T, p) = 0. \tag{9.163}$$

That is, the temperature and pressure must be those corresponding to the inversion point, T_i and p_i, for the maximum efficiency to be achieved in liquefaction.

If the virial equation of state is used to the first order in p, we obtain the fraction ϵ in a rather simple form

$$\epsilon = \frac{(T\frac{dB}{dT} - B)(p - p_b)}{\Delta h_v + \int_{T_b}^{T} c_p^{(g)} dT'} \approx \frac{(T\frac{dB}{dT} - B)(p - p_b)}{\Delta h_v + c_p^{(g)}(T_p - T_b)}, \tag{9.164}$$

where $c_{p0}^{(g)}$ is the temperature independent part of $c_p^{(g)}$. The second approximation will hold if the specific heat of the gas does not vary strongly with temperature in the range of temperature of interest.

9.9 Entropy Surface

The Gibbs relation

$$T \, dS = dE + p \, dV - \sum_{i=1}^{r} \mu_i \, dn_i,$$

implies that the entropy is a surface in the space of $(S, E, V, n_1, n_2, \ldots, n_r)$ in the case of a mixture, or, in the case of a closed system, the Gibbs relation

$$T \, dS = dE + p \, dV$$

implies that the entropy is a surface in the space of (S, E, V). The calculations which we have performed for thermodynamic quantities in Section 9.5 are in effect attempts at constructing such surfaces although the results obtained have not been cast in such forms. Such a surface contains all the necessary constitutive information on the system of which the entropy surface is representative, and, as we shall see, such surfaces can indeed generate all the thermodynamic constitutive relations when appropriate derivatives are taken from them. Since constructing entropy surfaces in the general case is as difficult as integrating an arbitrary differential form, we obviously have to specialize to particular cases if we wish to explicitly demonstrate the method. Here we shall consider the cases of an ideal gas and a van der Waals gas under the assumption that the heat capacities are independent of temperature.

It is worthwhile noting that in the axiomatic theory[6] of thermodynamics the existence of entropy is taken as an axiom following from the thermodynamic laws, and a theory of thermodynamics is formulated axiomatically. Such an approach is mathematically appealing, but it tends to remove us from the physical aspects important to the true understanding of thermodynamics, especially, of irreversible processes.

9.9.1 *Ideal Gas*

Let us collect the results for the internal energy and the entropy which we have calculated for the ideal gas:

$$S = C_v \ln T + R \ln V + S_0 - (C_v \ln T_0 + R \ln V_0), \qquad (9.165)$$

$$E = C_v T + E_0 - C_v T_0. \qquad (9.166)$$

If S and E are determined such that

$$S_0 = C_v \ln T_0 + R \ln V_0, \qquad (9.167)$$

$$E_0 = C_v T_0, \qquad (9.168)$$

then S and E are given by

$$S = C_v \ln T + R \ln V, \qquad (9.169)$$

$$E = C_v T. \qquad (9.170)$$

By eliminating T from Eqs. (9.169) and (9.170), we find

$$S = C_v \ln \left(\frac{E V^{R/C_v}}{C_v} \right) \qquad (9.171)$$

for the entropy surface in the space spanned by (S, E, V). This entropy contains all the information on the system (i.e., ideal gas) since it can yield all the thermodynamic quantities for the ideal gas.

---

[6]For example, see H. B. Callen, *Thermodynamics* (Wiley, New York, 1960).

In thermodynamics the temperature may be expressed such that

$$\frac{1}{T} = \left(\frac{\partial S}{\partial E}\right)_V. \tag{9.172}$$

The derivative on the right-hand side can be calculated from Eq. (9.171):

$$\left(\frac{\partial S}{\partial E}\right)_v = \frac{C_v}{E}.$$

Combining it with Eq. (9.172) yields Eq. (9.170). Analogously, the pressure may be expressed by the relation

$$\frac{p}{T} = \left(\frac{\partial S}{\partial V}\right)_E. \tag{9.173}$$

Since from Eq. (9.169)

$$\left(\frac{\partial S}{\partial V}\right)_E = \frac{R}{V},$$

we recover the equation of state for a mole of an ideal gas from Eq. (9.173):

$$pV = RT. \tag{9.174}$$

This is another constitutive equation postulated for calculating Eq. (9.169). These simple calculations imply the following: $1/T$ and p/T—the constitutive relations—are simply the tangents to the entropy surface. Nevertheless, however clean and appealing mathematically this point may be, the physical meanings of these tangents and their historical and experimental origins are much deeper than the mathematical meanings.

By combining the aforementioned results we also obtain from Eq. (9.169)

$$dS = \frac{1}{T}dE + \frac{p}{T}dV, \tag{9.175}$$

namely, the equilibrium Gibbs relation. This is the differential form which, with the help of the constitutive equations (Eqs. (9.170) and (9.174)), has yielded the entropy surface (Eq. (9.169)) in the first place. Other thermodynamic quantities follow from either the set of Eqs. (9.169), (9.172), and (9.173) or the set of Eqs. (9.170), (9.174), and (9.175). These two sets represent two different viewpoints to the same thermodynamic quantities.

9.9.2   *van der Waals Gas*

If the heat capacity C_v is independent of T then the molal energy E and entropy S of a van der Waals gas are given by

$$E = -\frac{a}{V} + C_v T + (E_0 - C_v T_0) + \frac{a}{V_0}, \tag{9.176}$$

$$S = C_v \ln T + R \ln(V - b) + [S_0 - C_v \ln T_0 - R \ln(V_0 - b)], \tag{9.177}$$

where E_0 and S_0 are the molal energy and entropy of the gas in the hypo-thetical state of an ideal gas at 1 atm and T_0. We select these integration constants such that

$$E_0 = C_v T_0, \quad S_0 = C_v \ln T_0 + R \ln(V_0 - b). \tag{9.178}$$

Then E and S are given by

$$E = -\frac{a}{V} + C_v T, \quad S = C_v \ln T + R \ln(V - b). \tag{9.179}$$

Elimination of T from the two equations in Eq. (9.179) yields the entropy surface for the van der Waals gas:

$$S = C_v \ln\left[\frac{(V - b)^{R/C_v}}{C_v} \left(E + \frac{a}{V} \right) \right]. \tag{9.180}$$

By following the same procedure as for the ideal gas considered earlier, this surface can be shown to yield the constitutive equations

$$E = -\frac{a}{V} + C_v T, \quad p = \frac{RT}{V - b} - \frac{a}{V^2}. \tag{9.181}$$

The verification of these results is left to the reader as an exercise. If the specific heat is not assumed to be independent of T then the entropy surface can be a more complicated function of E. In any event, the entropy surface (Eq. (9.180)) contains in it all the constitutive information on the system under the assumption of constant C_v. In this sense, the entropy is a storage of the constitutive information on the system in hand. For this reason and its statistical interpretation the concept of entropy holds a considerable importance in information science. However, it should be kept in mind that the entropy in the present work is defined only for reversible processes or systems in equilibrium. For irreversible processes the calortropy may be regarded as a storage of information.

Problems

(1) A mole of a liquid is reversibly heated from T_m to its boiling point T_b at 1 atm pressure and then the vapor is reversibly compressed and heated at the same time to pressure p and temperature T. Calculate the enthalpy change for the reversible process. The following data are available for the substance:

$c_p(l) = a + bT + cT^2$,
$c_p(v) = c_p^* + c_{p1}T + c_{p2}T^2$,
$pv = RT + BP$ for the vapor,
$\Delta h_v = $ heat of vaporization at $T = T_b$ and $p = 1\,\text{atm}$.

(2) Verify the results in Eq. (9.181) for the van der Waals fluid.

(3) Utilize the following data at 0°C to calculate the fugacity of carbon monoxide at 50, 100, 400, and 1000 atm by graphical integration based on the expression for fugacity.

p atm	pv/RT
25	0.9890
50	0.9792
100	0.9741
200	1.0196
400	1.2482
800	1.8075
1000	2.0819

(4) Calculate the efficiency of liquefaction by using the van der Waals model for the second virial coefficient $B = b - a/RT$.

(5) Using the pv/RT values for 50 atm and 400 atm given in problem 3, calculate the van der Waals constants a and b for carbon monoxide. From these determine the fugacities at various pressures and compare the results with those obtained in problem 3.

(6) By using the approximation $v \simeq RT/p$ in Eq. (9.153) and solving the resulting quadratic equation for T, show that the inversion temperature is given by

$$T_i = \frac{a}{Rb}\left[1 \pm \left(1 - \frac{3b^2}{a}p\right)^{\frac{1}{2}}\right],$$

which is real for all p satisfying the condition

$$p \leq \frac{a}{3b^2},$$

the same condition as Eq. (9.157).

Chapter 10

Canonical Equation of State

In the previous chapter on real fluids, we have shown that the van der Waals equation of state is capable of qualitatively describing the thermodynamic properties of fluids, although its description of critical and subcritical properties is inadequate in accuracy. Nevertheless, it is qualitatively better than the virial equation of state in its capability. It thus indicates that it somehow contains the basic elements required for correct description of fluid properties over the whole range of density and temperature. This attractiveness of the van der Waals equation of state has been enduring motivations for numerous attempts[1] at improving it in the literature.

In this chapter,[2] we present the canonical (i.e., generic van der Waals) equation of state that fulfills the desired aim of satisfactorily extending[3] the van der Waals equation of state. Here we develop the algorithms and formulas for the critical properties of fluids by means of which we can phenomenologically relate the critical characteristics of fluids to the parameters in the canonical equation of state. The critical and subcritical thermodynamic data, such as the critical parameters and the critical pressure–density data, spinodal and coexistence curves, isothermal compressibility, and specific heat in the subcritical neighborhood of the critical point, are shown closely related to the generic van der Waals parameters in the canonical equation of state—a phenomenological model. Each of these characteristic

---

[1]See, for example, J. J. van Laar, *Proc. Sci. Sec. Kon. ned. Akad. Wiesensch.* **14**, 278, 428, 563, 711 (1912); R. Planck, *Forsch. Geb. Ingenieures.* **7**, 161 (1936); R. Planck and L. Riedel, *Ing. Arch.* **16**, 255 (1947/48); H. D. Baehr, *Forsch. Geb. Ingenieures.* **29**, 143 (1963).

[2]This chapter is based on an unpublished research note on the subject by B. C. Eu (unpublished).

[3]See B. C. Eu and K. Rah, *Phys. Rev. E* **63**, 031303 (2001); B. C. Eu, *J. Chem. Phys.* **114**, 10899.(2001); K. Rah and B. C. Eu, *J. Phys. Chem. B* **107** 4382 (2003).

properties makes it possible to determine the empirical parameters in the model. It is thus shown that with suitable choices of parameter values the model would be capable of describing the thermodynamics of critical and subcritical fluids, appropriately generalizing the original van der Waals equation of state theory. The present chapter therefore shows how to fashion the equation of state in a van der Waals-like form, so that the experimental thermodynamic properties of fluids are better described than by the van der Waals equation of state, which is found defective in some aspects.

10.1   Canonical Equation of State

The genesis of the canonical equation of state is in a statistical mechanics study[4] of equation of state of real fluids and, in particular, of the Lennard–Jones (LJ) fluid, whose interaction potential energy consists of repulsive and attractive contributions. For such fluids it can be shown that the statistical mechanical virial equation of state can be split into the repulsive and attractive contributions, which can then be rearranged to a form reminiscent of the van der Waals equation of state

$$[p + A(\rho, \beta)\,\rho^2]\,[1 - B(\rho, \beta)\,\rho] = \rho\beta^{-1}, \qquad (10.1)$$

where p, ρ, and β are, respectively, the pressure, density, inverse temperature $\beta = 1/k_B T$ with T denoting the temperature and k_B the Boltzmann constant, and $A(\rho, \beta)$ and $B(\rho, \beta)$ are constants independent of p, which we call the generic van der Waals (GvdW) parameters. Equation (10.1) not only retains the useful features of the van der Waals equation of state but also removes defects of the latter in describing the thermodynamic properties of real fluids, particularly, in the critical and subcritical regimes.

According to the statistical mechanics study of Eq. (10.1) the parameters $A(\rho, \beta)$ and $B(\rho, \beta)$ are generally functions of ρ and β, but as the density and β diminish to zero, they become the van der Waals constants a' and b' of the van der Waals equation of state

$$(p + a'\rho^2)(1 - b'\rho) = \rho\beta^{-1}, \qquad (10.2)$$

that is,

$$\lim_{\rho \to 0, T \to \infty} A(\rho, \beta) = a', \qquad \lim_{\rho \to 0, T \to \infty} B(\rho, \beta) = b'. \qquad (10.3)$$

---

[4]See Footnotes 2 and 3 and also K. Rah and B. C. Eu, *J. Chem. Phys.* **115**, 2634 (2003).

Because of this limiting behavior of A and B the equation of state (Eq. (10.1)) was initially called the generic van der Waals equation of state. Since the form of equation of state (Eq. (10.1)) is generic to any fluid characterized by an interaction potential energy consisting of attractive and repulsive potential energies the terminology canonical equation of state is quite appropriate for fluids obeying such potentials of interaction. In fact, the van der Waals equation of state may be regarded as a special case of the canonical form.

By using phenomenological forms for $A(\rho, \beta)$ and $B(\rho, \beta)$ and calculating the critical isotherm, it will be shown that the canonical equation of state is capable of describing the critical isotherm correctly. Furthermore, its utility was also shown for developing a molecular theory of transport coefficients in the liquid density regime with the help of the modified free volume (MFV) theory,[5] since the canonical form (Eq. (10.1)) provides a well-defined formula for the mean free volume

$$v_f = \rho^{-1}[1 - B(\rho, \beta)\,\rho],$$

and A and B can be rigorously calculated from their statistical mechanical representations by using the pair correlation function of the fluid. Such an expression for the mean free volume has played a crucial role in the calculation of diffusion coefficients of liquids in the MFV theory of diffusion in liquids.

The statistical mechanics of the GvdW equation of state indicates that A and B are not only dependent on density and temperature, but also discontinuous in the subcritical regime, displaying a non-analytic behavior characteristic of the equation of state for a real fluid in the subcritical regime. Relying on the non-analytic behavior of A and B exhibited in the statistical mechanical study, we may now phenomenologically assume empirical non-analytic formulas—irrational functions of density and temperature— for them and then demonstrate that the equation of state thus constructed can be indeed capable of yielding the critical isotherm for pressure of the fluid. Turning this procedure around, we now formulate algorithms to construct a canonical equation of state so as to account for various thermodynamic properties and thus relate them to some characteristic aspects of the GvdW parameters; for example, such as the virial coefficients, Boyle temperature, Joule–Thomson coefficient and inversion temperature, various

---

[5]See K. Rah and B. C. Eu, *J. Chem. Phys.* **115**, 2634 (2001); K. Rah and B. C. Eu, *Phys. Rev. Lett.* **88**, 065901 (2002); K. Rah and B. C. Eu, *J. Chem. Phys.* **116**, 7967 (2002). See also B. C. Eu, *Transport Coefficients of Fluids* (Springer, Heidelberg, 2006).

critical properties of fluids—the critical isotherm, liquid–vapor coexistence curve, spinodal curve, isothermal compressibility, and specific heat in the neighborhood of the critical point. In this manner, the canonical equation of state is assured of summarizing as broadly as possible the fluid properties around the critical state and beyond.

The critical properties we are discussing in this chapter have been considered in the literature on the basis of the renormalization group theory, but the present empirical approach using the canonical equation of state is quite low-browed mathematically and algebraic, and thus can be easily grasped, although it appears complicated at first glance. It does not require anything other than a pair of constitutive equations for pressure and chemical potential based on the canonical equation of state, but involves only simple calculus and algebraic equations.

10.2 Reduced Variables

For the purpose of carrying out the desired analysis it is convenient to use non-dimensionalized variables. We define the following reduced (dimensionless) variables:

$$p^* = \frac{pv_0}{\varepsilon}, \qquad T^* = \frac{k_B T}{\varepsilon}, \qquad \eta = v_0 \rho,$$

$$A^* = \frac{A}{\varepsilon v_0}, \qquad B^* = \frac{B}{v_0}, \tag{10.4}$$

where ε is the potential well depth and σ is the size parameter, which may be identified with the Lennard–Jones (LJ) potential parameters, and $v_0 \equiv \pi \sigma^3 / 6$, the volume of a sphere of diameter σ. The critical parameters are denoted by p_c, η_c, and T_c. These reduced variables will be used when we examine the thermodynamic properties away from the critical region.

For studying the critical and subcritical behavior of the fluid it is convenient to scale p^*, η, and T^* with the reduced critical parameters p_c^*, η_c^*, and T_c^*. With the definitions of scaled reduced variables

$$\psi = \frac{p^*}{p_c^*} = \frac{p}{p_c}, \qquad y = \frac{\eta}{\eta_c}, \qquad \theta = \frac{T^*}{T_c^*} = \frac{T}{T_c}, \tag{10.5}$$

we define relative reduced variables

$$\phi = \psi - 1, \qquad x \equiv y - 1, \qquad t \equiv \theta - 1. \tag{10.6}$$

These relative reduced variables vanish at the critical point, (p_c, η_c, T_c). The reduced critical parameters p_c^*, η_c, and T_c^* will be expressed in terms of parameters making up A^* and B^*. We also define the dimensionless parameters

$$\nu \equiv B_c^* \eta_c, \quad \zeta \equiv \frac{A_c^* \eta_c^2}{p_c^*}, \quad \tau \equiv \frac{\eta_c T_c^*}{p_c^*}, \tag{10.7}$$

where A_c^* and B_c^* are the reduced values of A^* and B^* at the critical point. They are constants independent of density and temperature and may be related to the van der Waals parameters a' and b' in the van der Waals equation of state. The critical parameters then can be expressed in terms of these dimensionless parameters and A_c^* and B_c^*:

$$\eta_c = \frac{\nu}{B_c^*}, \quad p_c^* = \frac{\nu^2 A_c^*}{\zeta B_c^{*2}}, \quad T_c^* = \frac{\tau \nu A_c^*}{\zeta B_c^*}. \tag{10.8}$$

The dimensionless parameter τ, being independent of A_c^* and B_c^*, is an important quantity that enables us to gauge the quality of models for the GvdW parameters A^* and B^*, as will be shown later with more explicit, but simpler models for them—the quadratic model discussed later. The critical parameters given in Eq. (10.8) should be compared with the corresponding results in the van der Waals theory:

$$(\eta_c)_{vdw} = \frac{1}{3b}, \quad (p_c^*)_{vdw} = \frac{a}{27b^2}, \quad (T_c^*)_{vdw} = \frac{8a}{27b}, \tag{10.9}$$

where

$$a = a'/\varepsilon v_0, \quad b = b'/v_0.$$

Note that these parameters a and b are dimensionless in the present scheme of reducing variables and parameters. Since A_c^* and B_c^* are qualitatively comparable, respectively, to the reduced van der Waals constants a and b we see that in the low density limit

$$\nu \to \frac{1}{3}, \quad \frac{\nu^2}{\zeta} \to \frac{1}{27}, \quad \frac{\tau \nu}{\zeta} \to \frac{8}{27}$$

or

$$\nu \to \frac{1}{3}, \quad \zeta \to 3, \quad \tau \to \frac{8}{3}. \tag{10.10}$$

The structural similarity of the critical parameters in the van der Waals theory and the present canonical equation of state theory is one of the attractiveness of the latter because with relatively simple models for A^* and B^* the canonical equation of state can be shown capable of accounting for the critical phenomena with a relatively unsophisticated mathematical method as the van der Waals theory does, and also because the thermodynamics of fluids may be built around the van der Waals theory, which is capable of qualitatively describing most features of real fluids in the entire density range, as demonstrated in the previous chapter.

10.3 Reduced Canonical Equation of State

Upon using the aforementioned reduced variables and parameters ζ, ν, and τ, the canonical equation of state (Eq. (10.1)) can be reduced with respect to the critical parameters and expressed in the non-dimensional form

$$\left[\phi + 1 + \zeta(1+x)^2 \mathcal{A}(x,t)\right] \left[1 - \nu(1+x)\mathcal{B}(x,t)\right] = \tau(1+t)(1+x),$$
(10.11)

where

$$\mathcal{A}(x,t) = \frac{A^*}{A_c^*}, \qquad \mathcal{B}(x,t) = \frac{B^*}{B_c^*}.$$
(10.12)

In the limit where Eq. (10.10) holds, $\mathcal{A} = 1$ and $\mathcal{B} = 1$ and Eq. (10.11) then becomes the non-dimensionalized (in fact, corresponding state) van der Waals equation of state. If $\mathcal{A}$ and $\mathcal{B}$ are functions of x and t that are independent of material parameters then Eq. (10.11) would also represent the law of corresponding states. However, that will not be generally the case because $\mathcal{A}$ and $\mathcal{B}$ will not be free from (reduced) material parameters, which then may not be universal.

As mentioned earlier, in the statistical mechanics study of $\mathcal{A}$ and $\mathcal{B}$ it was found that $\mathcal{A}$ and $\mathcal{B}$ become discontinuous in the subcritical regime and there is a region of density in which they do not exist. The breakdown point of $\mathcal{A}$ and $\mathcal{B}$ will be denoted by z_{sl} and z_{sv}, where $z_{sl} = \eta_1/\eta_c$ is the reduced breakdown density η_1 on the liquid branch, whereas $z_{sv} = \eta_2/\eta_c$ is the reduced breakdown density η_2 on the vapor branch ($\eta_1 > \eta_2$). Note that η_1 and η_2 are different from the liquid–vapor coexistence densities η_l and η_v, respectively. As the critical point is approached, η_1 and η_2 tend to the critical density. Therefore

$$\lim_{t \to 0} z_{sk} = 1 \quad (k = l, v).$$
(10.13)

That is, $z_{sk}(t = 0) = 1$ coincides with the density at the critical point— the critical density. These reduced breakdown densities are functions of temperature only. It is convenient to define

$$x_{sk} = z_{sk} - 1 \quad (k = l, v),$$

which then has the limit

$$\lim_{t \to 0} x_{sk} = 0.$$
(10.14)

These breakdown densities will, in fact, turn out to be the spinodal densities at t, and such points will be mathematically defined as the points at which the first four density derivatives of pressure vanish in a manner similar to the density derivatives of pressure at the critical point. We remark that the fluid is thermodynamically unstable in the density interval $x_{sv} < x < x_{sl}$.

10.4   Models for the GvdW Parameters

10.4.1   *Subcritical Regime*

We have mentioned earlier that the GvdW parameters are discontinuous with respect to density for $t < 0$ (the subcritical regime of temperature) and do not exist in the interval $x_{sv} < x < x_{sl}$. Outside this interval they are continuous functions of density and temperature. In the subcritical regime of temperature they can be also irrational functions of t. Any phenomenological forms for the GvdW parameters should display these features so as to correctly describe the experimentally observed data on the thermophysical properties. The following forms consisting of series of integer and fractional powers in $(x - x_{sk})$ meet the requirements mentioned:

$$\mathcal{A} = \sum_{i\geq 0} a_i^{(k)}(t)(x - x_{sk})^i + \sum_{i\geq 0} \alpha_{nai}^{(k)}(t)(x - x_{sk})^{f+i}|x - x_{sk}|^{1+\delta}, \quad (10.15)$$

$$\mathcal{B} = \sum_{i\geq 0} b_i^{(k)}(t)(x - x_{sl})^i + \sum_{i\geq 0} \beta_{nai}^{(k)}(t)(x - x_{sk})^{f+i}|x - x_{sk}|^{1+\delta}, \quad (10.16)$$

where the coefficients $a_i^{(k)}(t)$, $\alpha_{nai}^{(k)}(t)$, $b_i^{(k)}(t)$, and $\beta_{nai}^{(k)}(t)$ $(k = l, v)$ are temperature-dependent and generally non-analytic with respect to t and the exponent f is an integer whereas δ is fractional ($\delta < 1$). In the quadratic model we will later use, $f = 3$ and the value of i in the first terms on the right is $i = 2$ whereas it is $i = 0$ in the non-analytic terms—the second terms on the right. The fractional number δ will be fixed from the critical isotherm. In this model the spinodal curve x_{sk} is an integral part of the equation of state.

The coefficients are chosen such that the liquid and vapor branches of the coefficients coincide with each other at the critical temperature, namely,

$$a_i^{(l)}(0) = a_i^{(v)}(0) \equiv a_i \neq 0, \qquad \alpha_{nai}^{(l)}(0) = \alpha_{nai}^{(v)}(0) \equiv \alpha_{nai} \neq 0,$$

$$b_i^{(l)}(0) = b_i^{(v)}(0) \equiv b_i \neq 0, \qquad \beta_{nai}^{(l)}(0) = \beta_{nai}^{(v)}(0) \equiv \beta_{nai} \neq 0, \quad (10.17)$$

$$\lim_{t\to 0} a_0^{(k)}(t) = 1 \quad \lim_{t\to 0} b_0^{(k)}(t) = 1 \quad (k = l, v),$$

and the fractional exponent $(4+\delta)\,(\delta = 0.3 \sim 0.5)$ is chosen in anticipation of the critical pressure–density isotherm discussed later. These properties suggest that in the subcritical regime near the critical point we may look for $a_i^{(k)}(t)$ and $b_i^{(k)}(t)$ $(k = l, v;\ i = 0, 1, 2)$ in the forms

$$a_i^{(k)}(t) = a_i + \widehat{a}_i^{(k)}(t) = a_i + \mathfrak{a}_i^{(k)}|t|^{\varepsilon_i}\left[1 + O(|t|^{1+\varepsilon_i})\right],$$

$$b_i^{(k)}(t) = b_i + \widehat{b}_i^{(k)}(t) = b_i + \mathfrak{b}_i^{(k)}|t|^{\varepsilon_i}\left[1 + O(|t|^{1+\varepsilon_i})\right] \quad (i = 0, 1, 2),$$

$$\tag{10.18}$$

where $\mathfrak{a}_i^{(k)}$ and $\mathfrak{b}_i^{(k)}$ $(k = l, v)$ are constants independent of x and t, and ε_i are fractional numbers, probably less than unity. Their values will be fixed empirically. Therefore

$$\widehat{a}_i^{(k)}(0) = \widehat{b}_i^{(k)}(0) = 0 \quad (i \geq 0). \tag{10.19}$$

Similarly, the coefficients of the irrational non-analytic terms must also be in the forms

$$\alpha_{na}^{(k)}(t) = \alpha_{na} + \widehat{\alpha}_{na}^{(k)}(t) = \alpha_{na} + \mathfrak{a}_{na}^{(k)}|t|^{\varepsilon_n},$$

$$\beta_{na}^{(k)}(t) = \beta_{na} + \widehat{\beta}_{na}^{(k)}(t) = \beta_{na} + \mathfrak{b}_{na}^{(k)}|t|^{\varepsilon_n}, \tag{10.20}$$

where ε_n is a fractional number, which may be less than unity. Its value also will be determined empirically. Therefore

$$\widehat{\alpha}_{na}^{(k)}(t), \widehat{\beta}_{na}^{(k)}(t) \to 0 \quad \text{as } t \to 0. \tag{10.21}$$

With this model for the t dependence of the parameters and with the exponents $\varepsilon_i\,[\varepsilon = \min(\varepsilon_0, \varepsilon_1, \ldots)]$ and ε_n appropriately chosen in comparison of the theory with experiment, the equation of state can be used to calculate the critical properties of fluids in the subcritical regime, as will be shown later. We have taken the same exponents for both for $\mathcal{A}$ and $\mathcal{B}$ because their statistical mechanical formulas are given in terms of an identical pair correlation function for the liquid (l) and vapor (v) branches—in which the temperature dependence is entirely vested—because they are symmetric in the neighborhood of the critical point. As will be shown later, the parameters $\mathfrak{a}_i^{(k)}$, $\mathfrak{b}_i^{(k)}$, $\mathfrak{a}_{na}^{(k)}$, and $\mathfrak{b}_{na}^{(k)}$ $(i = 0, 1;\ k = l, v)$ are not arbitrary, but constrained by the compatibility with experiment and by the stability conditions. *Our basic point in this chapter is that the GvdW parameters are directly related to certain specific and characteristic features of thermophysical data of the fluid under consideration, especially, in the neighborhood of the critical point.*

10.4.2 *Supercritical Regime*

Owing to the aforementioned properties of $a_i^{(k)}(t)$, $b_i^{(k)}(t)$, $\alpha_{nai}^{(k)}(t)$, and $\beta_{nai}^{(k)}(t)$ and the fact that at the critical point

$$x_{sl} = x_{sv} = 0,$$

the expansions for $\mathcal{A}$ and $\mathcal{B}$ in Eqs. (10.15) and (10.16) reduce, on the critical isotherm, to the forms

$$\mathcal{A} = 1 + \sum_{i \geq 1} a_i x^i + \sum_{i \geq 0} \alpha_{nai} x^{3+i} |x|^{1+\delta} \quad (t = 0), \qquad (10.22)$$

$$\mathcal{B} = 1 + \sum_{i \geq 1} b_i x^i + \sum_{i \geq 0} \beta_{nai} x^{3+i} |x|^{1+\delta} \quad (t = 0). \qquad (10.23)$$

These models can be extended into the supercritical regime of $t > 0$, in which case a_i, b_i, α_{nai}, and β_{nai} may have to be made t-dependent, to discuss the thermophysical properties of the fluid in the supercritical regime. It should be noted that $\mathcal{A}$ and $\mathcal{B}$ are analytic for $t > 0$ and $x \neq 0$ despite the presence of the irrational terms.

10.5 Reduced Chemical Potential

The chemical potential μ is reduced as follows:

$$\varpi(t, x) = \frac{\zeta B_c^*}{\nu A_c^* \varepsilon} \mu(x, t) = \frac{\tau \mu(x, t)}{T_c^* \varepsilon}. \qquad (10.24)$$

This reduced chemical potential is described by the reduced Gibbs relation for ϖ

$$d\varpi = -\hat{s}\, dt + \frac{1}{(1+x)}\, d\psi \quad [v = 1/y = 1/(1+x)], \qquad (10.25)$$

where the molar entropy $\bar{s}$ is scaled as follows:

$$\hat{s} = \frac{\tau}{k_B} \bar{s}. \qquad (10.26)$$

Now consider an isothermal process $(t, \psi_0) = (t, \phi_0 + 1) \rightarrow (t, \psi) = (t, \phi + 1)$. Then

$$\varpi(t, \phi) = \varpi(t, \phi_0) + \int_{\psi_0}^{\psi} \frac{d\psi}{1 + x}. \qquad (10.27)$$

The reference reduced pressure ψ_0 will be more explicitly specified later. The integral on the right of Eq. (10.27) is now calculated with the reduced equation of state (Eq. (10.11)).

By using Eq. (10.11) we express the integral for the chemical potential in terms of the GvdW parameters $\mathcal{A}$ and $\mathcal{B}$. First, the integral in Eq. (10.27) may be recast as follows:

$$K_k = \int_{\psi_0}^{\psi} \frac{d\psi}{1+x} = \int_{x_0}^{x} dx \left(\frac{\partial \psi}{\partial x}\right)_t \frac{1}{1+x}. \tag{10.28}$$

Performing integration by parts and using the reduced canonical equation of state (Eq. (10.11)), we obtain

$$K_k = \frac{1+\phi(x,t)}{1+x} - \frac{1+\phi(x_0,t)}{1+x_0} - \zeta \int_{x_0}^{x} dx \, \mathcal{A}_k(x,t)$$
$$+ \tau(1+t) \int_{x_0}^{x} \frac{dx}{(1+x)[1 - \nu(1+x)\mathcal{B}_k(x,t)]}. \tag{10.29}$$

Here the subscript k refers to the liquid (l) or vapor (v) branch; the liquid and vapor branches of the equation of state are different in the subcritical regime, but in the supercritical regime the equation of state involves a single set of continuous $\mathcal{A}$ and $\mathcal{B}$ representative of the gaseous states of the fluid. The expression for K_k is rearranged to the form which we find most convenient for the questions addressed in this chapter:

$$K_k = \frac{\phi(x,t)}{1+x} - \frac{\phi(x_0,t)}{1+x_0} + \int_{x_0}^{x} dx \frac{\Pi(x,t)}{(1+x)^2[1 - \nu(1+x)\mathcal{B}_k]}, \tag{10.30}$$

with the definition of $\Pi(x,t)$

$$\Pi(x,t) = \tau(1+x)(1+t)$$
$$- \left[1 + \zeta(1+x)^2 \mathcal{A}(x,t)\right][1 - \nu(1+x)\mathcal{B}(x,t)], \tag{10.31}$$

which is related to the reduced pressure as is evident from Eq. (10.59) below. Then we will be able to take advantage of the mathematical results developed for the equation of state when we deduce the mathematical properties of the chemical potential. The reduced chemical potential now is expressible in the form

$$\varpi(t,\phi) = \varpi_0(t) + \varpi_{ex}(x,t), \tag{10.32}$$

where the excess chemical potential is written as

$$\varpi_{ex}(x,t) = \frac{\phi(x,t)}{1+x} + \int_{x_0}^{x} dx \frac{\Pi(x,t)}{(1+x)^2[1 - \nu(1+x)\mathcal{B}]}. \tag{10.33}$$

With $\phi_0 = \phi(x_0, t)$, the reference chemical potential is defined by

$$\varpi_0(t) = \lim_{\phi \to \phi_0, x \to x_0} \left[\varpi(t, \phi) - \frac{\phi(x, t)}{1 + x} \right]. \qquad (10.34)$$

The chemical potential $\varpi(t, \phi)$ presented is formal in the sense that the integral is not performed as yet. We may specify the reference state more explicitly: we find it convenient to take $x_0 \equiv x_c = 0$ for both $k = l, v$ or $x_0 = x_{sk} (k = l, v)$, especially, when the excess chemical potential $\varpi_{ex}(x, t)$ is expressed in series of $(x - x_{sk})$. We will return to this question after we have discussed the definition of critical point.

10.6 Specific Heat

The isochoric specific heat may be calculated from the internal energy $\mathcal{E}$, which may be calculated in the same manner as for the chemical potential by following the well-known procedure in thermodynamics, namely, by varying the internal energy from state (x_0, t_0) to state (x, t). Then the specific heat is calculated with the derivative

$$\hat{c}_v(x, t) = \left(\frac{\partial \mathcal{E}}{\partial t} \right)_v, \qquad (10.35)$$

which can be eventually expressible in terms of the equation of state. It is also possible to obtain $\hat{c}_v(x, t)$ by integrating the derivative

$$\left(\frac{\partial \hat{c}_v}{\partial x} \right)_t = -\frac{(1 + t)}{(1 + x)^2} \left(\frac{\partial^2 \psi}{\partial t^2} \right)_x \qquad (10.36)$$

over x, given the equation of state. Still another way of calculating $\hat{c}_v$ employs the reduced work function (Helmholtz free energy) F. It gives rise to the formula

$$\hat{c}_v(x, t) = -(1 + t) \left(\frac{\partial^2 F}{\partial t^2} \right)_x$$

$$= (1 + t) \left[\left(\frac{\partial^2 \psi}{\partial t^2} \right)_x - (1 + x) \left(\frac{\partial^2 \varpi}{\partial t^2} \right)_x \right]. \qquad (10.37)$$

It should be remarked that this formula can be also obtained by using the Gibbs–Duhem equation; see Problem 5 of Chapter 7. All these relations can be shown to yield eventually the same formula expressed in terms of the equation of state. However, we will find Relation (10.37) more convenient to use for our purpose of calculating the asymptotic behavior of the anomalous specific heat in the subcritical neighborhood of the critical point, because the temperature derivative is taken after integration over x.

By using Eq. (10.32) it is easy to show that the excess part $\widehat{c}_{vk}^{(ex)}$ of the specific heat above and beyond the specific heat corresponding to $\varpi_0(t)$ is given by the formula

$$\widehat{c}_{vk}^{(ex)} = -(1+t)(1+x)\left(\frac{\partial^2 I_k}{\partial t^2}\right)_x, \qquad (10.38)$$

where $I_k(x_k, t)$ is defined by the integral

$$I_k(x_k, t) = \int_{x_0}^{x_k} dx \frac{\Pi_k(x, t)}{(1+x)^2 [1 - \nu \mathcal{B}_k(x+1)]} \quad (k = l, v) \qquad (10.39)$$

with x_0 chosen appropriately and $\Pi_k(x, t)$ is $\Pi(x, t)$ with the GvdW parameters for the liquid or vapor branch: for $k = l$ or v

$$\Pi_k(x, t) = \tau (1+x)(1+t)$$
$$- \left[1 + \zeta(1+x)^2 \mathcal{A}_k(x, t)\right][1 - \nu(1+x)\mathcal{B}_k(x, t)],$$
$$(10.40)$$

$\mathcal{A}_k$ and $\mathcal{B}_k$ denoting the k branch of $\mathcal{A}$ and $\mathcal{B}$, respectively. As will be shown later, the behavior of $\widehat{c}_{vk}^{(ex)}$ puts further restriction on the t dependence of parameters $\mathcal{A}(t, x)$ and $\mathcal{B}(t, x)$, because of its characteristic singular behavior in the neighborhood of the critical point.

10.7   Virial and Joule–Thomson Coefficients

The virial coefficients provide useful information on fluid behaviors and their experimental data are available for many fluids in the literature, so are the Joule–Thomson coefficients. Since these coefficients can be related to the GvdW parameters, the experimental information on the virial and Joule–Thomson coefficients provide valuable and important information on the GvdW parameters, particularly, in the supercritical regime. We relate them to each other in the following. The canonical equation of state reduced with respect to the potential parameters, but not with respect to the critical parameters, is used in this section.

10.7.1   *Virial Coefficients*

Then the virial coefficients $C_i (i \geq 2)$ of the virial equation of state in the supercritical regime

$$\frac{p^* \beta^*}{\eta} = 1 + C_2 \eta + C_3 \eta^2 + \cdots + C_i \eta^{i-1} + \cdots \qquad (10.41)$$

are given by the formulas in terms of A^* and B^* and their derivatives:

$$C_2 = B^*(T^*, 0) - A^*(T^*, 0) \beta^*, \qquad (10.42)$$

$$C_3 = [B^*(T^*, 0)]^2 + \left(\frac{dB^*}{d\eta} \right)_{\eta=0} - \beta^* \left(\frac{dA^*}{d\eta} \right)_{\eta=0}, \qquad (10.43)$$

$$\cdots.$$

Here C_j are reduced virial coefficients $(j \geq 2)$. Since the second virial coefficient vanishes at the Boyle temperature it follows that at the Boyle temperature $T^* = T_B^*$

$$B^*(T_B^*, 0) = \beta^* A^*(T_B^*, 0). \qquad (10.44)$$

Since C_2 tends to negative infinity as $T^* \to 0$ the parameters A and B should be such that

$$\lim_{T^* \to 0} [B^*(T^*, 0) - \beta^* A^*(T^*, 0)] = -\infty. \qquad (10.45)$$

On the other hand,

$$\lim_{T^* \to \infty} C_2 = C_2^{(\infty)} > 0. \qquad (10.46)$$

Moreover,

$$\lim_{T^* \to T_c} C_2 = C_2^{(c)}. \qquad (10.47)$$

The third virial coefficient also behaves similarly to the second virial coefficient with respect to T^*. In particular, it is known experimentally that there exists temperature T_3^* at which

$$C_3(T_3^*) = 0 \qquad (10.48)$$

and

$$\lim_{T^* \to 0} C_3(T^*) = -\infty. \qquad (10.49)$$

The Boyle temperature T_B^* does not coincide with T_3^* at which the third virial coefficient vanishes. If by any chance $T_3^* = T_B^*$, then at that temperature the equation of state assumes an ideal gas equation of state to the order of η^3.

We may conjecture that such a temperature exists for the canonical equation of state. Denote it by T_{can}^*. Then at $T^* = T_{can}^*$ there holds the relation

$$\frac{B^*(T_{can}^*, \eta)}{1 - B^*(T_{can}^*, \eta)\eta} = \beta^* A^*(T_{can}^*, \eta). \tag{10.50}$$

At temperature T_{can}^* the equation of state then takes the form for an ideal gas. However, at present it is not known that such a temperature is experimentally identified. If it was, it should be possible to define the standard state of the fluid with it instead of the hypothetical ideal gas that is conventionally defined as the standard state. It is the state at which $T = T_{can}^*$ and the equation of state is that of an ideal gas.

10.7.2 *Joule–Thomson Coefficient*

Another important property of fluids is the Joule–Thomson coefficient (9.145), which in terms of the canonical equation of state may be expressed as

$$\mu_{JT} = (\eta c_p)^{-1} \left[\frac{\eta T^{*2} (B^* + B^{*\prime} \eta)}{p^* (1 - B^* \eta)^2} - \frac{\eta T^*}{p^*} (A^* + A^{*\prime} \eta) - 1 \right]. \tag{10.51}$$

Here the prime on A^* and B^* denotes the temperature derivative. The Joule–Thomson coefficient vanishes at the inversion temperature. Therefore at $T^* = T_{in}^*$, the inversion temperature, there holds the equation

$$p_{in}^* = \frac{\eta T_{in}^{*2}(B_{in}^* + B_{in}^{*\prime}\eta)}{(1 - B_i^* \eta)^2} - \eta T_{in}^*(A_{in}^* + A_{in}^{*\prime}\eta), \tag{10.52}$$

where $A_{in}^* = A^*(T_{in}^*, \eta)$, $A_{in}^{*\prime} = (\partial A^*/\partial T^*)_{p^*}|_{T^*=T_{in}^*}$, $B_{in}^* = B^*(T_{in}^*, \eta)$, and $B_{in}^{*\prime} = (\partial B^*/\partial T^*)_{p^*}|_{T^*=T_{in}^*}$. If p_{in}^* on the left is replaced with the reduced canonical equation of state at $T^* = T_{in}^*$, Eq. (10.52) becomes an algebraic equation for T_{in}^* and η. The relation such as Eq. (10.52) puts constraints on the mathematical expressions for the GvdW parameters that may be empirically assumed, namely, it enables us to determine the parameters to reproduce the inversion curve.

10.7.3   *Asymptotic Behavior of the Second Virial Coefficient*

The asymptotic behaviors of the virial coefficients also put additional constraints on the GvdW parameters. For example, since the second virial coefficient for the LJ fluid asymptotically behaves as shown in the previous chapter

$$C_2 = 4\sqrt{2}\Gamma\left(\frac{3}{4}\right) v_0 T^{*-1/4}[1 + O(T^{*-1})] \quad \text{as } T^* \to \infty \qquad (10.53)$$

$$= -16\sqrt{2\pi}v_0 e^{1/T^*}T^{*-\frac{3}{2}}[1 + O(T^*)] \quad \text{as } T^* \to 0 \qquad (10.54)$$

the limiting behaviors of the GvdW parameters must conform to these limiting behaviors to be consistent with the LJ fluid and experiment. These therefore are additional means to fix the GvdW parameters.

10.8   Stability Conditions of Mechanical and Material Equilibria

10.8.1   *Classical Definition of Critical Point*

The critical state is a singular state of the fluid, and its definition is closely associated with the thermodynamic stability of the critical state of the fluid. In the conventional approach, the critical point is defined by the vanishing first two density derivatives of pressure or the chemical potential, and the sign of the third density derivative determines the stability of the system at the critical point so defined: it is stable if the third derivative is negative; unstable if it is positive; and neutral if it is equal to zero.

Thus if the reduced canonical equation of state is used, the critical point $(x, t, \phi) = (0, 0, 0)$ is given by (ν, τ, ζ) (see Eq. (10.7)) obeying the three algebraic equations

$$(1 + \zeta)(1 - \nu) = \tau, \qquad (10.55)$$

$$\left(\frac{\partial^i \phi}{\partial x^i}\right)_{t=0, x=0} = 0 \quad \text{for } i = 1, 2. \qquad (10.56)$$

This definition is possible and necessary to extend. It is possible to see its necessity if we examine it with the canonical equation of state in which the GvdW parameters can depend on t as well as x in general. Then it would be conceivable to imagine that there arise some cases in which the third density derivative happens to be equal to zero.

10.8.2   *Extended Definition*

J. J. van Laar[6] considered for the first time the variability of the van der Waals constant b and its relation to the critical parameters, although van der Waals himself disfavored such an extension. As noted earlier, introduction of density- and temperature-dependent van der Waals parameters might necessitate extending the definition of the critical point since such parameters could give rise to the vanishing third density derivative. He thus requires the first four density derivatives of pressure to identically vanish at the critical point.

We elaborate on this point by considering the thermodynamic stability theory of the critical point. In the classical thermodynamic stability theory of the critical state, the third density derivative provides the stability criteria for the fluid at the critical point. If the van der Waals parameters are assumed to depend on density and temperature, it is clearly possible to have a situation in which the third density derivative might identically vanish at the critical point. In that event, the thermodynamic stability of the fluid is determined by the fifth density derivative with the fourth density derivative also vanishing as well. Thus the critical point is determined by the equation of state and the first four density derivatives of pressure vanishing at the critical state. *The system is then thermodynamically stable if the fifth density derivative is negative at the critical point; unstable if it is positive; and neutral if it is equal to zero.* The sign of the fifth density derivative, therefore, also puts an additional condition on the equation of state or its parameters. This kind of situation arises for the canonical equation of state as will be shown. In fact, this extension of the definition of critical point makes it possible to relate the parameters of the canonical equation of state to the critical properties of the fluid.

Since the GvdW parameters $\mathcal{A}$ and $\mathcal{B}$ are generally density- and temperature-dependent, it is natural to extend the classical definition of the critical point. As a matter of fact, the GvdW parameters can be given exact statistical mechanical representations, which can be indeed shown density- and temperature-dependent. Although this extension necessarily gives rise to a larger and more complicated set of algebraic equations to solve for various thermodynamic properties, it widens the scope of the ideas embodied by the van der Waals theory and makes the theory more flexible, thereby enabling us to more effectively treat the critical phenomena

---

[6]See J. J. van Laar in Footnote 1.

phenomenologically than the van der Waals equation of state, but also renders the basic mathematical structures of algebraic equations that determine some thermophysical properties simpler than otherwise. As a matter of fact, such a simplification can be rather attractive in practice. Moreover, the additional conditions significantly reduce the degrees of freedom for the parameters necessarily introduced in $\mathcal{A}$ and $\mathcal{B}$ in the phenomenological approach. Therefore we make the following proposition:

Proposition 1. *The critical point of the fluid described by the canonical (GvdW) equation of state is defined by requiring the first four density derivatives of pressure to vanish at the critical point.*

This proposition is in essence about the thermodynamic stability of mechanical equilibrium of the fluid at the critical temperature and density. Away from the critical point in the subcritical regime, there exists a locus of the limit of stability in density, and this locus is the spinodal curve. There are two densities x_{sl} and $x_{sv}(x_{sv} < x_{sl})$ within the interval of which the fluid is unstable. According to the statistical mechanical expressions for the GvdW parameters, which have been investigated by means of an integral equation theory and simulations[7] the GvdW parameters extend into the liquid–vapor coexistence regime to the points, where they become discontinuous. This locus of discontinuity in GvdW parameters $\mathcal{A}$ and $\mathcal{B}$ is known to coincide with the spinodal curve. Since $x_{sk} = 0$ at the critical point $t = 0$, it is reasonable to extend Proposition 1 to the subcritical regime and assume that it is also applicable at $x = x_{sk}$ for $t < 0$. Therefore we make the following proposition for the spinodal curve:

Proposition 2. *Along the locus $x_{sk}(t)$ of the limit of instability the first four density derivatives of ϕ at $t < 0$ vanish. The fluid at the spinodal state is thermodynamically stable if the fifth density derivative of ϕ is negative; unstable if it is positive; and neutral if it is equal to zero.*

These propositions not only provide additional conditions on the GvdW parameters $\mathcal{A}$ and $\mathcal{B}$ and their density derivatives, but also yield, as a function of t, the spinodal curve that appears as a parameter in the model for $\mathcal{A}$ and $\mathcal{B}$ we have chosen. This point will become evident as we develop the theory for the critical characteristics of fluids in the following section.

---

[7]K. Rah and B. C. Eu, *J. Chem. Phys.* **115**, 9370 (2001).

10.9   Critical Properties of Fluids

10.9.1   *Critical Point*

According to *Proposition* 1, the non-dimensional form of GvdW equation of state (Eq. (10.11)) at the critical point is given by

$$(1 + \zeta)(1 - \nu) = \tau, \tag{10.57}$$

and the first four density derivatives are equal to zero:

$$\left(\frac{\partial^i \phi}{\partial x^i}\right)_{t=0, x=0} = 0 \quad \text{for } i = 1, 2, \ldots, 4. \tag{10.58}$$

Note that the first equation (Eq. (10.57)) is equivalent to $\phi(x, t)|_{t=0, x=0} = 0$.

The reduced equation of state, Eq. (10.11), can be rearranged to the form

$$\phi[1 - \nu(1 + x)\,\mathcal{B}(x, t)] = \Pi(x, t). \tag{10.59}$$

Henceforth $\Pi(x, t)$ will be referred to as the effective reduced pressure.

Then, since

$$\left(\frac{\partial^i p^*}{\partial \eta^i}\right)_{T^*=T_c^*,\,\eta=\eta_c} = \left(\frac{\partial^i \phi}{\partial x^i}\right)_{t=0, x=0} \quad \text{for } i = 1, 2, \ldots, 4$$

and, furthermore,

$$\left(\frac{\partial^i \phi}{\partial x^i}\right)_{t=0, x=0} = \left(\frac{\partial^i \Pi}{\partial x^i}\right)_{t=0, x=0} = 0 \quad \text{for } i = 1, 2, \ldots, 4 \tag{10.60}$$

owing to the fact that $\phi = 0$ at the critical point, Eqs. (10.57) and (10.58) take the forms

$$(1 + \zeta)(1 - \nu) = \tau, \tag{10.61}$$

$$\left(\frac{\partial^i \Pi}{\partial x^i}\right)_{t=0, x=0} = 0 \quad \text{for } i = 1, 2, \ldots, 4. \tag{10.62}$$

These equations are algebraic equations for the reduced critical parameters (ν, ζ, τ) and, when solved, yield (ν, ζ, τ) in terms of parameters making up $\mathcal{A}$ and $\mathcal{B}$. They are simpler to handle than Eqs. (10.57) and (10.58). Since there are five algebraic equations for the parameters appearing in the models for $\mathcal{A}(x, t)$ and $\mathcal{B}(x, t)$, the number of degrees of freedom is reduced by 5 for the parameters making up $\mathcal{A}$ and $\mathcal{B}$.

The first equation of this set is the reduced equation of state at the critical point $(\phi, t, x) = (0, 0, 0)$. If the density derivatives (Eq. (10.62)) are explicitly worked out at the critical point, we have the set of algebraic equations for the parameters:

$$P_0 \equiv \tau - (1 + \zeta)(1 - \nu) = 0, \tag{10.63}$$

$$1!P_1 \equiv \tau - \zeta(2 + a_1)(1 - \nu) + \nu(1 + \zeta)(1 + b_1) = 0, \tag{10.64}$$

$$2!P_2 \equiv -2\zeta(1 - \nu)(1 + 2a_1 + a_2) + 2\nu\zeta(2 + a_1)(1 + b_1)$$
$$+ 2\nu(1 + \zeta)(b_1 + b_2) = 0, \tag{10.65}$$

$$3!P_3 \equiv -6(\zeta a_1 + 2\zeta a_2 + \zeta a_3)(1 - \nu) + 6\zeta\nu(1 + 2a_1 + a_2)(1 + b_1)$$
$$+ 6\zeta\nu(2 + a_1)(b_1 + b_2) + 6\nu(1 + \zeta)(b_2 + b_3) = 0, \tag{10.66}$$

$$4!P_4 \equiv -24(1 - \nu)\zeta(a_2 + 2a_3 + a_4)$$
$$- 576\zeta^2\nu^2(a_1 + 2a_2 + a_3)(1 + 2a_1 + a_2)(1 + b_1)(b_1 + b_2)$$
$$+ 24\zeta\nu(2 + a_1)(b_2 + b_3) + 24\nu(1 + \zeta)(b_3 + b_4) = 0. \tag{10.67}$$

It should be noted that these algebraic equations are exact for the models chosen, Eqs. (10.15) and (10.16).

This set of algebraic equations, Eqs. (10.63)–(10.67), determine the reduced parameters (τ, ζ, ν) in terms of a_i and $b_i (i \le 4)$. Since the critical parameters are available experimentally, these parameters can be determined consistently with the experimental data, thus facilitating the determination of a_i and b_i from experiment. It is important to see that a_i and $b_i (i \le 4)$ are determined from the experimental values for the reduced critical parameters (τ, ζ, ν) such that Conditions (10.63)–(10.67) are satisfied. Such a set of (τ, ζ, ν) and a_i and $b_i (i \le 4)$ now assures the desired behavior for the critical isotherm for pressure as will be shown in the following.

10.9.2 *Critical Isotherm for Pressure*

The critical isotherm for ϕ now can be calculated if the effective reduced pressure $\Pi(x, t)$ is expanded around $x = 0$ on the critical isotherm $t = 0$. To obtain the idea of what sort of form $\Pi(x, t)$ would take, if the models for $\mathcal{A}$ and $\mathcal{B}$, for example, Eqs. (10.15) and (10.16), for the GvdW parameters are employed then we may express it in the form

$$\Pi(x, t) = \sum_{i \ge 0} L^{(i)}(x_{sk}, t)(x - x_{sk})^i$$
$$+ \sum_{i \ge 0} N^{(i)}(x_{sk}, t)(x - x_{sk})^{3+i}|x - x_{sk}|^{1+\delta}$$
$$+ \sum_{i \ge 0} M^{(i)}(x_{sk}, t)(x - x_{sk})^{6+i}|x - x_{sk}|^{2+2\delta}, \tag{10.68}$$

where $L^{(i)}(x_{sk}, t)$, $N^{(i)}(x_{sk}, t)$, and $M^{(i)}(x_{sk}, t)$ can be explicitly found in terms of $\mathcal{A}$ and $\mathcal{B}$ by using models (10.15) and (10.16). This in fact is a fairly general expansion arising from the required behavior of the GvdW parameters in the subcritical regime. In view of P_i defined earlier it is convenient to express $L^{(i)}(x_{sk}, t)$ in the form

$$L^{(i)}(x_{sk}, t) = P_i + \widehat{L}^{(i)}(x_{sk}, t), \tag{10.69}$$

with $\widehat{L}^{(i)}(x_{sk}, t)$ denoting that part of $L^{(i)}$ depending on t away from the critical temperature. Similarly, $N^{(i)}(x_{sk}, t)$ can be expressed in the form

$$N^{(i)}(x_{sk}, t) = Q_i + \widehat{N}^{(i)}(x_{sk}, t) \tag{10.70}$$

where

$$Q_i = N^{(i)}(x_{sk}, t)|_{t=0} \tag{10.71}$$

and $\widehat{N}^{(i)}(x_{sk}, t)$ is the t-dependent part. We note that

$$Q_0 = \nu(1 + \zeta)\beta_{na0} - \zeta(1 - \nu)\alpha_{na0}, \tag{10.72}$$

which will be used later. The coefficient $M^{(i)}(x_{sk}, t)$ can be similarly expressed. The t-dependent parts will be proportional to t^ε if the power series models (Eqs. (10.18) and (10.19)) are chosen.

Since the first four density derivatives of $\Pi(x, t)$ vanish at the critical point according to Proposition 1 we see that the power of the leading non-vanishing term in the expansion of $\Pi(x, t)$ near $x = 0$ should be at least larger than 4. Therefore on use of Eq. (10.59) we obtain

$$\phi \simeq \frac{\Pi(x, 0)}{1 - \nu} = \frac{Q_0}{1 - \nu} x^{4+\delta}[1 + O(x)] \quad (\delta > 0). \tag{10.73}$$

Thus, Eq. (10.73) is one of the necessary conditions for $\mathcal{A}$ and $\mathcal{B}$ to account for the experimental data on critical phenomena. It in fact enables us to vest the critical isotherm in the non-analytic term in $\mathcal{A}$ and $\mathcal{B}$ and choose the exponent δ on the basis of experiment: namely, $\delta = 0.3$–0.5. This indicates to us which part of the canonical equation of state is responsible for the critical isotherm for pressure. *This is the first important result for the canonical equation of state assumed.* Other subcritical characteristics of fluids can be vested in the t-dependent parts of $\mathcal{A}$ and $\mathcal{B}$, as will be shown subsequently.

10.9.3 *Spinodal Curve*

To calculate the spinodal curve as a function of t it is convenient to recast the conditions of Proposition 2

$$\left(\frac{\partial^i \phi}{\partial x^i} \right)_{x=x_{sk}} = 0 \quad (i = 1, \ldots, 4) \tag{10.74}$$

into another more convenient form by using Eq. (10.59). On use of this particular form of equation of state, we obtain in place of Eq. (10.74) the set of equations

$$-\nu \left[B_k^{(i-1)} + (1 + x_{sk}) B_k^{(i)} \right] \Pi_k(x_{sk}, t) = \Pi_k^{(i)} \left[1 - \nu(1 + x_{sk}) B_k^{(0)} \right], \tag{10.75}$$

where $i = 1, \ldots, 4$, $k = l, v$, and various symbols are:

$$B_k^{(i)}(t) = \left(\frac{\partial^i B}{\partial x^i} \right)_{x=x_{sk}}, \tag{10.76}$$

$$\Pi^{(i)}(x_{sk}, t) = \left(\frac{\partial^i \Pi}{\partial x^i} \right)_{x=x_{sk}}, \tag{10.77}$$

$$\phi_0 = \phi(x_{sk}, t). \tag{10.78}$$

It can be shown that Eq. (10.75) is quartic with respect to x_{sk} for all i. Thus the set can be written as

$$\varphi_{i4}^{(k)} x_{sk}^4 + \varphi_{i3}^{(k)} x_{sk}^3 + \varphi_{i2}^{(k)} x_{sk}^2 + \varphi_{i1}^{(k)} x_{sk} + \varphi_{i0}^{(k)} = 0 \tag{10.79}$$

for $k = l, v$ and $i = 1, \ldots, 4$. The coefficients $\varphi_{ij}^{(k)}$ in Eq. (10.79) can be worked out explicitly; they are presented in Appendix B. This set of four algebraic equations can be reduced to a single quadratic equation in x_{sk}

$$D_2^{(k)} x_{sk}^2 + D_1^{(k)} x_{sk} + D_0^{(k)} = 0, \tag{10.80}$$

where

$$D_2^{(k)} = \frac{1}{\Delta_1} \left(\varphi_{24}^{(k)} \varphi_{12}^{(k)} - \varphi_{14}^{(k)} \varphi_{22}^{(k)} \right) - \frac{1}{\Delta_2} \left(\varphi_{44}^{(k)} \varphi_{32}^{(k)} - \varphi_{34}^{(k)} \varphi_{42}^{(k)} \right),$$

$$D_1^{(k)} = \frac{1}{\Delta_1} \left(\varphi_{24}^{(k)} \varphi_{11}^{(k)} - \varphi_{14}^{(k)} \varphi_{21}^{(k)} \right) - \frac{1}{\Delta_2} \left(\varphi_{44}^{(k)} \varphi_{31}^{(k)} - \varphi_{34}^{(k)} \varphi_{41}^{(k)} \right), \tag{10.81}$$

$$D_0^{(k)} = \frac{1}{\Delta_1} \left(\varphi_{24}^{(k)} \varphi_{10}^{(k)} - \varphi_{14}^{(k)} \varphi_{20}^{(k)} \right) - \frac{1}{\Delta_2} \left(\varphi_{44}^{(k)} \varphi_{30}^{(k)} - \varphi_{34}^{(k)} \varphi_{40}^{(k)} \right),$$

with the definitions

$$\Delta_1 = \varphi_{24}^{(k)} \varphi_{13}^{(k)} - \varphi_{14}^{(k)} \varphi_{23}^{(k)}, \tag{10.82}$$

$$\Delta_2 = \varphi_{44}^{(k)} \varphi_{33}^{(k)} - \varphi_{34}^{(k)} \varphi_{43}^{(k)}. \tag{10.83}$$

The spinodal curve is then given by a solution of the quadratic equation (Eq. (10.80)):

$$x_{sk} = -\frac{D_0^{(k)}}{D_1^{(k)}} \cdot \frac{2}{\left(1 + \sqrt{1 - 4D_0^{(k)} D_2^{(k)}/D_1^{(k)2}}\right)}, \tag{10.84}$$

for which we have chosen the positive sign in the denominator. The important factor indicating the t-dependence of x_{sk} is the numerator.

It can be shown that $\varphi_{ij}^{(k)}$ can be decomposed into t-independent and t-dependent parts

$$\varphi_{ij}^{(k)}(t) = \omega_{ij} + \widehat{\varphi}_{ij}^{(k)}(t). \tag{10.85}$$

The t-independent part ω_{ij} consists of τ, ζ, ν, a_i, and $b_i (i \geq 0)$. More specifically,

$$\omega_{i0} = 0, \tag{10.86}$$

$$\omega_{ij} \neq 0 \quad (0 \leq i \leq 4; \ 1 \leq j \leq 4). \tag{10.87}$$

Therefore $\varphi_{i0}^{(k)}(t)$ appearing in D_0 do not contain the t-independent part. This means that in the leading order in t the numerator D_0 is a linear combination of $\widehat{\varphi}_{i0}^{(k)}(t)$ and hence x_{sk} is a linear combination of $\widehat{a}_i^{(k)}(t)$ and $\widehat{b}_i^{(k)}(t)$ in the leading order in t:

$$x_{sk} = -\sum_{i \geq 0} \left[s_{ai}^{(k)} \widehat{a}_i^{(k)}(t) + s_{bi}^{(k)} \widehat{b}_i^{(k)}(t) \right] + O\left(\widehat{a}^{(k)} \widehat{b}^{(k)} \right), \tag{10.88}$$

where $s_{ai}^{(k)}$ and $s_{bi}^{(k)}$ are constants made up of parameters $\nu, \tau, \zeta, a_1, \ldots,$ $a_4, b_1, \ldots, b_4$ related to the critical point and hence known by now; parameters $\nu, \tau, \zeta, a_1, a_2, b_1,$ and b_2 are explicitly determined in the quadratic model discussed later. These coefficients can be worked out more explicitly from the formulas for $\varphi_{ij}^{(k)}(t)$ given in Appendix B.

In any case, we now have a relatively simple algorithm to calculate the spinodal curve in terms of the model assumed. Therefore the spinodal curve may be written as

$$x_{sk} = |t|^{\varepsilon} f_k(t), \tag{10.89}$$

where the exponent ε is given by

$$\varepsilon = \min(\varepsilon_0, \varepsilon_1, \varepsilon_2, \ldots)$$

and the function $f_k(t)$ may be written as

$$f_k(t) = f_0^{(k)}[1 + O\left(|t|^\varepsilon, |t|^{\varepsilon_1}, \ldots\right)]. \tag{10.90}$$

Thus we see that the leading t-dependence is determined by that part of the GvdW parameters corresponding to the exponent ε. This means that the t-dependent part of the GvdW parameters is directly associated with the spinodal curve that is the locus of stability of the fluid. *We now have acquired the second important result for the model.*

To work out the coefficients $\varphi_{ij}^{(k)}$ it is necessary to calculate $\Pi^{(i)}(x_{sk}, t)$, the ith derivative of $\Pi(x,t)$ at $x = x_{sk}$. They are found to be cubic polynomials of x_{sk}:

$$\Pi^{(i)}(x_{sk}, t) = \Pi_0^{(i)} + \Pi_1^{(i)} x_{sk} + \Pi_2^{(i)} x_{sk}^2 + \Pi_3^{(i)} x_{sk}^3 \quad (0 \le i \le 4). \tag{10.91}$$

The coefficients $\Pi_j^{(i)}$ in this polynomial consist of t-independent and t-dependent parts as follows:

$$\Pi_j^{(i)} = P_{ij} + \widehat{\Pi}_j^{(i)}(t), \tag{10.92}$$

where P_{ij} is the t-independent part made up of the parameters determining the critical point and $\widehat{\Pi}_j^{(i)}(t)$ is the t-dependent part determined by the t-dependent part of the GvdW parameters, namely, $\widehat{a}_i(t)$ and $\widehat{b}_i(t)$. In particular, we note that

$$P_{i0} = P_i = 0 \tag{10.93}$$

by virtue of Proposition 1; Eq. (10.63). This means that

$$\Pi_0^{(i)} = \widehat{\Pi}_0^{(i)}(t) \quad (0 \le i \le 4). \tag{10.94}$$

These coefficients, although somewhat complicated but straightforward to obtain from the definition of $\Pi^{(i)}(x, t)$, are listed in Appendix B. It is useful to note that by virtue of Eqs. (10.93) and (10.94) we obtain Eq. (10.86) that gives rise to the conclusion (Eq. (10.88)) for x_{sk} given earlier.

10.9.4 *Excess Chemical Potential*

We have proposed Proposition 2 as a subcritical extension of the stability theory at the critical point. The proposition defines the stability of the fluid along the line of limit of stability — the spinodal curve. Since the derivatives of ϕ can be transcribed to those of chemical potentials the proposition can be phrased in terms of excess chemical potential ϖ_{ex}. Therefore, equivalently to Proposition 2 for the derivatives of ϕ, we have the conditions for the spinodal curve in the form

$$\left(\frac{\partial^i \varpi_{ex}}{\partial x^i}\right)_{x=x_{sk}} = 0 \quad (i = 1, \ldots, 4; \; k = l, v). \qquad (10.95)$$

Along the spinodal curve we also have the excess chemical potential $\varpi_{ex}(x, t)$ given by

$$\varpi_{ex}(x_{sk}, t) = \frac{\phi(x_{sk}, t)}{1 + x_{sk}} + \int_0^{x_{sk}} dx \frac{\Pi(x, t)}{(1 + x)^2 [1 - \nu(1 + x)\mathcal{B}]}, \qquad (10.96)$$

where the k branch (i.e., either the liquid or vapor branch) should be taken for $\mathcal{A}$ and $\mathcal{B}$ in the quantities involved. *The spinodal state is stable if the fifth density derivative of ϖ_{ex} is negative; unstable if positive; and neutral if equal to zero.* In fact, since the fluid is unstable at the spinodal state the fifth density derivative of ϖ_{ex} should be positive and, consequently, parameters associated with the t-dependent part of the GvdW parameters should be such that the derivative in question is positive at $x = x_{sk}$ for $t < 0$.

It is now necessary to examine the behaviors of Eq. (10.95) around x_{sk} to deduce some of thermodynamic properties in the neighborhood of x_{sk}. First, Eq. (10.95) can be written more explicitly:

$$0 = D_k(x_{sk}, t)\,\Pi(x_{sk}, t) + \int_0^{x_{sk}} dx \frac{D_k(x, t)}{(1 + x)}\Pi(x, t), \qquad (10.97)$$

$$0 = \left(D'_k + \frac{D_k}{1 + x}\right)_{x=x_{sk}} \Pi^{(0)} + D_k(x_{sk})\Pi^{(1)}, \qquad (10.98)$$

$$0 = \left(D'_k + \frac{D_k}{1 + x}\right)'_{x=x_{sk}} \Pi^{(0)} + \left(2D'_k + \frac{D_k}{1 + x}\right)_{x=x_{sk}} \Pi^{(1)}$$
$$+ D_k(x_{sk})\,\widehat{\Pi}^{(2)}, \qquad (10.99)$$

$$0 = \left(D'_k + \frac{D_k}{1 + x}\right)''_{x=x_{sk}} \Pi^{(0)} + \left(3D'_k + \frac{2D_k}{1 + x}\right)'_{x=x_{sk}} \Pi^{(1)}$$
$$+ \left(3D'_k + \frac{D_k}{1 + x}\right)_{x=x_{sk}} \Pi^{(2)} + D_k(x_{sk})\,\Pi^{(3)}, \qquad (10.100)$$

$$0 = \left(D_k' + \frac{D_k}{1+x} \right)''' \Pi^{(0)} \bigg|_{x=x_{sk}} + \left(4D_k' + \frac{3D_k}{1+x} \right)'' \Pi^{(1)} \bigg|_{x=x_{sk}}$$

$$+ \left(6D_k' + \frac{3D_k}{1+x} \right)' \Pi^{(2)} \bigg|_{x=x_{sk}} + \left(4D_k' + \frac{D_k}{1+x} \right) \Pi^{(3)} \bigg|_{x=x_{sk}}$$

$$+ D_k(x_{sk}) \Pi^{(4)}, \tag{10.101}$$

where the primes denote the density derivatives and

$$D_k(x,t) = \frac{1}{(1+x)[1 - \nu(1+x)\,\mathcal{B}_k(x,t)]} \quad (k = l, v). \tag{10.102}$$

Since the integral in Eq. (10.96) is simply $I_k(x_{sk}, t)$ (see Eq. (10.39)), we begin the analysis with it. We are interested in the asymptotic behavior of the integral in x_{sk} in the neighborhood of $t = 0$. For this purpose $D_k(x,t)/(1+x)$ is expanded in series of $(x - x_{sk})$:

$$\frac{D_k(x,t)}{1+x} = d_0^{(k)} \left[1 + d_1^{(k)}(x - x_{sk}) + \cdots \right], \tag{10.103}$$

where the coefficients $d_0^{(k)}$, $d_1^{(k)}$, etc. are independent of x; for example,

$$d_0^{(k)} = \frac{1}{(1+x_{sk})^2 \left[1 - \nu(1+x_{sk})\,\mathcal{B}_k^{(0)}(t) \right]}. \tag{10.104}$$

Note that in fact

$$\mathcal{B}_k^{(0)}(t) = b_0^{(k)}(t) \tag{10.105}$$

in the model taken. Then since Π is a series in $x - x_{sk}$, the integral can be evaluated. It is given by the form

$$I_k(x_{sk}, t) = d_0^{(k)} \left[-\Pi^{(0)} x_{sk} + \frac{1}{2} \left(\Pi^{(1)} + d_1^{(k)} \Pi^{(0)} \right) x_{sk}^2 + \cdots \right]$$

$$+ I_n^{(k)}(x_{sk}, t), \tag{10.106}$$

where $I_n^{(k)}(x_{sk}, t)$ represents the contributions to $I_k(x_{sk}, t)$ that are led by terms of $O(x_{sk}^{5+\delta})$ or higher:

$$I_n^{(k)}(x_{sk}, t) = -d_0^{(k)} \frac{N^{(0)}(x_{sk}, t)}{5+\delta} x_{sk}^{5+\delta} + \cdots. \tag{10.107}$$

For this result we have used the following: For $x_k > x > x_{sk}$, we obtain it in the form

$$
I_n^{(k)}(x, t) = \sum_{i \geq 0}^{5} \frac{\left[Q_i + \widehat{N}^{(i)}(x_{sk}, t)\right]}{5 + \delta} \left[(x - x_{sk})^{5+i+\delta} - x_{sk}^{5+\delta+i}\right]
$$

$$
+ d_0^{(k)} \sum_{i=5}^{7} \frac{\Pi^{(i)}(x_{sk}, t)}{1 + i} \left[(x - x_{sk})^{i+1} - (-x_{sk})^{1+i}\right]
$$

$$
+ \sum_{i \geq 0}^{3} \frac{\left[Q_{i+6} + \widehat{M}^{(i)}(x_{sk}, t)\right]}{9 + 2\delta} \left[(x_{sk} - x)^{9+2\delta+i} - x_{sk}^{9+2\delta+i}\right].
$$

$$(10.108)$$

On the other hand, for $x_k < x < x_{sk}$ the irrational terms are different, and we obtain

$$
I_n^{(k)}(x, t) = d_0^{(k)} \sum_{i \geq 0}^{5} \frac{\left[Q_i + \widehat{N}^{(i)}(x_{sk}, t)\right]}{5 + \delta} \left[(x_{sk} - x)^{5+\delta+i} - (-x_{sk})^{5+\delta+i}\right]
$$

$$
+ d_0^{(k)} \sum_{i \geq 5} \frac{\Pi^{(i)}(x_{sk}, t)}{1 + i} \left[(x - x_{sk})^{i+1} - (-x_{sk})^{1+i}\right]
$$

$$
+ d_0^{(k)} \sum_{i \geq 0} \frac{\left[Q_{i+6} + \widehat{M}^{(i)}(x_{sk}, t)\right]}{9 + 2\delta}
$$

$$
\times \left[(x_{sk} - x)^{9+2\delta+i} - (-x_{sk})^{9+2\delta+i}\right].
$$

$$(10.109)$$

The spinodal density x_{sk} is, in fact, proportional to $|t|^\varepsilon$ in the leading order; see Eq. (10.89). For calculation of the isochoric specific heat, the integral $I_k(x, t)$ is needed as will be seen.

By using the expansion (Eq. (10.103)) we find the leading order terms for the integral. Use of Eq. (10.106) in Eq. (10.97) provides us with an inhomogeneous linear algebraic set for $\Pi^{(i)}$ $(i = 0, 1, \ldots, 4)$ with the inhomogeneous term led by $x_{sk}^{5+\delta}$ arising from the irrational term in $\mathcal{A}$ and $\mathcal{B}$. We emphasize that $\Pi^{(i)}$ in Eq. (10.106) depend on t. With this result for I_k and on solving the set Eqs. (10.97)–(10.101) for $\Pi^{(i)}$, we find

$$
\Pi^{(i)}(x_{sk}, t) + O(x_{sk}^{5+\delta}) = 0 \ (i = 0, 1, \ldots, 4), \qquad (10.110)
$$

that is, $\Pi^{(i)}(x_{sk}, t)$ are equal to zero to $O(x_{sk}^{5+\delta})$.

Since we are interested in the asymptotic behavior near the critical point, the leading terms of the polynomial equations obtained will be of main interest. Therefore, for our purpose in this work we have the following lemma·

Lemma 1. *The leading order contribution to the integral $I_k(x,t)$ necessary in the asymptotic analysis is given by the form*

$$I_k(x,t) = \frac{d_0^{(k)} Q_0}{5+\delta}\left[(x - x_{sk})^{5+\delta} - x_{sk}^{5+\delta}\right] + \cdots \quad \text{for } x_k > x > x_{sk}$$

(10.111)

or

$$I_k(x,t) = \frac{d_0^{(k)} Q_0}{5+\delta}\left[(x_{sk} - x)^{5+\delta} - (-x_{sk})^{5+\delta}\right] + \cdots \quad \text{for } x_k < x < x_{sk}.$$

(10.112)

Lemma 2. *The excess chemical potential $\varpi_{ex}(x,t)$ is then given by*

$$\varpi_{ex}(x,t) = \frac{D_k(x,t)}{1+x}\Pi(x,t) + I_k(x,t),$$

(10.113)

where $\Pi(x,t)$ is given by Eq. (10.68) and $I_k(x,t)$ by Eq. (10.111) or Eq. (10.112).

All thermodynamic properties for the subcritical regime in the neighborhood of $t = 0$ can be deduced with this chemical potential and $\Pi(x,t)$, for which the parameters are determined subject to conditions deduced by applying Propositions 1 and 2. We make use of the results above in the analysis for the coexistence curve, excess heat capacity, and so on.

10.9.5 Liquid–Vapor Coexistence Curve

10.9.5.1 Equilibrium Conditions

The liquid–vapor coexistence curve is determined by the pressure and material equilibrium conditions. The pressure equilibrium condition in the present notation is given by

$$\frac{\Pi(x_l,t)}{1 - \nu\mathcal{B}_l(x_l + 1)} = \frac{\Pi(x_v,t)}{1 - \nu\mathcal{B}_v(x_v + 1)},$$

(10.114)

whereas the material equilibrium condition is

$$\varpi_l(x_l,t) = \varpi_v(x_v,t),$$

(10.115)

where the reduced chemical potential $\varpi(x,t)$ is given by Eq. (10.32) with Eq. (10.33). The GvdW parameters

$$\mathcal{A}_k = \mathcal{A}(x_k,t), \quad \mathcal{B}_k = \mathcal{B}(x_k,t) \quad (k = l, v)$$

(10.116)

are for the liquid (l) and vapor (v) branch, respectively. Then the material equilibrium condition may be written as

$$\frac{\Pi(x_l,t)}{(1+x_l)[1-\nu(x_l+1)\mathcal{B}_l]} - \frac{\Pi(x_v,t)}{(1+x_v)[1-\nu(x_v+1)\mathcal{B}_v]}$$
$$= I_v(x_v,t) - I_l(x_l,t). \qquad (10.117)$$

Using Eq. (10.114) in Eq. (10.117), this equation may be recast into the form

$$(x_v - x_l)\, \Pi(x_l,t) - [I_v(x_v,t) - I_l(x_l,t)](1+x_l)(1+x_v)$$
$$\times [1 - \nu(x_l+1)\,\mathcal{B}_l] = 0. \qquad (10.118)$$

Lemma 1 for the integral $I_k(x_k,t)$ asymptotically evaluated in the subcritical regime in the neighborhood of $t = 0$ will be used in this equation.

10.9.5.2 *Coexistence Curve*

The calculation of the coexistence curve requires solving the pair of equilibrium conditions (Eqs. (10.114) and (10.117)) for x_l and x_v. For this purpose, transform the variables:

$$x_l = R + \frac{1}{2}u, \quad x_v = R - \frac{1}{2}u \qquad (10.119)$$

and expand first $\mathcal{A}_k$ and $\mathcal{B}_k$ in series of R and u to obtain

$$\mathcal{A}_k = \alpha_0^{(k)} + \alpha_R^{(k)}R + \alpha_u^{(k)}u + O(Ru),$$
$$\mathcal{B}_k = \beta_0^{(k)} + \beta_R^{(k)}R + \beta_u^{(k)}u + O(Ru), \qquad (10.120)$$

where expansion coefficients are given by the formulas

$$\alpha_0^{(k)} = a_0^{(l)} - a_1^{(l)}x_{sl} + a_2^{(l)}x_{sl}^2 - a_3^{(l)}x_{sl}^3 + \cdots$$
$$\alpha_R^{(k)} = a_1^{(l)} - 2a_2^{(l)}x_{sl} + 3a_3^{(l)}x_{sl}^2 + \cdots,$$
$$\alpha_u^{(k)} = \frac{1}{2}a_1 - a_2 x_{sl} + \frac{3}{2}a_3^{(l)}x_{sl}^2 + \cdots, \qquad (10.121)$$
$$\beta_0^{(k)} = b_0^{(k)} - b_1^{(k)}x_{sk} + b_2^{(k)}x_{sk}^2 - b_3^{(k)}x_{sk}^3 + \cdots,$$
$$\beta_R^{(k)} = b_1^{(k)} - 2b_2^{(k)}x_{sk} + 3b_3^{(k)}Rx_{sk}^2 + \cdots,$$
$$\beta_u^{(k)} = \frac{1}{2}b_1^{(l)} - b_2^{(l)}x_{sl} + \frac{3}{2}b_3^{(l)}x_{sl}^2 + \cdots \qquad (k = l,v). \qquad (10.122)$$

The ellipsis represents terms of parameters $a_i^{(k)}$ and $b_i^{(k)}$ for $i \geq 4$. These series stem from the expansions for $\mathcal{A}_k$ and $\mathcal{B}_k$—the models (10.15) and

(10.16). Therefore we find

$$\Pi_k(x_k, l) = T_0^{(k)} + T_R^{(k)} R + T_u^{(k)} u + O(Ru) \tag{10.123}$$

with the t-dependent coefficients defined by

$$T_R^{(k)} = \tau + \nu \left(\beta_0^{(k)} + \beta_R^{(k)} \right) \left(\zeta \alpha_0^{(k)} + 1 \right) \tag{10.124}$$
$$+ \zeta \left(\nu \beta_0^{(k)} - 1 \right) \left(2\alpha_0^{(k)} + \alpha_R^{(k)} \right) + \tau t,$$

$$T_u^{(k)} = \frac{1}{2}\tau + \zeta \left(\nu \beta_0^{(k)} - 1 \right) \left(\alpha_0^{(k)} + \alpha_u^{(k)} \right) \tag{10.125}$$
$$+ \nu \left(\frac{1}{2}\beta_0^{(k)} + \beta_u^{(k)} \right) \left(\zeta \alpha_0^{(k)} + 1 \right) + \frac{1}{2}\tau t,$$

$$T_0^{(k)} = \tau + \left(\zeta \alpha_0^{(k)} + 1 \right) \left(\nu \beta_0^{(k)} - 1 \right) + \tau t. \tag{10.126}$$

The integral $I_k(x_k, t)$ are also expanded in series of R and u. Since

$$x_{sk}^{5+\delta} g(x_k) = -(5 + \delta) x_{sk}^{4+\delta} x_k + O(x_k^2), \tag{10.127}$$

we find the integral along the isochore $x = x_k$ is given by[8]

$$I_k(x_k) = \left(\tau\nu - \zeta d_{na}^{(k)} \right) \beta_{na0}^{(k)}(t) x_{sk}^{4+\delta} x_k + O(x_k^2) \quad (k = l, v), \tag{10.128}$$

where

$$d_{na}^{(k)}(t) = \frac{\alpha_{na0}^{(k)}(t)}{\beta_{na0}^{(k)}(t)}. \tag{10.129}$$

Therefore, we find

$$I_l(x_l) - I_v(x_v) = \mu_R R + \mu_u u + \cdots \tag{10.130}$$

where

$$\mu_R = \left[\tau\nu - \zeta d_{na}^{(l)}(t) \right] \beta_{na0}^{(l)}(t) x_{sl}^{4+\delta}(1 - \vartheta_{vl}), \tag{10.131}$$

$$\mu_u = \frac{1}{2} \left[\tau\nu - \zeta d_{na}^{(l)}(t) \right] \beta_{na0}^{(l)}(t) x_{sl}^{4+\delta}(1 + \vartheta_{vl}), \tag{10.132}$$

with ϑ_{vl} defined by the ratio

$$\vartheta_{vl} = \frac{\left[\tau\nu - \zeta d_{na}^{(v)}(t) \right] \beta_{na0}^{(v)}(t) x_{sv}^{4+\delta}}{\left[\tau\nu - \zeta d_{na}^{(l)}(t) \right] \beta_{na0}^{(l)}(t) x_{sl}^{4+\delta}} \approx \frac{\left[\tau\nu - \zeta d_{na}^{(v)}(0) \right] \beta_{na0}^{(v)}(0) x_{sv}^{4+\delta}}{\left[\tau\nu - \zeta d_{na}^{(l)}(0) \right] \beta_{na0}^{(l)}(0) x_{sl}^{4+\delta}}. \tag{10.133}$$

---

[8]We remark that along the isochore $x = x_{sk}$ the result for the integral $I_k(x)$ would have been simpler than for the isochore $x = x_k$. Nevertheless, the isochore $x = x_k$ is chosen here because the value of the integral along $x = x_k$ is required to calculate the specific heat along the isochore mentioned, owing to the experimental data measured along the isochore.

It should be noted that μ_R and μ_u are made up of quantities associated with the non-analytic terms in $\mathcal{A}$ and $\mathcal{B}$ and $\beta_{na0}^{(k)}(t) = \beta_{na}^{(k)}(t) = \beta_{na} + \widehat{\beta}_{na}(t)$ with $\widehat{\beta}_{na}(0) = 0$. Thus we see that Eqs. (10.114) and (10.118) give rise to the following pair of algebraic equations

$$\Lambda_{1R}R + \Lambda_{1u}u = -T_0^{(v)}\left(\nu\beta_0^{(l)} - 1\right) + O(Ru), \tag{10.134}$$

$$\Lambda_{2R}R + \Lambda_{2u}u = 0 + O(Ru), \tag{10.135}$$

where the coefficients are defined by

$$\Lambda_{1R} = T_R^{(v)}\left(\nu\beta_0^{(l)} - 1\right) + \nu T_0^{(v)}\left(\beta_0^{(l)} + \beta_R^{(l)}\right), \tag{10.136}$$

$$\Lambda_{1u} = T_u^{(v)}\left(\nu\beta_0^{(l)} - 1\right) + \nu T_0^{(v)}\left(\frac{1}{2}\beta_0^{(l)} + \beta_u^{(l)}\right), \tag{10.137}$$

$$\Lambda_{2R} = \left(\nu\beta_0^{((l)} - 1\right)\mu_R, \tag{10.138}$$

$$\Lambda_{2u} = T_0^{(l)} + \left(\nu\beta_0^{((l)} - 1\right)\mu_u. \tag{10.139}$$

Thus the leading order of the set is linear in R and u. This set is solved to the linear order to obtain the coexistence curve and the rectilinear diameter:

$$R = -\frac{T_0^{(v)}\left(\nu\beta_0^{(l)} - 1\right)}{\det|\Lambda|}\Lambda_{2u}, \tag{10.140}$$

$$u = \frac{T_0^{(v)}\left(\nu\beta_0^{(l)} - 1\right)}{\det|\Lambda|}\Lambda_{2R}. \tag{10.141}$$

We now observe that these solutions are intimately related to $x_{sk}^{4+\delta}$ and also to the linear combinations of $a_i^{(k)}(t)$ and $b_i^{(k)}(t)$. We also emphasize that these results are rigorous asymptotic formulas holding in the neighborhood of the critical point and that, as will be shown, the t dependence of the coexistence curve (i.e., u) is given by $|t|^{(7+\delta)\varepsilon}$ in the leading order, where ε is the smallest of the exponents of the t dependence of the spinodal curve. It should be remembered that the exponent $(4 + \delta)$ is that of the critical isotherm and hence the exponent $(7 + \delta)\varepsilon$ is a composite of the exponents of the critical isotherm and spinodal curve, reflecting its physical origin.

To learn more details of these results let us explore the coefficients. Since

$$\beta_0^{(k)} = 1 - b_1 x_{sk} + b_2 x_{sk}^2 + \widehat{b}_0^{(k)} - \widehat{b}_1^{(k)} x_{sk} + \widehat{b}_2^{(k)} x_{sk}^2 + \cdots,$$

$$\alpha_0^{(k)} = 1 + \widehat{a}_0^{(k)} + b_2 x_{sk}^2 + \widehat{b}_0^{(k)} - \widehat{b}_1^{(k)} x_{sk} + \widehat{b}_2^{(k)} x_{sk}^2 + \cdots, \tag{10.142}$$

and

$$T_0^{(v)} = \zeta(\nu - 1)\,\widehat{a}_0^{(v)} + \nu(1+\zeta)\,\widehat{b}_0^{(v)} + \zeta\nu\widehat{a}_0^{(v)}\widehat{b}_0^{(v)} + \tau t + \cdots, \qquad (10.143)$$

for which we have made use of Eq. (10.63), the factors in the numerators of the solutions for R and u are found in the forms

$$T_0^{(v)}\left(\nu\beta_0^{(l)} - 1\right) = \zeta\nu\,(\nu-1)\,\widehat{a}_0^{(v)}\widehat{b}_0^{(l)} + \nu^2(1+\zeta)\,\widehat{b}_0^{(l)}\widehat{b}_0^{(v)} + \cdots, \qquad (10.144)$$

$$\Lambda_{2R} = \nu\widehat{b}_0^{(l)}\left(\tau\nu - \zeta d_{na}^{(l)}\right)\beta_{na}^{(l)}(t)\,x_{sl}^{4+\delta} + \cdots, \qquad (10.145)$$

$$\Lambda_{2u} = \frac{1}{2}\left[\zeta(\nu-1)\,\widehat{a}_0^{(l)} + \nu(1+\zeta)\,\widehat{b}_0^{(l)}\right]$$
$$+ \nu(1+\zeta)(\tau\nu - \zeta d_{na})\,\beta_{na}\widehat{b}_0^{(v)}x_{sl}^{4+\delta} + \cdots, \qquad (10.146)$$

where the ellipsis stands for higher order terms. This means that R and u are asymptotically

$$R \sim |t|^{3\varepsilon}, \qquad (10.147)$$

$$u \sim |t|^{(7+\delta)\varepsilon} \qquad (10.148)$$

in the regime of $|t| \ll 1$. These are, respectively, the asymptotic behavior of the rectilinear diameter and of the coexistence curve in the neighborhood of the critical point in the model employed. We see that they are composites of the exponents for the spinodal curve and the critical isotherm. The value of ε may be chosen from the information on the excess specific heat near the critical point.

10.9.6 *Excess Heat Capacity*

To calculate the excess heat capacity anomaly the formal result presented for the excess chemical potential ϖ_{ex} given in Eqs. (10.33) and (10.32) is preferable to use. In the following we examine the leading order excess heat capacity which exhibits a singular behavior in the critical region.

To obtain the asymptotic behavior of the excess heat capacity in the neighborhood of the critical point in the subcritical regime we need to consider only the integral $I_k(x,t)$, which can be shown from Lemma 1 (Eq. (10.111) or Eq. (10.112)) to be expressible in the form

$$I_k(x,t) = \frac{1}{(5+\delta)}\left[\tau\nu - \zeta\frac{\alpha_{na0}^{(k)}(t)}{\beta_{na0}^{(k)}(t)}\right]g(x)\,\beta_{na0}^{(k)}(t)\,x_{sk}^{5+\delta} + \cdots, \qquad (10.149)$$

where

$$g(x) = \left(\frac{x}{x_{sl}} - 1\right)^{5+\delta} - 1 \quad \text{for liquid} \tag{10.150}$$

$$= \left(1 - \frac{x}{x_{sv}}\right)^{5+\delta} - 1 \quad \text{for vapor.} \tag{10.151}$$

Since $\alpha_{na0}^{(k)}(t)$ and $\beta_{na0}^{(k)}(t)$ remain adjustable, we are free to further assume that $\alpha_{na0}^{(k)}(t)$ and $\beta_{na0}^{(k)}(t)$ are such that

$$\frac{\alpha_{na0}^{(k)}(t)}{\beta_{na0}^{(k)}(t)} = d_{na} \text{ independent of } t + O(|t|^{\varepsilon_n}). \tag{10.152}$$

Again, the argument for this can be made based on the statistical mechanics that the GvdW parameters $\mathcal{A}$ and $\mathcal{B}$ share the same pair distribution function and hence $\alpha_{na0}^{(k)}(t)$ and $\beta_{na0}^{(k)}(t)$ should have a common t dependence, at least, in the neighborhood of $t = 0$ and hence scale to a constant. Since we are interested in x along the isochore $x = x_k$ ($k = l, v$) and since x_k/x_{sk} is a constant independent of t, the t-dependence of the excess chemical potential in the order considered is mainly vested in $\beta_{na0}^{(k)}(t) x_{sk}^{5+\delta}$. Since we may take

$$\beta_{na0}^{(k)}(t) = \beta_{na0} + \mathbf{b}_{na0}^{(k)}|t|^{\varepsilon_n} \tag{10.153}$$

with $0 < \varepsilon_n < 1$ and $x_{sk} \sim t^{\epsilon}(t > 0)$ we find to the leading order

$$\frac{\partial^2}{\partial t^2} x_{sk}^{5+\delta} \sim x_{sk}^{(7+\delta)\epsilon - 2}.$$

This shows that in the neighborhood of $t \lesssim 0$ we have the temperature dependence

$$-(1+t)(1+x_k)\frac{\partial^2}{\partial t^2} I_k(x,t) \sim t^{-\alpha'}, \tag{10.154}$$

where

$$\alpha' = 2 - (7+\delta)\epsilon, \tag{10.155}$$

and the excess heat capacity in the neighborhood of the critical point therefore is given by

$$\widehat{c}_{vk}^{(ex)} \sim |t|^{-\alpha'} \quad (k = l, v). \tag{10.156}$$

The exponent α' is made up of the exponents characterizing the critical isotherm, coexistence curve, and spinodal curve. Therefore their values may be chosen to optimize collectively the behaviors of the properties mentioned, in addition to the excess specific heat and the isothermal compressibility discussed below

Tentatively, with the values for ϵ and δ taken earlier, respectively, for the spinodal and coexistence curves and the critical isotherm in this work, it follows

$$\alpha' = 2 - (7 + 0.3)\, 0.25 = 0.18. \tag{10.157}$$

This value of α' for $\hat{c}_{vk}^{(ex)}$ along the isochore $x = x_k$ is consistent with the experimental observation on the behavior of the specific heat along the isochore $x = x_k$ $(k = l, v)$. Thus we have shown that the specific heat anomaly, from the standpoint of the canonical equation of state, arises from the combination of the exponents of non-analytic irrational terms in $\mathcal{A}_k^{(i)}(t)$ and $\mathcal{B}_k^{(i)}(t)$ making up the GvdW parameters $\mathcal{A}$ and $\mathcal{B}$. Furthermore, it also provides information on the parameters $\mathcal{A}_k^{(i)}(t)$ and $\mathcal{B}_k^{(i)}(t)$; in fact, their mutual relations provided by Propositions 1 and 2, which are algebraic equations of parameters and their relations, drastically reduce the number of free parameters to just only a few.

10.9.7   *Isothermal Compressibility*

The isothermal compressibility κ_T along the critical isochore or along the isochores x_k $(k = l, v)$ is experimentally known to behave[9]

$$\frac{\kappa_T}{\kappa_I} = Ct^{-\gamma} \quad (\rho = \rho_c,\ t > 0) \tag{10.158}$$

$$= C\,(-t)^{-\gamma'} \quad (\rho = \rho_l(T)\ \text{or}\ \rho = \rho_g(T),\ t < 0). \tag{10.159}$$

By using the equation of state near $t = 0$, we find that along the isochore at $x = x_k$

$$\kappa_T \sim x_k^{-2}|x_k|^{-1-\delta} + \cdots. \tag{10.160}$$

To determine the t dependence of κ_T it is necessary to know the t-dependence of x_k, namely, the liquid–vapor coexistence curve, which has been already determined in the previous section. Using the result for x_k presented earlier, the t-dependence of the isothermal compressibility in the neighborhood of $t = 0$ is deduced to be

$$\kappa_T \sim |t|^{-\gamma'} \tag{10.161}$$

---

[9]P. Heller, *Rep. Prog. Phys.* **30**, 731 (1967); J. V. Sengers and J. M. H. Levelt Sengers, in *Progress in Liquid Physics*, C. A. Croxton, ed. (Wiley, New York, 1978), pp. 103–174; P. A. Egelstaff and J. W. Ring, in *Physics of Simple Liquids*, H. N. V. Temperley, J. S. Rowlinson and G. S. Rushbrooke, eds. (North-Holland, Amsterdam, 1968), pp. 255–297.

where $\gamma' = 3(3 + \delta)\,\varepsilon$. Therefore, if we choose $\varepsilon = 0.25$ as we have for the coexistence curve we find γ' within the range of experimental observation, namely, $\gamma' = 2.48$. This value is about two times too large, but it should be noted that the estimate of the exponent for the coexistence curve itself given here is too rough; a more accurate value, however, takes numerical evaluations of various quantities involved that we have avoided to see how the exponents arise within the range of experimental values for them. Moreover, as mentioned earlier, the values of the exponents should be optimized for the whole set of properties together. The point we would like to emphasize here is that the exponents stem from the t dependence of the GvdW parameters $\mathcal{A}$ and $\mathcal{B}$. Note that the exponent γ' is a composite number stemming from the critical isotherm, the spinodal curve, and the coexistence curve, which ultimately originates from the t-dependence of $\mathcal{A}$ and $\mathcal{B}$ away from the critical temperature. This is the principal point we would like to make with the model for the canonical equation of state studied here.

10.10   Quadratic Model

In this section, we implement the algorithm developed to calculate the critical behavior of fluids by using particular models for the GvdW parameters of the canonical equation of state, namely, the quadratic model in which the number of terms in A and B is limited to $i \leq 2$ for the terms regular in x and to $i = 0$ for the non-analytic terms. This model obviously renders various coefficients simpler. More specifically, by the quadratic model we mean the pair of the formulas for $\mathcal{A}$ and $\mathcal{B}$ in the following two subsections:

10.10.1   *Subcritical Regime*

In the subcritical regime

$$\mathcal{C} = \sum_{i=0}^{2} \mathcal{C}_i^{(k)}(t)(x - x_{sk})^i + \mathcal{C}_{na}^{(k)}(t)(x - x_{sk})^3 |x - x_{sk}|^{1+\delta} \quad t \leq 0, \quad (10.162)$$

where $\mathcal{C} = (\mathcal{A}, \mathcal{B})$ and the coefficients are:

$$\mathcal{C}_i^{(k)}(t) = \left[a_i^{(k)}(t), b_i^{(k)}(t); \; 0 \leq i \leq 2 \right],$$

$$\mathcal{C}_{na}^{(k)}(t) = \left[\alpha_{na}^{(k)}(t), \beta_{na}^{(k)}(t) \right] \quad (k = l, v).$$

The exponent $\delta < 1$ (e.g., $\delta = 0.3$–0.5) is a parameter already familiar to us here.

These coefficients are temperature-dependent and non-analytic (discontinuous) with respect to t. They are assumed to take the form

$$C_i^{(k)}(t) = C_i + \widehat{C}_i^{(k)}(t) = C_i + \mathfrak{c}_i^{(k)}|t|^{\varepsilon_i} \quad (i = 0, 1, 2), \tag{10.163}$$

where $C_i = (a_i, b_i)$, $\widehat{C}_i^{(k)} = \left(\widehat{a}_i^{(k)}, \widehat{b}_i^{(k)}\right)$ and $\mathfrak{c}_i^{(k)} = \left(\mathfrak{a}_i^{(k)}, \mathfrak{b}_i^{(k)}\right)$ $(k = l, v)$ are constants, and $\varepsilon_i(\varepsilon_0 \leq \varepsilon_1 \leq \varepsilon_2)$ are fractional numbers less than unity. Therefore

$$\widehat{C}_i^{(k)}(0) = 0 \quad (i = 0, 1, 2). \tag{10.164}$$

The liquid and vapor branches of the coefficients are sought to coincide with each other at the critical temperature, namely,

$$C_i^{(l)}(0) = C_i^{(v)}(0) \equiv C_i \neq 0 \ (i = 1, 2), \quad \lim_{t \to 0} C_0^{(k)}(t) = 1. \tag{10.165}$$

Similarly, the coefficients of the irrational non-analytic terms must also be in the forms

$$C_{na}^{(k)}(t) = C_{na} + \widehat{C}_{na}^{(k)}(t) = C_{na} + \mathfrak{c}_{na}^{(k)}|t|^{\varepsilon_n}, \tag{10.166}$$

where $C_{na} = (\alpha_{na}, \beta_{na})$ and $\mathfrak{c}_{na}^{(k)} = \left(\mathfrak{a}_{na}^{(k)}, \mathfrak{b}_{na}^{(k)}\right)$, and it is assumed that $\varepsilon_n \leq \varepsilon_2$ and is also a fractional number less than unity. Therefore

$$\widehat{C}_{na}^{(k)}(t) \to 0 \quad \text{as } t \to 0. \tag{10.167}$$

The irrational terms multiplied by $\alpha_{na}^{(k)}(t)$ and $\beta_{na}^{(k)}(t)$ $(k = l, v)$ are the non-analytic parts, which give rise to the non-analytic (in fact, discontinuous) behavior of the equation of state with respect to density, and the non-classical critical exponent for the pressure–density relation different from the mean field (van der Waals) theory values. This is the principal reason for assuming the non-analytic term with a fractional exponent for the density dependence. The parameters $\alpha_{na}^{(k)}$ and $\beta_{na}^{(k)}$ may depend on temperature t only, but are constants on the critical isotherm in the same sense as for $a_i^{(k)}$ and $b_i^{(k)}$ in Eq. (10.165), namely,

$$\alpha_{na}^{(l)}(0) = \alpha_{na}^{(v)}(0) \equiv \alpha_{na}, \quad \beta_{na}^{(l)}(0) = \beta_{na}^{(v)}(0) \equiv \beta_{na}. \tag{10.168}$$

With this model for the t dependence of the parameters and with the exponents $\varepsilon_i(\epsilon \equiv \varepsilon_0)$ and ε_n appropriately chosen in comparison of the theory with experiment (principally on the critical exponents), the equation of state can be used to calculate the critical properties of fluids in the subcritical regime. We have taken the same exponents ε_i and ε_n for both vapor (v) and liquid (l) branches as an approximation near the critical point because the vapor and liquid branches of the experimental coexistence curve is practically symmetric in the close neighborhood of the critical point. As will be explicitly seen, the parameters $\mathfrak{c}_i^{(k)} = \left(\mathfrak{a}_i^{(k)}, \mathfrak{b}_i^{(k)} : i = 0, 1, 2; \; k = l, v \right)$ and $\mathfrak{c}_{na}^{(k)} = \left(\mathfrak{a}_{na}^{(k)}, \mathfrak{b}_{na}^{(k)} : k = l, v \right)$ are not arbitrary, but constrained by some conditions given by Propositions 1 and 2, which remove the degree of freedom for the parameters except for only a few. Therefore there are not so many free parameters left, as one might get a contrary impression at first glance. Put differently, the quadratic model is an empirical model constructed to ensure to reproduce the critical characteristics of fluids with a least number of adjustable parameters by imposing Propositions 1 and 2.

10.10.2 *Supercritical Regime*

Owing to these properties of $\mathcal{C}_i^{(k)}$ and $\mathcal{C}_{na}^{(k)}$ and the fact that the spinodal curve coincides with the coexistence curve at the critical point, that is,

$$x_{sl} = x_{sv} = 0,$$

the expansions for $\mathcal{C}$ in Eq. (10.162) reduce, on the critical isotherm or in the supercritical regime, to the forms

$$\mathcal{C} = 1 + \sum_{i=1}^{2} \mathcal{C}_i x^i + \mathcal{C}_{na} x^3 |x|^{1+\delta} \quad (t = 0). \tag{10.169}$$

The parameters $\mathcal{C}_c = (A_c^*, B_c^*)$ of the quadratic model are related to the van der Waals parameters a and b in the following sense: since $x \to -1$ as $\eta \to 0$ it follows that, if T_c is such that $\exp(\varepsilon/k_B T_c) - 1 \ll 1$—in other words, $T_c^* \gg 1$, then

$$a = A_c^* \left[1 + \sum_{i=1}^{2} (-1)^i a_i - \alpha_{na} \right], \quad b = B_c^* \left[1 + \sum_{i=1}^{2} (-1)^i b_i - \beta_{na} \right]. \tag{10.170}$$

This means that A_c^* and B_c^* do not coincide with the reduced van der Waals parameters a and b, respectively, unless a_i, b_i, α_{na}, and β_{na} all vanish.

10.10.3 *Critical Point*

In the quadratic model, five of the seven parameters a_1, a_2, b_1, b_2, τ, ζ, and ν are determined in terms of two free parameters from the information on the critical isotherm ($t = 0$) with the help of Proposition 1. As a consequence, with an appropriate choice for the exponent δ so that the critical $\phi - x$ relation (reduced critical isotherm) agrees with experiment, the parameters a_1, b_1, τ, ζ, and ν are determined by suitably choosing a_2 and b_2 such that the critical parameters agree with experiment. The remaining two parameters are in fact constrained by the stability condition—namely, the fifth derivative—that should be either positive or negative or equal to zero, depending on the stability of the critical state. Therefore they are not completely arbitrary. The so-determined parameters and the parameters necessary for the critical isotherm are summarized in Table 10.1. The critical isotherm calculated in the quadratic model is found in excellent agreement with experiment as shown in Fig. 10.1.

It is interesting to note that the critical parameters are given by the parameters of the regular, rational power terms in the model whereas the critical isotherm is determined by the parameters making up the irrational part of the model. Thus we see that in this model the irrational part is essential to account for the characteristic behavior of the critical isotherm of the fluid. It has been already shown that the same irrational term, in coupling with the (irrational) t-dependence of $a_i^{(k)}(t)$ and $b_i^{(k)}(t)$ ($k = l, v$), is closely associated with the critical behaviors of other thermodynamic quantities, such as the isothermal compressibility, heat capacity, etc. This feature is seen unchanged in the case of the quadratic model. Thus the important point we would like to make with the canonical equation of state postulated is that specific parts of the equation of state are associated with characteristic properties of thermodynamic properties, subcritical and supercritical, of the fluid of interest.

Table 10.1 Various parameter values at $t = 0$ in the vdW theory and the GvdW theory in the quadratic model.

a_i	vdW	GvdW	Parameters	vdW	GvdW
a_0	1	1	ν	1/3	0.535
b_0	1	1	ζ	3	6.424
a_1	0	−0.336	τ	8/3	3.448
b_1	0	−0.618	α_{na}	0	0.01
a_2	0	−0.360	β_{na}	0	−0.015
b_2	0	0.0436			

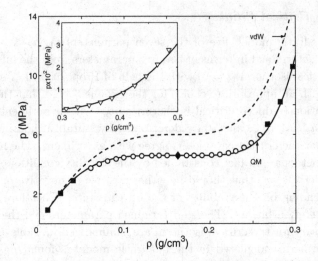

Fig. 10.1 The critical isotherm predicted by the quadratic model for methane (solid curve) and compared with the van der Waals equation of state prediction (broken curve). The symbols are experimental data. The filled diamond represents the critical point. Reproduced with permission from K. Rah and B. C. Eu, *J. Phys. Chem. B* **107**, 4382 (2003) © 2003 American Chemical Society.

10.10.4 *Critical Isotherms*

It is easy to show that the critical isotherm is in the same form as before:

$$\phi(x,0) = \left[\frac{\nu(1+\zeta)\,\beta_{na} - \zeta(1-\nu)\,\alpha_{na}}{1-\nu b_0} \right] x^{4+\delta}[1+O(x)]. \tag{10.171}$$

With the values of the parameters given in Table 10.1 the critical isotherm is in excellent agreement with experiment.

10.10.5 *Spinodal Curve*

In the quadratic model, since to the leading order in $|t|^{\varepsilon}$ and so on

$$\begin{aligned}
\varphi_{24}^{(k)}\varphi_{13}^{(k)} - \varphi_{14}^{(k)}\varphi_{23}^{(k)} &= \Delta_1 + O(|t|^{\varepsilon}, \ldots), \\
\varphi_{44}^{(k)}\varphi_{33}^{(k)} - \varphi_{34}^{(k)}\varphi_{43}^{(k)} &= \Delta_2 + O(|t|^{\varepsilon}, \ldots), \\
\Delta_1 &= \omega_{24}\omega_{13} - \omega_{14}\omega_{23}, \\
\Delta_2 &= \omega_{44}\omega_{33} - \omega_{34}\omega_{43},
\end{aligned} \tag{10.172}$$

we find

$$D_0^{(k)} = \frac{\omega_{24}}{\Delta_1} \widehat{\varphi}_{10}^{(k)}(t) - \frac{\omega_{14}}{\Delta_1} \widehat{\varphi}_{20}^{(k)}(t) - \frac{\omega_{44}}{\Delta_2} \widehat{\varphi}_{30}^{(k)}(t) + \frac{\omega_{34}}{\Delta_2} \widehat{\varphi}_{40}^{(k)}(t)$$

$$= \sum_{i=0}^{2} s_{ai} \widehat{a}_i^{(k)}(t) + \sum_{i=0}^{2} s_{bi} \widehat{b}_i^{(k)}(t) + O\left(|t|^{2\varepsilon}, \dots\right), \qquad (10.173)$$

$$D_1^{(k)} = D_{10}^{(k)} + O(|t|^{\varepsilon}, \dots), \qquad (10.174)$$

$$D_2^{(k)} = D_{20}^{(k)} + O(|t|^{\varepsilon}, \dots), \qquad (10.175)$$

where s_{ai}, s_{bi}, $D_{10}^{(k)}$, and $D_{20}^{(k)}$ are constants consisting of ω_{ij}; they can be calculated explicitly in terms of the parameters determined and hence known from the critical point data by now. Then to the leading order in t the spinodal curve is given by

$$x_{sk} = -\sum_{i=0}^{2} \frac{2}{D_{10}^{(k)}} \left[s_{ai} \widehat{a}_i^{(k)}(t) + s_{bi} \widehat{b}_i^{(k)}(t) \right] + \text{higher order terms.} \quad (10.176)$$

Together with the information on the critical point data, the critical isotherm and coexistence curve can be calculated by using the procedure described earlier. Therefore the information on the spinodal curve facilitates the determination of the quadratic model. We remark that if a more accurate result is desired of x_{sk} one should simply calculate x_{sk} in the quadratic model without approximations. With the spinodal curve thus calculated and GvdW parameters determined, we can calculate the other thermodynamic quantities as already described in Sec. 10.9. The expressions for other critical properties do not get simpler than those already presented. For the reason of space we do not discuss their version for the quadratic model here.

10.11   Concluding Remarks

The van der Waals theory provides a fairly simple form of equation of state which accounts for most of fluid behaviors in a qualitatively correct manner. It, however, has sometimes glaring defects that should be removed so as to acquire an improvement if quantitatively correct results are desired, especially, with regard to the critical properties. Because of the attractive features of the van der Waals theory, we have taken an approach in which an improved theory is built around the van der Waals theory, incorporating the virtuous parts of the van der Waals theory into the improved theory. The canonical equation of state is designed to achieve the aim. In the canonical equation of state approach, we phenomenologically elucidate the manner in which the critical properties and other thermodynamic properties of fluids are connected to the GvdW model with regard to its density and temperature dependences. Thus a set of thermodynamic properties of a fluid is specifically determined in association with the GvdW parameters. Once such a GvdW equation of state is constructed for a class of fluids we can use it to study the thermodynamics of all fluids of the class and expect to correctly reproduce, at least, various aspects of critical phenomena phenomenologically. Because of the statistical mechanical representations of GvdW parameters $\mathcal{A}$ and $\mathcal{B}$, the statistical mechanical basis of the parameters so determined can be studied and elucidated systematically from the molecular theory viewpoint. We believe that therein lies the promising potential for the canonical equation of state.

The model for $\mathcal{A}$ and $\mathcal{B}$ consists of analytical and non-analytical (irrational) parts, the former being made up of a finite order polynomial of density, and the latter factor proportional to density with fractional exponents—thus non-analytic. The temperature-dependent coefficients are also non-analytic with respect to t and closely related to various critical exponents for the coexistence and spinodal curves, isothermal compressibility, excess heat capacity, and critical isotherm. Since the quadratic model yields an excellent result for the critical isotherm for reduced pressure, an empirical model for $\mathcal{A}$ and $\mathcal{B}$ is constructed to account for the critical behavior of fluids in the subcritical regime near $t = 0$ on the basis of the model that correctly yields the critical isotherm. We have therewith shown that $\mathcal{A}$ and $\mathcal{B}$ indeed can determine the temperature dependence of the spinodal and coexistence curves through coefficients $a_i^{(k)}(t)$ and $b_i^{(k)}(t)$ $(k = l, v; \ i = 0, 1, 2)$; see Eqs. (10.80), (10.84), (10.140), and (10.141). These are exact results in the model assumed. However, the determination of the liquid–vapor coexistence curve and other related thermodynamic properties cannot be obtained exactly in a simple form for the model. If their behaviors are desired in the neighborhood of the critical point, asymptotic formulas that exhibits a qualitatively correct critical behavior of the fluid near the critical point can be extracted only as approximations, provided that the exponents are suitably chosen. This way, we have also shown that the isothermal compressibility and the specific heat also diverges at the critical point in a manner in accord with the experimental observation; see Eq. (10.161).

The divergences of isothermal compressibility and isochoric excess heat capacity essentially owe their origin to the non-analytic fractional power term of density as well as the irrational t dependences of $a_i^{(k)}(t)$ and $b_i^{(k)}(t)$. Moreover, their divergence is also closely related to the behavior of the coexistence curve or the spinodal curve. Therefore it, in fact, indicates the internal consistency of the models (10.15) and (10.16) as well as the quadratic model in accounting for the critical properties. An interesting conclusion of the present study of the canonical equation of state is that the critical properties of fluids provide not only a way to construct a model for a canonical form of equation of state, but also a set of algebraic relations among the parameters of the model in a self-consistent manner that is reminiscent of the scaling theory relations. And therein lies the attractiveness of the present canonical equation of state approach.

The molecular theoretic origin of the non-analytic behaviors of the GvdW parameters must be sought through the statistical mechanics theory, but that is beyond the scope of a phenomenological model for the equation of state described in this chapter. The present phenomenological equation of state is, in fact, built upon the salient points of the statistical mechanical studies and empirical observations of the critical phenomena.

For a global thermodynamic description over a wider range of density and temperature of subcritical fluids the present asymptotic analysis in the subcritical neighborhood of the critical point would not be sufficient, but a more detailed numerical analysis with the full models assumed for $a_i^{(k)}(t)$, $b_i^{(k)}(t)$, $\alpha_{na}^{(k)}(t)$, and $\beta_{na}^{(k)}(t)$ would be necessary. The algorithm presented already presents itself for such numerical analysis.

Chapter 11

Thermodynamics of Real Gas Mixtures

The foregoing formalisms for calculating thermodynamic functions hold for pure substances and, in particular, for real fluids. The most important result among the host of thermodynamic functions is the formula for chemical potentials, since we can calculate other thermodynamic functions from them by either taking suitable derivatives or combining the derivatives thereof. The formalisms for calculating thermodynamic functions for mixtures proceed similarly regardless of the states of aggregation of the substance involved. However, we shall specialize to the case of real gas mixtures in this chapter in order to equip us with a formalism, so that gas phase chemical equilibria can be treated in the subsequent chapter.

11.1 Chemical Potentials for Mixtures

We have seen that the chemical potential of a pure gas may be given in a form similar to that for the ideal gas, if the fugacity is introduced, namely, it is written in a form reminiscent of the ideal gas form

$$\mu(T, p) = \mu^*(T) + RT \ln p_f.$$

(See Eq. (9.131)). Therefore, in the case of a real gas mixture it is convenient to look for the formula for chemical potentials in a mathematically similar form

$$\mu_i(T, p) = \mu_i^*(T) + RT \ln p_{fi}, \qquad (11.1)$$

where p_{fi} is the fugacity for component i in the mixture and $\mu_i^*(T)$ is the chemical potential of i when $p_{fi} = 1$. Let us now find the precise meaning of the fugacity p_{fi} and its mathematical expansion.

Since the partial molar volume may be written as

$$\bar{v}_i = \left(\frac{\partial \mu_i}{\partial p}\right)_{T,x} \tag{11.2}$$

(see Eq. (7.31)), if we look for μ_i in the form of Eq. (11.1), we must have the relation

$$\bar{v}_i = RT\left(\frac{\partial \ln p_{fi}}{\partial p}\right)_{T,x}, \tag{11.3}$$

where p is the total pressure and $x = \{x_i\}$. Since chemical potentials are logarithmically singular at $p = 0$ as we have found out in the case of a single component system, it is convenient to subtract the singular part from $\bar{v}_i$. Therefore we cast Eq. (11.3) in a slightly different, but more suitable, form as below:

$$RT\left[\frac{\partial}{\partial p}\ln\left(\frac{p_{fi}}{x_i p}\right)\right]_{T,x} = \bar{v}_i - \frac{RT}{p}. \tag{11.4}$$

On integrating it at constant T we obtain

$$RT \ln p_{fi} = RT\left[\ln\left(\frac{p_{fi}}{x_i p}\right)\right]_{p=0} + RT \ln(x_i p) + \int_0^p dp\left(\bar{v}_i - \frac{RT}{p}\right). \tag{11.5}$$

It is now necessary to calculate the first term on the right-hand side of Eq. (11.5). It can be rewritten as

$$\left[\ln\left(\frac{p_{fi}}{x_i p}\right)\right]_{p=0} = \lim_{p \to 0}\left[\ln\left(\frac{p_{fi}}{p_i}\right) + \ln\left(\frac{p_i}{x_i p}\right)\right]. \tag{11.6}$$

Since the Gibbs–Dalton law holds for the ideal gas, we have the limit

$$\lim_{p \to 0}\left(\frac{p_i}{x_i p}\right) = 1. \tag{11.7}$$

The existence of the limit of (p_{fi}/p_i) can be established[1] if the mixture is put in equilibrium with component i across a semipermeable membrane which is permeable to component i only. Let us consider the following situation. A mixture of real gases is in equilibrium with a pure gas of the ith species. The membrane is also assumed to be diathermal and deformable, so that the two systems are in thermal and mechanical equilibrium as well as in material equilibrium with respect to component i. We designate the mixture phase A, and the pure component system phase B, and affix

---

[1]See J. G. Kirkwood and I. Oppenheim, *Chemical Thermodynamics* (McGraw-Hill, New York, 1960).

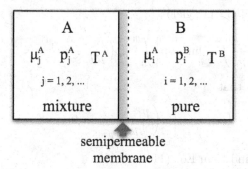

Fig. 11.1 Equilibrium between a mixture and pure species i across a semipermeable membrane.

a superscript A or B to thermodynamic quantities of the two phases in order to distinguish them. It then follows from the equilibrium conditions established for heterogeneous systems in Chapter 8 that

$$\mu_i^A = \mu_i^B \equiv \mu_i,$$
$$T^A = T^B \equiv T, \qquad (11.8)$$
$$p_i^A = p_i^B \equiv p_i.$$

According to the analysis made of pure (single-component) systems in Chapter 9, we have for the chemical potential in phase B

$$
\begin{aligned}
\mu_i^B(T, p_i^B) &= \mu_i^*(T) + RT \ln p_{fi}(T, p_i^B) \\
&= \mu_i^*(T) + RT \ln\left[p_i^B \exp \int_0^{p_i^B} dp\left(\frac{v}{RT} - p^{-1}\right)\right], \qquad (11.9)
\end{aligned}
$$

which, in view of the equilibrium conditions listed in Eq. (11.8), may be written as

$$\mu_i(T, p_i) = \mu_i^*(T) + RT \ln\left[p_i \exp \int_0^{p_i} dp\left(\frac{v}{RT} - p^{-1}\right)\right]. \qquad (11.10)$$

We have made use of the fugacity expression for the pure component i. Note that $\mu_i^*(T)$ in Eq. (11.9) is the chemical potential of i when $p_{fi} = 1$ or the chemical potential of the hypothetical pure ideal gas of component i at 1 atm pressure. Therefore, if we determine the fugacity p_{fi} such that the values for $\mu_i^*(T)$ in Eqs. (11.1) and (11.9) coincide then comparison of

Eqs. (11.10) and (11.11) yields

$$\lim_{p \to 0} \left(\frac{p_{fi}}{p_i} \right) = \lim_{p \to 0} \exp\left[\int_0^{p_i} dp \left(\frac{v}{RT} - p^{-1} \right) \right] = 1. \qquad (11.11)$$

This establishes that

$$\left[\ln\left(\frac{p_{fi}}{x_i p} \right) \right]_{p=0} = 0$$

and we finally find from Eq. (11.5)

$$p_{fi}(T,p) = x_i p \exp\left[\int_0^p dp' \left(\frac{\bar{v}_i}{RT} - \frac{1}{p'} \right) \right] \qquad (11.12)$$

for fugacities of the components of the mixture. We clearly see that it is a generalization of the fugacity formula (Eq. (9.124)) for a single component to which it reduces as x_i tends to unity.

Chemical potentials are sometimes given in terms of the ideal gas and excess chemical potentials:

$$\mu_i(T,p) = \mu_i^*(T) + RT \ln(x_i p) + \mu_i^{(ex)}, \qquad (11.13)$$

where

$$\mu_i^{(ex)} = \int_0^p dp' \left(\bar{v}_i - \frac{RT}{p'} \right). \qquad (11.14)$$

Since $p\bar{v}_i = RT$ for the ideal gas, for which the excess chemical potential obviously vanishes, we recover from Eq. (11.14) the chemical potential of the ideal gas i from Eq. (11.14):

$$\mu_i(T,p) = \mu_i^*(T) + RT \ln(x_i p).$$

For many common gases the volume change on mixing is negligible. In this case, the partial molar volume $\bar{v}_i$ of gas i in the mixture is almost the same as the molar volume of the pure gas. That is,

$$\Delta v_i = \bar{v}_i - v_i \simeq 0. \qquad (11.15)$$

When this approximation is applicable, the partial molar volume in Eq. (11.12) may be replaced with the molar volume and we obtain

$$p_{fi} \simeq x_i p_{fi}^0, \qquad (11.16)$$

where

$$p^0_{fi} = p \exp\left[\int_0^p dp'\left(\frac{v_i}{RT} - p'^{-1} \right) \right],$$ (11.17)

which is the fugacity of pure gas i. In this approximation the fugacity of component i in a mixture is the product of the mole fraction and the fugacity of pure gas i. This approximate relation holds with a fair accuracy up to about 100 atm pressure for common gases. Equation (11.16) represents the Lewis–Randall rule.[2]

According to statistical mechanics the virial equation of state for a mixture may be written in the form

$$pv = RT + B_2(T)p + B_3(T)p^2 + \cdots,$$ (11.18)

where

$$v = \frac{V}{n} = \frac{V}{n_1 + n_2 + \cdots + n_c},$$

$$B_2 = \sum_{i=1}^c \sum_{j=1}^c B_{ij}(T)x_i x_j,$$ (11.19)

$$B_3 = \sum_{i=1}^c \sum_{j=1}^c \sum_{k=1}^c B_{ijk}(T)x_i x_j x_k, \text{ etc.}$$

with x_i and so on denoting mole fractions. The second virial coefficients[3] are generally symmetric with respect to the indices:

$$B_{ij} = B_{ji}.$$ (11.20)

Using the equation of state (11.18), we may calculate the fugacity of a mixture. For the purpose it is first necessary to calculate the partial molar

---

[2] G. N. Lewis and M. Randall, *Thermodynamics*, pp. 225–227 (McGraw Hill, New York, 1923); *J. Am. Chem. Soc.* **50**, 1522 (1928); *ibid.* **59**, 2733 (1937).

[3] According to the statistical mechanics of simple fluids the second virial coefficient B_{ij} is given in terms of the intermolecular potential energy $u_{ij}(r)$ between two molecules of species i and j:

$$B_{ij}(T) = -2\pi \int_0^\infty dr\, r^2 \left\{ \exp\left[-\frac{u_{ij(r)}}{k_B T} \right] - 1 \right\},$$

where k_B is the Boltzmann constant. Since $u_{ij} = u_{ji}$, the symmetry property of B_{ij} in Eq. (11.20) is a consequence of the symmetry of the potential energies with respect to interchange of particle indices.

volume $\bar{v}_i$. Since from the virial equation of state (Eq. (11.18))

$$V = \sum_{i=1}^{c} n_i \frac{RT}{p} + \sum_{i=1}^{c} \sum_{j=1}^{c} n_i n_j B_{ij}(T) n^{-1}$$

$$+ p \sum_{i=1}^{c} \sum_{j=1}^{c} \sum_{k=1}^{c} n_i n_j n_k B_{ijk}(T) n^{-2} + \cdots, \tag{11.21}$$

we obtain on differentiation with respect to n_i

$$\bar{v}_i = \frac{RT}{p} + \sum_{j=1}^{c} \sum_{k=1}^{c} x_j x_k (2B_{ij} - B_{jk}) \tag{11.22}$$

$$+ p \sum_{j=1}^{c} \sum_{k=1}^{c} \sum_{l=1}^{c} x_j x_k x_l (B_{ijk} + B_{jil} + B_{jki} - 2B_{jkl}) + \cdots.$$

When this is substituted into Eq. (11.12), there follows an approximate fugacity

$$p_{fi} = x_i p \exp\left[\frac{p}{RT} \sum_{j=1}^{c} \sum_{k=1}^{c} x_j x_k (2B_{ij} - B_{jk}) \right], \tag{11.23}$$

for which we have ignored the term linear in p in Eq. (11.22). If the van der Waals theory is applied to the mixture the second virial coefficients are found to be in the form

$$B_{ij} = b_{ij} - \frac{a_{ij}}{RT}, \tag{11.24}$$

where b_{ij} are known to be related to the sum of molecular radii of components i and j, and a_{ij} to the strength of the van der Waals attraction between the molecules of components i and j.

It is sometimes a fair approximation to calculate B_{ij} as the arithmetic mean of the second virial coefficients of pure gases i and j, namely,

$$B_{ij} = \tfrac{1}{2}(B_{ii} + B_{jj}), \tag{11.25}$$

where B_{ii} and B_{jj} are, respectively, the second virial coefficient of pure gases i and j. In this approximation the fugacity takes a rather simple form:

$$p_{fi} = x_i p \exp\left(\frac{B_{ii}p}{RT} \right). \tag{11.26}$$

Since we may write the virial equation of state for pure gas i in the form

$$v_i = \frac{RT}{p} + B_{ii}(T) + O(p),$$

where v_i is the molar volume of the gas, by combining it with Eq. (11.22), we obtain the volume change on mixing

$$\bar{v}_i - v_i = \sum_{j=1}^{c} \sum_{k=1}^{c} x_j x_k (2B_{ij} - B_{jk} - B_{ii}) + O(p). \tag{11.27}$$

When the approximation (Eq. (11.25)) is used for the second virial coefficients in this equation, there follows

$$\bar{v}_i - v_i - \sum_{j=1}^{c} \sum_{k=1}^{c} \tfrac{1}{2} x_j x_k (B_{jj} \quad B_{kk})$$

$$= 0 \tag{11.28}$$

and therefore, to the same order of approximation, the volume is unchanged on mixing

$$\Delta v = \sum_{i=1}^{c} n_i (\bar{v}_i - v_i) = 0. \tag{11.29}$$

It is useful to recall the Lewis–Randall rule which assumes the validity of a negligible change in volume on mixing. The rule therefore is seen to be related to the combination rule (Eq. (11.25)) and the virial equation of state truncated at the first order in p.

11.2 Entropy of Mixing

Once the chemical potentials are calculated for a mixture other thermodynamic functions may be easily calculated from the chemical potentials. We describe the procedure here in the case of a gas. Since the partial molar entropy of component i may be written as

$$\bar{s}_i = -\left(\frac{\partial \mu_i}{\partial T}\right)_{p,x}, \tag{11.30}$$

differentiation of Eq. (11.13) yields the partial molar entropy

$$\bar{s}_i = \bar{s}_i^*(T) - R\ln(x_i p) - \int_0^p dp' \left[\left(\frac{\partial \bar{v}_i}{\partial T}\right)_{p',x} - \frac{R}{p'}\right], \tag{11.31}$$

where

$$\bar{s}_i^* = -\left(\frac{\partial \mu_i^*}{\partial T}\right)_{p,x}. \tag{11.32}$$

Its meaning is similar to the entropy $s^*(T)$ of a pure gas; it is the partial molar entropy of a hypothetical pure ideal gas at 1 atm pressure or alternatively

$$\bar{s}_i(T) = \lim_{p\to 0}[\bar{s}_i(T,p) + R\ln(x_i p)]$$

$$= s_i^0 + \int_{T_1}^{T} dT\, c_{pi}^*(T)$$

$$= s_i^*(T). \tag{11.33}$$

If the system is composed of component i alone, the partial molar entropy of component i is equal to its molar entropy which may be written as

$$s_i(T,p) = s_i^*(T) - R\ln p - \int_0^p dp'\left[\left(\frac{\partial v}{\partial T}\right)_p - \frac{R}{p'}\right]. \tag{11.34}$$

This is the molar entropy of pure real gas i.

Let us imagine the following process: $n_1, n_2, \ldots, n_c$ moles of real gases are reversibly[4] combined at constant pressure and temperature. The partial molar entropies are then expected to be different from the molar entropies and therefore there will be a change in the overall entropy of the system on mixing. This change in entropy is called the entropy of mixing and is calculated as follows:

$$\Delta s = \sum_{i=1}^{c} x_i(\bar{s}_i - s_i) \tag{11.35}$$

$$= -R\sum_{i=1}^{c} x_i \ln x_i + \sum_{i=1}^{c} x_i \int_0^p dp'\left[\left(\frac{\partial v_i}{\partial T}\right)_{p'} - \left(\frac{\partial \bar{v}_i}{\partial T}\right)_{p',x}\right](p').$$

Equation (11.33) is made use of in this calculation. Since for ideal gases

$$\bar{v}_i = v_i,$$

---

[4]Strictly speaking, mixing various components cannot be regarded as a reversible process, because it cannot be reversed in practice. Here it means equilibrium is maintained over the course of the process.

it follows that

$$\left(\frac{\partial \bar{v}_i}{\partial T}\right)_{p,x} = \left(\frac{\partial v_i}{\partial T}\right)_p,$$

and therefore the entropy of mixing takes the form

$$\Delta s = -R \sum_{i=1}^{c} x_i \ln x_i. \tag{11.36}$$

This is the entropy of mixing for ideal gases. As is obvious from the formula, the entropy of mixing is always positive for ideal gases.

We see that the second term on the right-hand side of Eq. (11.35) is the nonideality correction to the ideal gas entropy of mixing. The nonideality correction to the entropy of mixing can be estimated by making use of Eq. (11.27). Differentiating Eq. (11.27) with respect to T and substituting the result into the integral in Eq. (11.35), we obtain the nonideality correction

$$\Delta s_{real} = p \sum_{i=1}^{c} \sum_{j=1}^{c} x_i x_j \left(\frac{dB_{ii}}{dT} - \frac{dB_{ij}}{dT}\right). \tag{11.37}$$

Note that this vanishes when the combination rule (Eq. (11.25)) is used for evaluating the derivatives of the second virial coefficients in Eq. (11.37). The Δs_{real} generally vanishes if the Lewis–Randall rule holds. If the van der Waals model is used for B_{ii} and B_{ij} then

$$\Delta s_{real} = \frac{p}{RT^2} \sum_{i=1}^{c} \sum_{j=1}^{c} x_i x_j (a_{ii} - a_{ij}). \tag{11.38}$$

If the combination rule is used for a_{ij} then still

$$\Delta s_{real} = 0,$$

and the entropy of mixing again has the same form as an ideal mixture. The entropy of mixing obtained in this section, for example, Eqs. (11.36), (11.37), and (11.38), permits the following remark. In physical chemistry textbooks it is often mentioned that the entropy of mixing is consistent with the requirement of the second law of thermodynamics since $\Delta s \geq 0$. This is obviously true in the case of an ideal gas mixture, but not assured if the gas mixture is real since, for example, if $a_{ij} \geq a_{ii}$ and if p and temperature are such that Δs_{real} is larger than Δs_{ideal} defined by Eqs. (11.36) then it is possible that

$$\Delta s < 0.$$

The ideal gas formula Δs_{ideal} for the entropy of mixing is isomorphic to the statistical mechanical entropy formula if x_i is taken to mean the distribution function of finding species i, but such an interpretation is over-extending the notion of probability in the phase space of particles. Recall also the remarks made in Chapter 4 with regard to the second law of thermodynamics and $\Delta s \geq 0$ for an isolated system.

11.3   Heat of Mixing

The partial molar enthalpy may be calculated in terms of μ_i and $\bar{s}_i$ just calculated, if we make use of the formula

$$\bar{h}_i = \mu_i + T\bar{s}_i.$$

Upon substitution of the expressions for the chemical potential and the partial molar entropy we easily find the partial molar enthalpies of a mixture

$$\bar{h}_i(T, p) = h_i^*(T) + \int_0^p dp' \left[\bar{v}_i - T\left(\frac{\partial \bar{v}_i}{\partial T}\right) \right]_{p', x}, \tag{11.39}$$

where

$$h_i^*(T) = \mu_i^*(T) + Ts_i^*(T). \tag{11.40}$$

Therefore we again see that

$$h_i^*(T) = \lim_{p \to 0} \bar{h}_i(T, p) \tag{11.41}$$

and hence the molar enthalpy of pure ideal gas i. As a consequence, we see that the second term on the right-hand side of Eq. (11.39) is the nonideality correction for the partial molar enthalpies. It is the generalization of the nonideality correction for the molar enthalpy of a pure gas.

The molar enthalpy was calculated for pure gases, and for pure gas i it may be written in the form

$$h_i = h_i^*(T) + \int_0^p dp' \left[v_i - T\left(\frac{\partial v_i}{\partial T}\right)_{p'} \right], \tag{11.42}$$

where v_i is the molar volume of gas i.

The heat of mixing is defined as the sum of all enthalpy changes for all substances upon formation of a mixture of $x_1, x_2, \ldots, x_c$ mole fractions from the c pure constituents of the mixture:

$$\Delta h = \sum_{i=1}^{c} x_i (\bar{h}_i - h_i). \tag{11.43}$$

It is then easily calculated with Eqs. (11.39) and (11.42). We obtain

$$\Delta h = \sum_{i=1}^{c} x_i \int_0^p dp' \left[\Delta \bar{v}_i - T \left(\frac{\partial \Delta \bar{v}_i}{\partial T} \right)_{T,x} \right], \tag{11.44}$$

where $\Delta v_i = \bar{v}_i - v_i$. Since the molar volumes do not change on mixing if the gases are ideal, $\Delta v_i = 0$ as was already discussed previously, and we find that there is no heat of mixing for ideal gases:

$$\Delta h = 0.$$

For real gases, if we make use of Eq. (11.27) there follows to the first order in p the expression

$$\Delta h = p \sum_{i=1}^{c} \sum_{j=1}^{c} x_i x_j \left[B_{ij} - B_{ii} - T \left(\frac{dB_{ij}}{dT} - \frac{dB_{ii}}{dT} \right) \right]. \tag{11.45}$$

This equation indicates that the heat of mixing will be equal to zero if the virial equation of state is truncated at the first order in p and if the combination rule (Eq. (11.25)) holds for the second virial coefficients. In general, there will be no heat of mixing for gases for which the Lewis–Randall rule holds.

11.4 Activity and Activity Coefficient

Let us assume that by some method we have determined the change in the chemical potential of a gaseous component when the concentrations are varied at constant temperature and pressure. This determination amounts to measuring the ratio of the fugacities, since

$$\Delta \mu_i = \mu_i - \mu_i^0 = RT \ln \left(\frac{p_{fi}}{p_{fi}^{(0)}} \right), \tag{11.46}$$

where $p_{fi}^{(0)}$ is the fugacity of i at the state corresponding to μ_i^0. It is common practice in thermodynamics to determine and use such a ratio of fugacities when it is difficult to obtain the numerical value of either one of the fugacities. It is proven convenient to work with the ratio of the fugacity p_{fi} of a substance in a given state to its fugacity $p_{fi}^{(0)}$ in some other state chosen as a standard state. We call this relative fugacity the activity and will denote it by a_i:

$$a_i = \frac{p_{fi}}{p_{fi}^{(0)}}. \tag{11.47}$$

Therefore the activity is equal to unity at the standard state. The chemical potential is then written

$$\mu_i = \mu_i^0 + RT \ln a_i, \tag{11.48}$$

where μ_i^0 is the chemical potential of i at the standard state chosen.

There are a number of conventions adopted for the standard state. For gases the standard state is chosen with the one in which the fugacity at a given temperature is unity. It is also the state at which the heat capacity, enthalpy, and entropy are those of the gas at infinitely low pressure:

$$a_i = p_{fi}, \tag{11.49}$$

that is, in such a limit

$$\lim_{p \to 0} \frac{a_i}{x_i p} = 1. \tag{11.50}$$

This choice of standard state for gases makes the standard chemical potential μ_i^0 coincide with $\mu_i^*(T)$:

$$\mu_i^0 = \mu_i^*(T),$$

where $\mu_i^*(T)$ is given in Eq. (11.1) and is the chemical potential of the hypothetical ideal gas i at 1 atm pressure.

If the activity coefficient f_i for gas i is introduced by the relation

$$a_i = f_i p_i = x_i f_i p \tag{11.51}$$

and if we define

$$\mu_i^0(T, p) = \mu_i^*(T) + RT \ln p, \tag{11.52}$$

where it must be noted that $\mu_i^0(T, p)$ is not the same as μ_i^0 in Eq. (11.48), then the chemical potential is given in the form

$$\mu_i = \mu_i^0(T, p) + RT \ln(x_i f_i). \tag{11.53}$$

For liquids and solids the standard chemical potentials are chosen such that

$$\mu_i^0 = \mu_i^l(T,p) \quad \text{for liquids } (i = 1, 2, \ldots, c)$$

$$= \mu_i^s(T,p) \quad \text{for solids } (i = 1, 2, \ldots, c),$$

(11.54)

where $\mu_i^l(T,p)$ and $\mu_i^s(T,p)$ are the chemical potential of pure liquid i and pure solid i, respectively. This will be referred to as Convention I on activity.

Sometimes, the following convention is adopted by distinguishing the solvent designated as component 1 and the solutes designated as components $2, 3, \ldots, c$:

$$\mu_1^0 = \mu_1^l(T,p) \quad \text{for the solvent,}$$

$$\mu_i^0 = \lim_{x_1 \to 1}(\mu_i - RT \ln x_i) \quad \text{for the solutes.}$$

(11.55)

This will be referred to as Convention II. Note that μ_i^0 for solutes $i = 2, 3, \ldots, c$ are defined as the chemical potentials in the limit of infinite dilution in this convention. One can choose one of the aforementioned conventions, depending on the situation and convenience. In any case, by introducing the activity coefficient f_i by the relation

$$a_i = x_i f_i,$$

(11.56)

we can express the chemical potentials of components in a solution in the form

$$\mu_i = \mu_i^0(T,p) + RT \ln(x_i f_i).$$

(11.57)

In this manner we can put the formalisms for gaseous mixtures and liquid or solid solutions on the same footing. The conventions adopted for the standard state give rise to the following conventions for the activity coefficients. For gases

$$\lim_{p \to 0} f_i = 1.$$

(11.58)

For liquids and solids

Convention I

$$\lim_{x_i \to 1} f_i = 1 \ (i = 1, 2, \ldots, c);$$

(11.59)

Convention II

$$\lim_{x_1 \to 1} f_i = 1 \ (i = 1, 2, \ldots, c). \tag{11.60}$$

Activities and activity coefficients introduced here will be used for discussions of thermodynamics of real substances in the chapters for real gas mixtures and real solutions in this work where various theories of determining activity and activity coefficients are developed.

11.5   Canonical Equation of State for a Mixture

Instead of the virial equation of state, the theory can be developed by means of the canonical equation of state suitably generalized for mixtures. It is only necessary to use the canonical equation of state for the equation of state in the formalism presented earlier in this chapter. The canonical equation of state for a c-component mixture may be written as

$$\left[p + A(n, T) \, n^2\right] \left[1 - nB(n, T)\right] = nk_B T, \tag{11.61}$$

where n is the number density of molecules and

$$A(n, T) = \sum_{i,j=1}^{c} A_{ij}(n, T) \, x_i x_j, \tag{11.62}$$

$$B(n, T) = \sum_{i,j=1}^{c} B_{ij}(n, T) \, x_i x_j, \tag{11.63}$$

with x_i denoting the mole fraction of species i

$$x_i = \frac{n_i}{n} = \frac{n_i}{\sum_{i=1}^{c} n_i}.$$

The coefficients $A_{ij}(n, T)$ and $B_{ij}(n, T)$ are phenomenological functions of n_i, n_j, and T. The statistical mechanical representations of the GvdW parameters A_{ij} and B_{ij} are

$$A_{ij}(n, T) = \frac{2\pi}{3} \int_{r_{ij}^{\ddagger}}^{\infty} dr r^3 u'_{ij}(r) \exp[-\beta u_{ij}(r)] \, g_{ij}(r), \tag{11.64}$$

$$B_{ij}(n, T) = -\frac{2\pi}{3} \beta \int_{r_{ij}^0}^{r_{ij}^{\ddagger}} dr r^3 u'_{ij}(r) \exp[-\beta u_{ij}(r)] \, g_{ij}(r)$$

$$\times \left[1 - \frac{2\pi}{3} \beta n \sum_{i,j=1}^{c} x_i x_j \int_{r_{ij}^0}^{r_{ij}^{\ddagger}} dr r^3 u'_{ij}(r) \exp[-\beta u_{ij}(r)] \, g_{ij}(r)\right]^{-1},$$

$$\tag{11.65}$$

where $\beta = 1/k_B T$, $u_{ij}(r)$ is the potential energy of the pair (i,j), $u'_{ij}(r)$ is the derivative of $u_{ij}(r)$, $g_{ij}(r)$ is the radial distribution function of the pair (i,j), $r^{\ddagger}_{ij}$ is the position at which $-r^3 u'_{ij}(r) \exp[-\beta u_{ij}(r)] g_{ij}(r)$ is a maximum, and r^0_{ij} is the hard core or the point below which the radial distribution function vanishes practically to zero; generally, $r^0_{ij} = 0$ for $u_{ij} \to \infty$ as $r \to 0$. It should be noted that $B_{ij}(n,T)$ in the canonical equation of state are, of course, different from the second virial coefficients appearing in the virial expansion used earlier in this chapter. In phenomenological thermodynamics these statistical mechanical representations, of course, have no place in the theory, but they are presented to help us cast the thermodynamic formulas in terms of the GvdW parameters introduced earlier.

We calculate partial molar volumes which we have found to be a building block of the thermodynamic solution theory. For this purpose and also for insight and completeness we make use of the virial equation of state expressed, for notational simplicity, in the form

$$Z = \beta p - n = \sum_{i,j=1}^{c} n_i n_j \mathfrak{Z}_{ij}, \qquad (11.66)$$

where the statistical mechanical form for $\mathfrak{Z}_{ij}$ is given by the formula

$$\mathfrak{Z}_{ij} = -\frac{2\pi}{3}\beta \int_0^\infty dr\, r^3 u'_{ij}(r) g_{ij}(r; \{n_k\}). \qquad (11.67)$$

Here $u'_{ij}(r) = \partial u_{ij}/\partial r$. Since the intermolecular potential energy is symmetric with respect to the interchange of indices i and i, it follows that $\mathfrak{Z}_{ij} = \mathfrak{Z}_{ji}$. The partial molar volume is then calculated from this form of compressibility factor Z. Differentiating with the specific volume $v = 1/n$ and using

$$\frac{\partial}{\partial n_k} = \frac{\partial x_k}{\partial n_k}\frac{\partial}{\partial x_k} = \frac{1}{n}(1 - x_k)\frac{\partial}{\partial x_k},$$

we obtain

$$\frac{\bar{v}_k}{v} = \sum_{i,j=1}^{c}\left[2\delta_{ik}x_j\mathfrak{Z}_{ij} + x_i x_j(1 - x_k)\left(\frac{\partial \mathfrak{Z}_{ij}}{\partial x_k}\right)_{T,p,n'}\right]. \qquad (11.68)$$

We now observe that if we split $\mathfrak{Z}_{ij}$ into two parts, attractive and repulsive, as in

$$\mathfrak{Z}_{ij} = \mathfrak{Z}_{ij}^{(A)} + \mathfrak{Z}_{ij}^{(B)},$$

where

$$3_{ij}^{(A)} = -\frac{2\pi}{3}\beta \int_{r_{ij}^{\ddagger}}^{\infty} dr\, r^3 u'_{ij}(r)\exp[-\beta u_{ij}(r)]\, g_{ij}(r), \qquad (11.69)$$

$$3_{ij}^{(B)} = -\frac{2\pi}{3}\beta \int_{r_{ij}^{0}}^{r_{ij}^{\ddagger}} dr\, r^3 u'_{ij}(r)\exp[-\beta u_{ij}(r)]\, g_{ij}(r). \qquad (11.70)$$

The partial molar volume in Eq. (11.68) then may be expressible in terms of the GvdW parameters for the mixture defined by

$$A_{ij} = -\beta^{-1}3_{ij}^{(A)}, \qquad (11.71)$$

$$B_{ij} = \frac{3_{ij}^{(B)}}{1 + n\sum_{k,l=1}^{c} x_i x_j 3_{kl}^{(B)}}. \qquad (11.72)$$

These definitions put the equation of state in the canonical form given in Eq. (11.61). Inverting the relation B_{ij} and $3_{ij}^{(B)}$ in Eq. (11.72) we find

$$3_{kl}^{(B)} = \frac{B_{kl}}{1 - n\sum_{ij}^{c} x_i x_j B_{ij}}. \qquad (11.73)$$

With these identifications of $3_{ij}^{(A)}$ and $3_{kl}^{(B)}$, the partial molar volume can be expressed in terms of GvdW parameters:

$$\frac{\bar{v}_k}{v} = \sum_{i,j=1}^{c} 2\delta_{ik}x_j\left(\beta A_{ij} - \frac{B_{ij}}{1 - n\sum_{l,m}^{c} x_l x_m B_{lm}}\right) + \sum_{i,j=1}^{c} x_i x_j(1 - x_k)$$

$$\times \left[\beta\left(\frac{\partial A_{ij}}{\partial x_k}\right)_{T,p,x'} - \frac{\partial}{\partial x_k}\left(\frac{B_{ij}}{1 - n\sum_{l,m}^{c} x_l x_m B_{lm}}\right)_{T,p,x'}\right]. \qquad (11.74)$$

From the statistical mechanical viewpoint there is little to gain by using Eq. (11.74), but since we can model the GvdW parameters by using the phenomenological canonical equation of state as shown in Chapter 10, the formula presented for $\bar{v}_k$ provides an approach alternative to that of the virial expansion for solutions, which is not particularly reliable for liquids. We note that for a study of subcritical fluids of a single component a quadratic model for the GvdW parameters was used, which were non-analytic and discontinuous functions of density and temperature; see Sec. 10.10, Chapter 10. It is expected that their mathematical properties would be similar even for mixtures.

Expanding the rational fraction term in Eq. (11.74) yields a virial expansion-like series

$$\frac{\overline{v}_k}{v} = \sum_{i,j=1}^{c} 2\delta_{ik}x_j\left(\beta A_{ij} - B_{ij} - n\sum_{lm}^{c} x_l x_m B_{ij}B_{lm} + \cdots\right) + \sum_{i,j=1}^{c} x_i x_j$$

$$(11.75)$$

$$\times (1 - x_k)\left[\frac{\partial}{\partial x_k}\left(\beta A_{ij} - B_{ij} - n\sum_{lm}^{c} x_l x_m B_{ij}B_{lm} + \cdots\right)\right]_{T,p,x'}.$$

If A_{ij} and B_{ij} are expanded in series of mole fractions

$$A_{ij} = A_{ij}^{(0)}(T) + \sum_{k=1}^{c} A_{ijk}^{(1)}(T)x_k + \sum_{k,l=1}^{c} A_{ijkl}^{(2)}(T)x_k x_l + \cdots, \qquad (11.76)$$

$$B_{ij} = B_{ij}^{(0)}(T) + \sum_{k=1}^{r} B_{ijk}^{(1)}(T)x_k + \sum_{k,l=1}^{c} B_{ijkl}^{(2)}(T)x_k x_l + \cdots, \qquad (11.77)$$

then to the leading order mole fractions in the supercritical regime the partial molar volume is given by the series

$$\frac{\overline{v}_k}{v} = \sum_{j=1}^{c} 2\left(\beta A_{kj}^{(0)} - B_{kj}^{(0)}\right)x_j - 2n\sum_{j,l,m=1}^{c} B_{kj}^{(0)}B_{lm}^{(0)}x_j x_l x_m + \cdots. \qquad (11.78)$$

The partial molar volumes[5] obtained can be made use of to deduce other thermodynamic quantities for a mixture such as the entropy and heat of mixing. The thermodynamics of critical phenomena can be similar to the formalism for the case of pure substance presented in the previous chapter, but because of the space limitation this subject is left to the reader as an exercise.

---

[5]We remark that molar and partial molar volumes discussed in this chapter are those of a uniform equilibrium fluid, but they are not necessarily equal to their nonequilibrium counterpart, for example, if the fluid is spatially nonuniform, since then the local partial molar volumes are not conserved over space and time owing to the nonequilibrium evolution of the fluid structure in space-time. Therefore, since dynamic molar volume is not necessarily the same as the equilibrium molar volume which is conserved, care must be exercised when molar and partial molar volumes are treated for nonequilibrium fluids. See B. C. Eu, *J. Chem. Phys.* **129**, 094502, 134509 (2008). See also J. S. Hunjan and B. C. Eu, *J. Chem. Phys.* **132**, 134510 (2010) for a molecular representation of molar volume.

Problems

(1) Calculate the entropy of mixing for a c-component mixture obeying the canonical equation of state.
(2) Calculate the heat of mixing for a c-component mixture obeying the canonical equation of state.
(3) Calculate the fugacity of a c-component mixture obeying the canonical equation of state to the leading order expansion Eq. (11.78) for the partial molar volume.

Chapter 12

Chemical Equilibria

In most of the discussions up to now we have excluded chemical reactions among the substances in a mixture. This assumption should be removed for chemically reacting fluids. In this chapter we consider chemical equilibria. Nonequilibrium phenomena in reacting systems will be deferred to the last chapter of this work. Since chemical reactions transform one or more species into other species through molecular interactions of reactants or products, there are additional relations required of the chemical potentials that are absent when chemical reactions are precluded, and equilibrium is accordingly modified. We first find these additional conditions on equilibrium. The discussion can be made more general by considering a heterogeneous system of ν phases where a number of coupled chemical reactions may occur as in the cases of many biological systems and catalysis. Since a general formulation also invites complications and often cumbersome notations without necessarily increasing our understanding of the fundamental points in question, we may first consider a simple case and then extend the result to a more general situation removing the restriction. With the general results so obtained we will also be able to consider a generalized phase rule as well.

12.1   A Single Reaction

We are interested in obtaining the equilibrium condition for a chemical reaction occurring in a closed homogeneous system composed of c components of $n_1, n_2, \ldots, n_c$ moles in a single phase, for example, the gas phase.

Let us assume that there is a reversible chemical reaction occurring in the system:

$$r_1\mathcal{R}_1 + r_2\mathcal{R}_2 + r_3\mathcal{R}_3 + \cdots \rightleftharpoons q_1\mathcal{Q}_1 + q_2\mathcal{Q}_2 + q_3\mathcal{Q}_3 + \cdots. \qquad (12.1)$$

The symbols $\mathcal{R}_i(i \geq 1)$ denote the reactants, and $\mathcal{Q}_i(i \geq 1)$ the products. The small letters r_i and q_i denote the stoichiometric coefficients of the reaction. Reaction (12.1) collectively represents a unimolecular, bimolecular, or trimolecular reaction, and so on. It is convenient to write the equation in the form

$$\sum_{i=1}^{c} \sigma_i X_i = 0, \qquad (12.2)$$

where X_i stand for the species and σ_i the stoichiometric coefficients *counted positive for the products and negative for the reactants*. Equation (12.2) is obtained from Eq. (12.1) by transferring the left-hand side to the right.

If the reactants are predominant in concentration relative to the products or if the conditions are favorable for the forward reaction, the reaction will proceed toward the right in Eq. (12.1), and vice versa. Eventually the forward and reverse reactions will be precisely balanced, when the system has reached chemical equilibrium. The condition for chemical equilibrium is the object we wish to investigate here.

For the purpose, it is useful to introduce the progress variable λ indicating the degree of the chemical reaction progressed. Let us denote by $dn_1, dn_2, \ldots, dn_c$ the molar changes for species arising from the chemical reaction. The stoichiometry of the reaction then demands that

$$\frac{dn_1}{\sigma_1} = \frac{dn_2}{\sigma_2} = \cdots = \frac{dn_c}{\sigma_c} = d\lambda. \qquad (12.3)$$

This equation basically defines the progress variable and in this definition we are counting the reaction as progressing to the right if $d\lambda > 0$, and to the left if $d\lambda < 0$. We will denote the molecular weights of compounds by M_i, $i = 1, 2, \ldots, c$. Since the total mass must be balanced before and after the reaction, as is demanded by the conservation of mass, there must hold the relation

$$\sum_{i=1}^{c} \sigma_i M_i = 0. \qquad (12.4)$$

This is the mass balance equation for the reaction. With this preparation we are now ready to consider the chemical equilibrium condition. Since the system is closed, for an arbitrary virtual variation of the internal energy

$$(\delta E)_{\Psi, V} \geq 0 \tag{12.5}$$

in accordance with the demand made by the second law of thermodynamics. Since for reversible processes $S = \Psi_e$ and thus the fundamental relation may be written as

$$dE = T \, dS - p \, dV + \sum_{i=1}^{c} \mu_i \, dn_i, \tag{12.6}$$

where n_i changes because of the chemical reaction, the equilibrium condition (Eq. (12.5)) now takes the form

$$\sum_{i=1}^{c} \mu_i \, \delta n_i > 0 \tag{12.7}$$

for arbitrary virtual variations in n_i, $i = 1, 2, \ldots, c$. Substitution of equations in Eq. (12.3) puts this inequality in the form

$$\delta \lambda \sum_{i=1}^{c} \mu_i \sigma_i \geq 0. \tag{12.8}$$

If $\delta \lambda > 0$ then

$$\sum_{i=1}^{c} \mu_i \sigma_i \geq 0,$$

but if $\delta \lambda < 0$ then

$$\sum_{i=1}^{c} \mu_i \sigma_i \leq 0.$$

However, since the variation in progress variable is arbitrary its sign is also arbitrary. The only way to satisfy the aforementioned two conditions simultaneously is then

$$\sum_{i=1}^{c} \mu_i \sigma_i = 0. \tag{12.9}$$

This is the chemical equilibrium condition we are looking for. Th. de Donder[1] introduced the affinity defined by

$$\mathcal{A} = -\sum_{i=1}^{c} \mu_i \sigma_i. \tag{12.10}$$

---

[1]Th. de Donder, *Bull. Acad. Roy. Belg. (Cl. Sc.)* (5) **7**, 197, 205 (1922).

The chemical equilibrium condition (Eq. (12.9)) then implies that the affinity is equal to zero at chemical equilibrium. The concept of affinity plays an important role in irreversible phenomena of chemical reactions. Its meaning can be better understood quite easily, if we simply write out Eq. (12.10) for the chemical Reaction (12.1). By using the convention we have adopted for σ_i, we find

$$\mathcal{A} = \sum_{j \in reac} r_j \mu_j - \sum_{i \in prod} q_i \mu_i, \qquad (12.11)$$

which shows that the affinity is the difference between the Gibbs free energies of the products and reactants of a chemical reaction. When the two Gibbs free energies are precisely balanced, the chemical equilibrium is reached and there is no longer a macroscopically discernible change in the concentrations of the reactants and the products at the given temperature and pressure.

12.2   Coupled Chemical Reactions

The equilibrium condition just obtained for a single chemical reaction occurring in a phase can be generalized to the case of a system of coupled chemical reactions occurring in a phase. Let us denote the coupled chemical reactions by the equations

$$\sum_{i=1}^{c} \sigma_{i\alpha} X_i = 0 \quad (\alpha = 1, 2, \dots, m), \qquad (12.12)$$

where $\sigma_{i\alpha}$ stands for the stoichiometric coefficient of the compound X_i in the αth chemical reaction. The same convention on the stoichiometric coefficients is used as for the single reaction considered earlier. Examples for coupled chemical reactions are:

$$\mathrm{NaH_2PO_4} \rightarrow \mathrm{Na^+} + \mathrm{H_2PO_4^-},$$
$$\mathrm{H_2PO_4^-} \rightarrow \mathrm{H^+} + \mathrm{HPO_4^{2-}},$$
$$\mathrm{HPO_4^{2-}} \rightarrow \mathrm{H^+} + \mathrm{PO_4^{3-}}.$$

Each chemical reaction may be followed in terms of progress variables λ_α $(\alpha = 1, 2, \dots, m)$. We may then write

$$dn_{i\alpha} = \sigma_{i\alpha} \, d\lambda_{i\alpha}. \qquad (12.13)$$

Since the total change in the number of moles for species i is

$$dn_i = \sum_{i=1}^{c} dn_{i\alpha} = \sum_{\alpha=1}^{m} \sigma_{i\alpha}\, d\lambda_\alpha, \qquad (12.14)$$

the Gibbs relation may be written in the form

$$dE = T\, dS - p\, dV + \sum_{i=1}^{c} \sum_{\alpha=1}^{m'} \sigma_{i\alpha} \mu_i d\lambda_\alpha. \qquad (12.15)$$

The equilibrium condition for the closed reacting system therefore is

$$\sum_{\alpha=1}^{m} \left[\sum_{i=1}^{c} \sigma_{i\alpha} \mu_i \right] \delta\lambda_\alpha \geq 0.$$

By following the same line of argument as for Eq. (12.9) we find the equilibrium condition

$$\mathcal{A}_\alpha = -\sum_{i=1}^{c} \sigma_{i\alpha} \mu_i = 0 \quad (\alpha = 1, 2, \ldots, m). \qquad (12.16)$$

That is, the affinities must vanish for all chemical reactions at chemical equilibrium. This is obviously a generalization of Eq. (12.7).

12.3 Chemical Reactions in a Multiphase System

The chemical equilibrium condition (Eq. (12.16)) which we have obtained in Sec. 12.2 holds in the case of a single phase. In nature, there can arise a situation in which chemical reactions occur in different phases which are not necessarily in equilibrium with each other. One typical example would be biological cells where various chemical reactions are known to occur in cytoplasm, membranes, and vesicles. Another more physical example would be a system of liquid solution–gaseous mixture, in which chemical reactions occur. We would like to extend to such systems the equilibrium conditions which we have obtained for non-reacting systems.

If there is a chemical reaction proceeding in a phase, there are two factors contributing to the concentration changes for the components involved: one stems from the chemical transformation of species, and the other arises from the permeation (diffusion) of species through the interfacial boundaries. These two factors will be denoted $d_i n_k$ and $d_c n_k$, the former standing for the transfer of matter across the interfacial boundaries, and the latter for the concentration change arising from the chemical reactions. We shall consider for the sake of simplicity the case where the same chemical reactions occur in every phase involved and each phase is in internal equilibrium with regard to temperature and pressure, apart from chemical reactions which are assumed to occur uniformly over the phases. It will be assumed that there are m chemical reactions.

In this case, because of the assumption of internal equilibrium, the Gibbs relation for phase a may be written as

$$dE^{(a)} = T^{(a)}dS^{(a)} - p_a\, dV^{(a)} + \sum_{k=1}^{c} \mu_k^{(a)} d_i n_k^{(a)} + \sum_{k=1}^{c} \mu_k^{(a)} d_c n_k^{(a)}, \quad (12.17)$$

where, on use of progress variables $\lambda_\alpha^{(a)}$ $(\alpha = 1, 2, \ldots, m;\ a = 1, 2, \ldots, m)$ for m chemical reactions, we may express the chemical reaction part of the concentration change as follows:

$$d_c n_k^{(a)} = \sum_{\alpha=1}^{m} d_c n_{k\alpha}^{(a)} = \sum_{\alpha=1}^{m} \sigma_{k\alpha} d\lambda_\alpha^{(a)}. \quad (12.18)$$

If Eq. (12.18) is substituted into the last term on the right-hand side of Eq. (12.17), it may be given a familiar form:

$$\sum_{k=1}^{c} \mu_k^{(a)} d_c n_k^{(a)} = \sum_{\alpha=1}^{m} \left[\sum_{k=1}^{c} \sigma_{k\alpha} \mu_k^{(a)} \right] d\lambda_\alpha^{(a)}$$

$$\equiv - \sum_{\alpha=1}^{m} \mathcal{A}_\alpha^{(a)} d\lambda_\alpha^{(a)}, \quad (12.19)$$

where

$$\mathcal{A}_\alpha^{(a)} = - \sum_{k=1}^{c} \sigma_{k\alpha} \mu_k^{(a)} \quad (12.20)$$

is the affinity of reaction α in phase a.

The equilibrium conditions for the reacting heterogeneous systems can be obtained from Eq. (12.17) which must satisfy the inequality

$$\sum_{a=1}^{\nu} dE^{(a)} \geq 0 \quad (12.21)$$

for the system to be in stable thermodynamic equilibrium. The procedure to obtain this is the same as for the non-reacting multiphases or the reacting single phase we have previously considered. The result is

$$T_1 = T_2 = \cdots = T_\nu,$$

$$p_1 = p_2 = \cdots = p_\nu,$$

$$\mu_1^{(1)} = \mu_1^{(2)} = \cdots = \mu_1^{(\nu)},$$
$$\mu_2^{(1)} = \mu_2^{(2)} = \cdots = \mu_2^{(\nu)},$$
$$\vdots$$
$$\mu_c^{(1)} = \mu_c^{(2)} = \cdots = \mu_c^{(\nu)},$$
$$\mathcal{A}_\alpha^{(a)} = -\sum_{k=1}^c \sigma_{k\alpha}\mu_k^{(a)} = -\sum_{k=1}^c \sigma_{k\alpha}\mu_k = 0 \quad (\alpha = 1, 2, \ldots, m).$$

$$(12.22)$$

Because of the material equilibrium conditions that equate the chemical potentials in the ν phases, the chemical potential $\mu_k^{(a)}$ may be replaced with the equilibrium chemical potential μ_k in the chemical equilibrium conditions which now read

$$\mathcal{A}_\alpha^{(a)} = -\sum_{k=1}^c \sigma_{k\alpha}\mu_k = 0 \quad (\alpha = 1, 2, \ldots, m). \qquad (12.23)$$

When compared with the equilibrium conditions for non-reacting multi-phase systems, there are m additional conditions related to chemical equilibrium which supply additional constraints on chemical potentials. The phase rule for the system under consideration is then easily found to be

$$f = c + 2 - \nu - m. \qquad (12.24)$$

This generalizes the phase rule for non-reacting multiphase systems

$$f = c + 2 - \nu, \qquad (12.25)$$

which was derived in Chapter 8.

12.4 Equilibrium Constant

The equilibrium conditions obtained earlier imply that the concentrations are no longer independent, but are constrained by one or more relations. Such constraining relations can be made more explicit, if chemical potentials previously calculated for a mixture in terms of fugacities or activities are made use of. By substituting the formula for chemical potentials into, for example, the equilibrium condition (Eq. (12.9)), we obtain

$$RT \sum_{i=1}^c \sigma_i \ln a_i = -\Delta G^0, \qquad (12.26)$$

where

$$\Delta G^0 = \sum_{i=1}^{c} \sigma_i \mu_i^0(T, p)$$
$$= (q_1 \mu_1^0 + q_2 \mu_2^0 + \cdots) - (r_1 \mu_1^0 + r_2 \mu_2^0 + \cdots)$$
$$= G_{prod}^0 - G_{reac}^0. \tag{12.27}$$

Equation (12.27) may be rearranged to the form

$$K(T, p) = \exp\left(-\frac{\Delta G^0}{RT}\right), \tag{12.28}$$

where $K(T, p)$ is called the equilibrium constant. It is defined by

$$K(T, p) = \prod_{i=1}^{c} a_i^{\sigma_i} = \frac{\prod_i a_i^{q_i}}{\prod_i a_i^{r_i}}$$
$$= \frac{a_{\mathcal{Q}_1}^{q_1} a_{\mathcal{Q}_2}^{q_2} a_{\mathcal{Q}_3}^{q_3} \cdots}{a_{\mathcal{R}_1}^{r_1} a_{\mathcal{R}_2}^{r_2} a_{\mathcal{R}_3}^{r_3} \cdots}. \tag{12.29}$$

If the system is ideal (e.g., at infinite dilution), the activity a_i is equal to mole fraction x_i and the equilibrium constant is given in terms of mole fractions:

$$K(t, p) = \prod_{i=1}^{c} x_i^{\sigma_i}. \tag{12.30}$$

Since for real gases

$$a_i = p_{fi}$$

by choice of the standard state, the equilibrium constant for a real gas chemical reaction takes the form

$$K(T) = \prod_{i=1}^{c} p_{fi}^{\sigma_i}, \tag{12.31}$$

where the equilibrium constant may be given in terms of the standard free energy change ΔG^*

$$K(T) = \exp\left(-\frac{\Delta G^*}{RT}\right) \tag{12.32}$$

with ΔG^* defined by

$$\Delta G^* = \sum_{i=1}^{c} \sigma_i \mu_i^*(T). \tag{12.33}$$

In this case, since the standard chemical potentials are dependent on the temperature only, the equilibrium constant is a function of temperature only. Since fugacity coefficients are unity for ideal gases, the fugacities are equal to the partial pressures and hence we obtain for an ideal gas reaction

$$K(T) = p_i^{\Delta\sigma} \prod_{i=1}^{c} x_i^{\sigma_i} \equiv K_0(T, p),$$

(12.34)

where $\Delta\sigma$ is the difference in stoichiometric coefficients for the reaction:

$$\Delta\sigma = \sum_{i=1}^{c} \sigma_i.$$

(12.35)

The temperature and pressure dependence of equilibrium constants is contained in ΔG^0. It is therefore possible to investigate the T and p dependence of $K(T)$ by studying the thermochemical data on ΔG^0. The standard free energy change ΔG^0 may be obtained from the data on ΔH^0 and ΔS^0. Since we may write for an isothermal process

$$\Delta G^0 = \Delta H^0 - T\,\Delta S^0,$$

(12.36)

the temperature dependence of $K(T)$ is often expressed in the form

$$K(T, p) = \exp\left(\frac{\Delta S^0}{R}\right) \exp\left(-\frac{\Delta H^0}{RT}\right),$$

(12.37)

where

$$\Delta H^0 = \sum_{i=1}^{c} \sigma_i h_i^0(T, p),$$

(12.38)

$$\Delta S^0 = \sum_{i=1}^{c} \sigma_i s_i^0(T, p).$$

(12.39)

Here ΔH^0 is called the heat of reaction and ΔS^0 the entropy change of reaction, and h_i^0 and s_i^0 are, respectively, the standard enthalpy and standard entropy of species i defined previously in Chapter 3, Section 3.7.

12.5 van't Hoff Equation

Chemical equilibria are influenced by the state of the system where chemical reactions occur. Especially, a change in temperature or pressure results in a shifting of chemical equilibrium. Here we consider the temperature dependence of equilibrium constants. To find it let us differentiate $K(T)$ with respect to T:

$$\left[\frac{\partial \ln K(T)}{\partial T}\right]_p = -\frac{1}{R}\sum_{i=1}^{c}\sigma_i\left(\frac{\partial}{\partial T}\frac{\mu_i^0}{T}\right)_p, \qquad (12.40)$$

for which we have used

$$\Delta G^0 = \sum_{i=1}^{c}\sigma_i\mu_i^0$$

in Eq. (12.28). Since

$$\left(\frac{\partial}{\partial T}\frac{\mu_i^0}{T}\right)_p = -\frac{h_i^0}{T^2},$$

by making use of Eq. (12.38) we may cast Eq. (12.40) in the form

$$\left[\frac{\partial \ln K(T)}{\partial T}\right]_p = \frac{\Delta H^0}{RT^2}. \qquad (12.41)$$

This is called the van't Hoff equation for chemical equilibrium constant. Here ΔH^0 is the heat of reaction. If the temperature dependence of ΔH^0 is known, it is possible to find the variation of $K(T)$ with respect to T. In the case of gaseous reactions, heats of reaction are measured in reference to the ideal gas state for all the components. It is the heat absorbed when the reaction proceeds in the forward direction. Note that h_i^0 is chosen for gases such that

$$h_i^0 = \mu_i^0(T,p) + Ts_i^0(T,p) = h_i^*(T), \qquad (12.42)$$

which is the enthalpy of i in the dilute ideal gas state. In the case of reactions in the liquid state ΔH^0 may be defined by either one of the following conventions:

(a) *the heat absorbed in the forward reaction when all the components are in their most stable pure state.*
(b) *the heat absorbed in the forward reaction when the solvent is pure and the other components (solutes) are in the hypothetical, infinitely dilute state.*

It is helpful to note that these conventions are rooted on the conventions taken for the standard states for thermodynamic functions, which we have adopted previously. The same standard states will be adopted when we develop the thermodynamics of solutions in subsequent chapters.

Let us pursue a little further the discussion of the temperature dependence of $K(T)$ in the case of gaseous reactions. We recall that

$$h_i^*(T) = h_{i0} + \int^T dT\, c_p^*(T)$$
$$= h_{i0} + c_{i0}^* T + \tfrac{1}{2} c_{i1}^* T^2 + \cdots$$

and

$$s_i^*(T) = s_{i0} + c_{i0}^* \ln T + c_{i1}^* T + \cdots.$$

Therefore the heat of reaction ΔH^0 may be obtained from the calorimetric data as indicated below:

$$\Delta H^0 = \Delta h^0 + \Delta c_0 T + \tfrac{1}{2}\Delta c_1 T^2 + \cdots, \qquad (12.43)$$

where

$$\Delta h^0 = \sum_{i=1}^c \sigma_i h_{i0},$$

$$\Delta c_0 = \sum_{i=1}^c \sigma_i c_{i0}^*, \quad \Delta c_1 = \sum_{i=1}^c \sigma_i c_{i1}^*, \text{ etc.}$$

Similarly, with definition

$$\Delta s^0 = \sum_{i=1}^c \sigma_i s_{i0},$$

there holds the relation

$$\Delta s^0 = \Delta s^0 + \Delta c_0 \ln T + \Delta c_1 T + \cdots. \qquad (12.44)$$

By using these relations we finally obtain the temperature dependence of $K(T)$

$$\ln K(T) = -\frac{\Delta h^0}{RT} + \frac{\Delta s^0}{R} - \frac{\Delta c_0}{R} + \frac{\Delta c_0}{R}\ln T + \frac{\Delta c_1}{2R}T + \cdots. \qquad (12.45)$$

On the other hand, by inserting the expansion (Eq. (12.43)) for ΔH^0 into Eq. (12.41) and integrating the equation we find

$$\ln K(T) = I - \frac{\Delta h^0}{RT} + \frac{\Delta c_0}{R}\ln T + \frac{\Delta c_1}{2R}T + \cdots, \qquad (12.46)$$

where I is the integration constant. Comparison of Eqs. (12.45) and (12.46) yields the integration constant in the form

$$I = \frac{1}{R}(\Delta s^0 - \Delta c_0). \qquad (12.47)$$

This result suggests that to obtain $K(T)$ at a temperature all we need is Δh^0, Δs^0, Δc^0, and so on.

If the integration constant I is eliminated with the equation for $K(T)$ at another temperature, say, T_1, then we find

$$\ln\left[\frac{K(T)}{K(T_1)}\right] = \frac{\Delta h^0}{R}\left(\frac{1}{T_1} - \frac{1}{T}\right) + \frac{\Delta c_0}{R}\ln\left(\frac{T}{T_1}\right) + \frac{\Delta c_1}{2R}(T - T_1) + \cdots.$$

(12.48)

This, of course, can be obtained directly from Eqs. (12.41) and (12.43) if Eq. (12.41) is integrated from T_1 to T.

12.6 Equilibrium Constant for Real Gases

Since chemical reactions do not necessarily occur between ideal gases or at sufficiently low pressures so that gases may be treated as ideal, it is generally useful to consider what roles the nonideality of gases play in chemical reactions. We now examine the nonideality correction for $K(T)$ in the case of gaseous reactions. For the purpose we return to Eq. (12.31). Recalling that chemical potentials for real gases may be given in terms of fugacities

$$p_{fi} = f_i p_i = x_i f_i p,$$

we separate out the ideal gas part and the nonideality correction of $K(T)$:

$$K(T) = K_0(T, p) \prod_{i=1}^{c} f_i^{\sigma_i},$$

(12.49)

where f_i is the fugacity coefficient of i

$$f_i = \exp\left[\int_0^p dp'\left(\frac{\bar{v}_i}{RT} - \frac{1}{p'}\right)\right]$$

(12.50)

and $K_0(T, p)$ is the ideal gas part of $K(T)$ and defined in Eq. (12.34). The nonideality correction may be more explicitly given if Eq. (12.50) is made use of:

$$K_{corr}(T, p) = \prod_{i=1}^{c} f_i^{\sigma_i} = \exp\left[\sum_{i=1}^{c} \sigma_i \int_0^p dp'\left(\frac{\bar{v}_i}{RT} - \frac{1}{p'}\right)\right].$$

(12.51)

Since according to Eq. (11.22) the partial molar volume $\bar{v}_i$ may be approximated, to the first order in p, by the formula

$$\bar{v}_i = \frac{RT}{p} + \sum_{j=1}^{c}\sum_{k=1}^{c} x_j x_k(2B_{ij} - B_{jk}),$$

substitution of this formula into Eq. (12.51) yields $K_{corr}(T,p)$ in the form

$$K_{corr}(T,p) = \exp\left[\sum_{i=1}^{c}\sum_{j=1}^{c}\sum_{k=1}^{c}\sigma_i x_j x_k (2B_{ij} - B_{jk})\frac{p}{RT}\right]. \qquad (12.52)$$

If we use the combination rule (Eq. (11.25)) for the second virial coefficients, the exponent in Eq. (12.52) can be considerably simplified. We thereby obtain

$$\sum_{i=1}^{c}\sum_{j=1}^{c}\sum_{k=1}^{c}\sigma_i x_j x_k (2B_{ij} - B_{jk}) = \sum_{i=1}^{c}\sigma_i B_{ii} \equiv \Delta B \qquad (12.53)$$

and hence

$$K(T,p) = K_0(T,p)\exp\left(\frac{p\Delta B}{RT}\right). \qquad (12.54)$$

This enables us to make a first-order correction to chemical equilibrium constants, when the gases participating in the chemical reaction are not ideal. It gives chemical equilibrium constants to a fair degree of accuracy up to a few tens of atmospheric pressure, but if the pressure is higher, the higher order terms in the virial equation of state must be included, or another more suitable equation of state must be used. It is straightforward to do so, if the theory of real fluids developed earlier is made use of.

Problems

(1) For chemical reaction $\frac{1}{2}N_2 + \frac{3}{2}H_2 = NH_3$ the equilibrium constant is known to be 0.0065 at 450°C. The following is available for ΔH^0 as a function of T:

$$\Delta H^0 = \left(-38.20 - 3.12 \times 10^{-2}T\right.$$
$$\left. + 1.54 \times 10^{-5}T^2 - 1.97 \times 10^{-9}T^3\right)\text{kJ mol}^{-1}.$$

Obtain the equilibrium constant as a function of T.

(2) Use the van der Waals model for the second virial coefficients B_{ii} of species in gas phase chemical reaction and express the nonlinearity correction for the equilibrium constant in terms of the van der Waals parameters.

(3) Two liquid phases are separated by a semipermeable membrane which permeates species 1 only. One phase is a solution of three species

where species 1 is the solvent and reacts with species 2 to form the third species (3) according to the reaction

$$\sigma_1 X_1 + \sigma_2 X_2 \rightleftharpoons \sigma_3 X_3.$$

Obtain the Gibbs phase rule and discuss the thermodynamics of this system, including chemical equilibrium.

Chapter 13

Thermodynamics of Solutions

The general principles developed in the previous chapters can be applied to study thermodynamic properties of liquid solutions. A liquid composed of more than one component is called a liquid solution. The predominant component is called the solvent, and the other components the solute. Henceforth, when the term solution is used in this chapter, we mean a liquid solution. We assume that the solution consists of c neutral components, which do not react with each other nor do they ionize into constituent ions. This assumption is easy to remove if the theory is developed as discussed in Chapter 12.

13.1 Chemical Potentials of Solutions

As in the case of gases, the quantity of central interest is the chemical potentials, since they are the storage of thermodynamic information of the system and other thermodynamic functions can be derived from them. For a c-component solution chemical potentials are generally functions of T, p, and $x_1, x_2, \ldots, x_{c-1}$, but their explicit forms are not completely known a priori. In effect, the aim of thermodynamics of solutions is just in finding the functional forms for chemical potentials through laboratory experiments and comparison of the results of experiments with a thermodynamic theory which one might develop for solutions of interest.

We have seen that in the thermodynamics of gases the ideal gas plays a special role and thermodynamic functions for real gases are calculated in reference to the thermodynamic functions of the ideal gas. Thus thermodynamic functions for real gases are invariably calculated such that they consist of the ideal gas part and the nonideality correction. We develop a theory of solutions in a way parallel to that of gases. We call a solution ideal in close analogy to gases, if the chemical potentials depend on the concentrations in the following manner:

$$\mu_i(T, p, x_1, \ldots, x_{c-1}) = \mu_i^0(T, p) + RT \ln x_i, \tag{13.1}$$

where $\mu_i^0(T, p)$ is the reference chemical potential of i whose value we may fix according to the conventions introduced in Chapters 9 and 11. Our experience with the ideal gas thermodynamics shows that thermodynamics will be simple for ideal solutions, but it will be only a limiting theory of a more realistic thermodynamics of real solutions in the ideal solution limit. Since Eq. (13.1) will generally deviate from the chemical potential for species i in the real solution, a correction must be made to account for the deviation arising from nonideal behaviors[1] of the constituents. It is generally introduced in the form of excess chemical potential

$$\mu_i^{(ex)} = \mu_i - \left[\mu_i^0(T, p) + RT \ln x_i\right]. \tag{13.2}$$

It is common practice to express the excess chemical potential in terms of activity coefficient f_i, which is defined as follows:

$$f_i = \exp\left(\frac{\mu_i^{(ex)}}{RT}\right) = x_i^{-1} \exp\left(\frac{\mu_i - \mu_i^0}{RT}\right). \tag{13.3}$$

The chemical potential then may be expressed in terms of the concentration and the activity coefficient f_i

$$\mu_i = \mu_i^0(T, p) + RT \ln(x_i f_i) \tag{13.4}$$

or, more simply,

$$\mu_i = \mu_i^0(T, p) + RT \ln a_i, \tag{13.5}$$

where a_i is the activity defined by

$$a_i = x_i f_i. \tag{13.6}$$

The reference values of activity coefficients are fixed in accordance with the standard states adopted for chemical potentials. We recall that there

---

[1]The ideal gas is often regarded as a gas where constituent molecules do not interact with each other. This notion of no interaction between the molecules is quite correct for ideal gases, but it is misleading in the case of ideal solutions, because the molecules do interact intimately with each other in a condensed phase. The definition of ideal solutions is based on the mathematical form of the constitutive equation for the chemical potential, for example, Eq. (13.1), where the information on molecular interactions in pure solvent is contained in the reference chemical potential $\mu_i^0(T, p)$.

are two modes of choosing the reference state, and the reference activity coefficients are determined by the limits

$$\lim_{x_i \to 1} f_i(x_1, x_2, \ldots, x_{c-1}) = 1 \quad (i = 1, 2, \ldots, c) \tag{13.7}$$

or by the limits

$$\lim_{x_1 \to 1} f_i(x_1, x_2, \ldots, x_{c-1}) = 1 \quad (i = 1, 2, \ldots, c), \tag{13.8}$$

depending on the convention for the standard state — a reference state — adopted. It is conventional to express concentrations by units other than mole fractions, such as molality or molarity, although it is most convenient to use mole fractions from the theoretical viewpoint. Since the actual values of activity coefficients can be different for different concentration units taken, it is necessary to distinguish them by different symbols. Since different concentration units are interrelated, the activity coefficients in different concentration units are also related to each other.

If the concentrations are expressed in molality m_i, we will write the chemical potential μ_i as

$$\mu_i = \mu_i^{(m)}(T, p) + RT \ln(\gamma_i m_i), \tag{13.9}$$

where γ_i is the activity coefficient and $\mu_i^{(m)}$ is the corresponding standard chemical potential. When chemical potentials are given in terms of molality, only Convention II applies, since the definition of molality necessarily designates the solvent. We therefore have

$$\mu_i^{(m)} = \mu_1^l(T, p),$$

$$\tag{13.10}$$

$$\mu_i^{(m)} = \lim_{x_i \to 0} (\mu_i - RT \ln m_i) \quad (i = 2, \ldots, c),$$

where μ_1^l is the chemical potential of pure liquid 1, and thus

$$\lim_{x_1 \to 1} \gamma_i = 1 \quad (i = 2, \ldots, c). \tag{13.11}$$

By definition, the molality is related to mole fraction x_i as follows:

$$x_i = \frac{m_i M_1}{1000 \left(1 + \frac{M_1}{1000} \sum_{j=2}^{c} m_j\right)}, \tag{13.12}$$

where M_1 is the molecular weight of the solvent and m_i is the molality of component i—namely, the mole numbers of the solutes in 1 kg of the solvent. Since the chemical potentials in the two concentration units must be the same

$$\mu_i^0 + RT \ln(x_i f_i) = \mu_i^{(m)} + RT \ln(\gamma_i m_i), \qquad (13.13)$$

if we set

$$\mu_i^{(m)} = \mu_i^0 + RT \ln\left(\frac{M_1}{1000}\right), \qquad (13.14)$$

we find by making use of Eq. (13.12) the relation between two activity coefficients f_i and γ_i:

$$\gamma_i = \frac{f_i}{1 + \frac{M_1}{1000} \sum_{j=2}^c m_j}. \qquad (13.15)$$

Therefore, if the solution is dilute, the activity coefficients γ_i and f_i are approximately equal:

$$\gamma_i \approx f_i.$$

If the molarity c_i is taken to represent the concentration, the chemical potential μ_i may be written as

$$\mu_i = \mu_i^{(c)}(T, p) + RT \ln(\alpha_i c_i), \qquad (13.16)$$

where α_i is the activity coefficient of component i in molarity units and $\mu_i^{(c)}$ is the corresponding standard chemical potential. Since there is no necessity of distinguishing the solvent and the solutes in this case, both Conventions I and II for the standard state apply, and we therefore have

$$\mu_i^{(c)} = \mu_i^l(T, p) \quad (i = 1, 2, \ldots, c). \qquad (13.17)$$

Consequently in Convention I

$$\lim_{x_1 \to 1} \alpha_i = 1 \quad (i = 1, 2, \ldots, c), \qquad (13.18)$$

and in Convention II

$$\mu_1^0 = \mu_1^l(T, p),$$
$$\qquad (13.19)$$
$$\mu_i^{(c)} = \lim_{x_1 \to 1} (\mu_i - RT \ln c_i) \quad (i = 2, \ldots, c).$$

The chemical potential must have the same value regardless of the concentration units in which it is expressed. Therefore there should hold the equation

$$\mu_i^{(c)} + RT \ln(\alpha_i c_i) = \mu_i^0 + RT \ln(x_i f_i).$$

Since c_i is related to x_i as follows

$$c_i = \frac{1000 n_i}{\sum_{j=1}^c \bar{v}_j n_j} = \frac{1000 x_i}{\sum_{j=1}^c \bar{v}_j x_j}, \tag{13.20}$$

if we set

$$\mu_i^{(c)} = \mu_i^0 + RT \ln\left(\frac{\bar{v}_1 n_1}{1000}\right), \tag{13.21}$$

there follows the relation between α_i and f_i:

$$f_i = \frac{\alpha_i}{1 + \sum_{j=2}^c (\bar{v}_j n_j / \bar{v}_1 n_1)}. \tag{13.22}$$

Therefore f_i coincides with α_i in the limit of infinite dilution, where the second term in the denominator becomes negligible.

13.2 Ideal Solutions

As an ideal gas mixture is an idealization of the real gas mixture at low pressure, an ideal solution is an idealization of real solutions. In this case, it is a fictitious representation of real solutions in the limit of infinite dilution.[2] Nevertheless, it is mathematically easy to handle and serves as a reference state of real solutions around which a thermodynamic theory of real solutions can be constructed. This approach makes it easier for us to draw analogy from the experience we have gained through our study of real gas mixtures and develop a theory quite parallel to the thermodynamics of real gas mixtures.

If Convention I is adopted for the reference state, ideal solutions may be defined as solutions whose chemical potentials are given by the formula

$$\mu_i = \mu_i^l(T, p) + RT \ln x_i \quad (i = 1, 2, \ldots, c). \tag{13.23}$$

They are sometimes defined as solutions for which Raoult's law holds. As will be seen shortly, these two definitions are equivalent. The expression

---

[2]This notion is a little limited. Generally, ideal solutions include solutions of liquids of structurally similar molecules interacting similarly, as will be seen later.

for chemical potentials is similar to the formula for chemical potentials of ideal gas mixtures except for the difference in the reference state.

If the chemical potentials for $(c-1)$ components in a c-component solution are given by Eq. (13.23), the chemical potential for the cth component must be also in the same form as Eq. (13.23). Therefore, if we have a binary solution, it is not possible to treat only one component as an ideal component obeying Eq. (13.23): both components must be treated on the same footing. To prove this statement we use the Gibbs–Duhem equation at constant T and p

$$\sum_{i=1}^{c} x_i d\mu_i = 0. \tag{13.24}$$

Since Eq. (13.23) holds for $(c-1)$ components we obtain from Eq. (13.23)

$$d\mu_i = \frac{RT}{x_i} dx_i \ (i = 1, 2, \ldots, c-1),$$

where T and p are kept constant. Substituting it into Eq. (13.24) and making use of

$$\sum_{i=1}^{c} dx_i = 0,$$

we obtain the differential equation

$$d\mu_c = \frac{RT}{x_c} dx_c.$$

Integration of this differential equation yields

$$\mu_c = C + RT \ln c,$$

where C is the integration constant. By using Convention II for chemical potentials we find

$$C = \mu_c^l(T, p)$$

and thus

$$\mu_c = \mu_c^l(T, p) + RT \ln x_c,$$

which proves the statement.

Ideal solutions have a number of characteristic properties that remind us of the properties of ideal gas mixtures. These properties in practice may be used collectively as an indicator of the ideality of a solution. We list them in the following.

(1) *An ideal solution has no heat of mixing.*
As in the case of gases, the heat of mixing for a solution is defined as

$$\Delta h = \sum_{i=1}^{c} x_i(\bar{h}_i - h_i^l), \tag{13.25}$$

where h_i^l is the molar enthalpy of pure component i. The partial molar enthalpy may be calculated with Eq. (13.23). Since

$$\bar{h}_i = -T^2\left(\frac{\partial}{\partial T}\frac{\mu_i}{T}\right)_{p,x}, \tag{13.26}$$

substitution of Eq. (13.23) yields

$$\bar{h}_i = -T^2\left(\frac{\partial}{\partial T}\frac{\mu_i^l}{T}\right)_p = h_i^l, \tag{13.27}$$

which implies that

$$\Delta h = 0. \tag{13.28}$$

(2) *There is no volume change on mixing for ideal solutions.*
The volume change of mixing is defined by

$$\Delta v = \sum_{i=1}^{c} x_i(\bar{v}_i - v_i^l), \tag{13.29}$$

where v_i^l is the molar volume of liquid component i. Since the partial molar volume is given by

$$\bar{v}_i = \left(\frac{\partial \mu_i}{\partial p}\right)_{T,x}, \tag{13.30}$$

substitution of Eq. (13.23) yields

$$\bar{v}_i = \left(\frac{\partial \mu_i^l}{\partial p}\right)_T = v_i^l, \tag{13.31}$$

which implies that

$$\Delta v = 0. \tag{13.32}$$

(3) *The entropy of mixing is in the same form as for an r-component ideal gas mixture.*

The entropy of mixing for a solution is defined by

$$\Delta s = \sum_{i=1}^{r} x_i \left(\bar{s}_i - s_i^l \right), \tag{13.33}$$

where s_i^l is the molar entropy of pure liquid component i. Since the partial molar entropy of component i is

$$\bar{s}_i = -\left(\frac{\partial \mu_i}{\partial T} \right)_{p,x}, \tag{13.34}$$

substitution of Eq. (13.23) into the right-hand side yields

$$\bar{s}_i = s_i^l - R \ln x_i, \tag{13.35}$$

where

$$\bar{s}_i = -\left(\frac{\partial \mu_i^l}{\partial T} \right)_p. \tag{13.36}$$

Therefore Eq. (13.35) implies

$$\Delta s = -R \sum_{i=1}^{c} x_i \ln x_i, \tag{13.37}$$

which is in exactly the same form as for the entropy of mixing for an r-component ideal gas mixture. Note that since $x_i \leq 1$ the entropy of mixing is always positive. That is, the entropy always increases on mixing two or more components. This conclusion can be drawn only for ideal solutions, because for real solutions the nonideality correction may be such that Δs may not be positive in the whole range of x_i. The non-positivity, in any case, does not mean that the second law is broken by the mixing process; the reason for this statement was given in Chapter 4.

13.3 Raoult's Law

The vapor pressure of a solution reflects the properties of the solution. In fact, the vapor pressure is in a well defined, simple relationship with the composition of the solution if the solution is ideal. This relation is the content of Raoult's law. If a solution is in equilibrium with its vapor at a given T and p, by the equilibrium condition (Eq. (8.22)) there holds the relation

$$\mu_i = \mu_i^v, \tag{13.38}$$

where μ_i^v is the chemical potential of species i in the vapor. It is generally given by

$$\mu_i^v = \mu_i^*(T) + RT \ln p_{fi}. \tag{13.39}$$

By substituting this equation and Eq. (13.23) into Eq. (13.38) and rearranging the terms, we obtain for the fugacity the following expression:

$$p_{fi} = x_i \exp\left[\frac{\mu_i^l(T,p) - \mu_i^*(T)}{RT}\right]. \tag{13.40}$$

If x_i is put equal to unity then the solution consists of a single component, and the vapor of the same component only. In that case, p_{fi} becomes the fugacity p_{fi}^0 of pure component i. That is, as x_i tends to unity, the left-hand side becomes the fugacity of pure vapor i:

$$\exp\left[\frac{\mu_i^l(T,p) - \mu_i^*(T)}{RT}\right] = p_{fi}^0. \tag{13.41}$$

Thus, when Eqs. (13.40) and (13.41) are combined, there follows the relation

$$p_{fi} = x_i p_{fi}^0. \tag{13.42}$$

This is Raoult's law. If the vapor is ideal then the fugacity is equal to the pressure, namely,

$$p_{fi} = p_i; \quad p_{fi}^0 = p_i^0, \tag{13.43}$$

where p_i^0 is the vapor pressure of pure component i, and consequently we obtain Raoult's law for the ideal vapor:

$$p_i = x_i p_i^0. \tag{13.44}$$

As an application of Raoult's law, we now consider the case of a binary ideal solution in equilibrium with its vapor. An example is a solution of 2-methyl propanol-1 and propanol-2. This system is almost ideal. Let us designate the solvent as component 1 and the solute as component 2. The mole fractions in the solution will be denoted by x_i, and the mole fractions in the vapor by y_i. We assume that the vapor is ideal. Then according to Raoult's law

$$p_i = p_i^0 x_i.$$

The total pressure is

$$p = p_1 + p_2.$$

If the Raoult's law is used, it is given by

$$p = p_1^0 + (p_2^0 - p_1^0)x_2. \tag{13.45}$$

Since according to the Gibbs–Dalton law

$$p_i = y_i p, \tag{13.46}$$

it is possible to express the total pressure in terms of the mole fraction of the solute in the vapor, y_2, as follows:

$$(p_2^0 - p_1^0)x_2 = \frac{(p_2^0 - p_1^0)x_2 p_2^0}{p_2^0} = \frac{(p_2^0 - p_1^0)p_2}{p_2^0}$$
$$= \frac{(p_2^0 - p_1^0)y_2 p}{p_2^0}.$$

Substituting it into Eq. (13.45) and solving the resulting equation for p, we obtain

$$p = \frac{p_1^0 p_2^0}{p_2^0 + (p_1^0 - p_2^0)y_2}. \tag{13.47}$$

A phase diagram for the two phase binary solution system is given in Fig. 13.1. The total vapor pressure p is linear with respect to x_2 according to Eq. (13.45), whereas it is concave upward with respect to y_2 as shown in Fig. 13.1. As the pressure is reduced to p_s from p, the solution attains an equilibrium with its vapor at the composition corresponding to p_v. As the pressure is further reduced below p', the solution is completely vaporized and the system attains a single phase consisting of vapor. This phase diagram serves as a theoretical and thermodynamic basis of fractional distillation of a liquid mixture. Suppose there is a two-component liquid solution at pressure p and concentration x_2, as shown in Fig. 13.1. As the pressure is lowered to p_s from some point, say, p, the state of two-phase equilibrium is reached at p_s at which point the vapor phase acquires the mole fraction y_2. Collect the vapor at y_2 and condense it to the solution at another concentration x_2' which is in equilibrium with the vapor phase at concentration y_2'. This process can be continued until the liquid phase acquires a desired composition. This process is called fractional distillation.

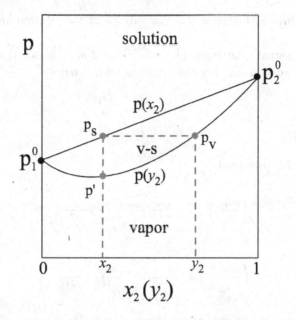

Fig. 13.1   A phase diagram for a two-component system in $p-x(y)$ plane.

13.4   Two-Component, Two-Phase Equilibrium Reconsidered

In Chapter 8 we have considered phase equilibria between two phases composed of a binary mixture and derived a set of differential equations comparable to the Clapeyron equation. In particular, we have derived a pair of differential equations for concentration dependences of vapor pressure. Integration of the differential equations for real solutions is rather complex and cannot be generally performed analytically, even if the chemical potentials are completely known as a function of concentrations. However, it is possible to integrate them analytically for ideal solutions. In this section, we consider a binary ideal solution which is in equilibrium with its vapor. Since the governing equations can be solved analytically, it makes a good illustrative example for application of the theory of two-phase equilibria of a binary mixture developed in Chapter 8. Therefore we take up the same subject discussed in the previous section and derive the vapor pressure equations (Eqs. (13.45) and (13.47)) by integrating Eq. (8.72). The object under consideration here is two phases of a binary solution and vapor which are in equilibrium at a given T. The vapor and the solution are assumed to be ideal.

Let us designate the vapor phase as phase 2 and the solution phase as phase 1. In order to simplify the notation we will write

$$x_1 = x_1^{(1)}, \quad x_2 = x_2^{(1)},$$
$$y_1 = x_1^{(2)}, \quad y_2 = x_2^{(2)}.$$

Equation (8.72) now reads

$$\left(\frac{\partial p}{\partial y_2}\right)_T = -\frac{\sum_{i=1}^2 x_i \phi_i^{(2)}}{\sum_{i=1}^2 x_i (\bar{v}_i^{(2)} - \bar{v}_i^{(1)})},$$

$$\left(\frac{\partial p}{\partial x_2}\right)_T = -\frac{\sum_{i=1}^2 y_i \phi_i^{(1)}}{\sum_{i=1}^2 y_i (\bar{v}_i^{(1)} - \bar{v}_i^{(2)})}.$$

(13.48)

Since the vapor is assumed ideal, the partial molar volumes of the vapor are equal to the molar volumes, which are given by

$$\bar{v}_1^{(2)} = \frac{RT}{p} \quad \text{and} \quad \bar{v}_2^{(2)} = \frac{RT}{p}.$$

Moreover, it is reasonable to assume

$$\bar{v}_i^{(2)} \gg \bar{v}_i^{(1)} \quad (i = 1, 2).$$

Under these approximations the denominators in Eq. (13.48) are very much simplified:

$$\sum_{i=1}^2 x_i(\bar{v}_i^{(2)} - \bar{v}_i^{(1)}) \simeq \sum_{i=1}^2 x_i \bar{v}_i^{(2)} = \frac{RT}{p}\sum_{i=1}^2 x_i = \frac{RT}{p},$$

(13.49)

$$\sum_{i=1}^2 y_i(\bar{v}_i^{(1)} - \bar{v}_i^{(2)}) \simeq -\sum_{i=1}^2 y_i \bar{v}_i^{(2)} = -\frac{RT}{p}\sum_{i=1}^2 y_i = -\frac{RT}{p}.$$

Since both the solution and vapor are ideal, the chemical potentials are those of ideal solutions and gases. By using such chemical potentials and the definition (Eq. (8.59)) for $\phi_i^{(\alpha)}$ we obtain

$$\phi_1^{(1)} = -\frac{RT}{x_1}, \quad \phi_2^{(1)} = \frac{RT}{x_2},$$

$$\phi_1^{(2)} = -\frac{RT}{y_1}, \quad \phi_2^{(2)} = \frac{RT}{y_2}.$$

(13.50)

When these results are used, the numerators in Eq. (13.48) take the following simple forms:

$$\sum_{i=1}^{2} x_i \phi_i^{(2)} = \frac{RT(x_2 - y_2)}{y_1 y_2},$$

(13.51)

$$\sum_{i=1}^{2} y_i \phi_i^{(1)} = -\frac{RT(x_2 - y_2)}{x_1 x_2}.$$

On substitution of Eqs. (13.49) and (13.50) into Eq. (13.48) there follows a pair of relatively simple pressure–concentration equations

$$\left(\frac{\partial p}{\partial x_2}\right)_T = -p\frac{x_2 - y_2}{x_1 x_2},$$

(13.52)

$$\left(\frac{\partial p}{\partial y_2}\right)_T = -p\frac{x_2 - y_2}{y_1 y_2}.$$

When these two equations are added side by side, there follows a partial differential equation for pressure

$$x_1 x_2 \left(\frac{\partial p}{\partial x_2}\right)_T - y_1 y_2 \left(\frac{\partial p}{\partial y_2}\right)_T = 0.$$

(13.53)

The characteristic equations[3] for this partial differential equation are

$$\frac{dx_2}{dt} = x_1 x_2,$$

$$\frac{dy_2}{dt} = -y_1 y_2,$$

(13.54)

$$\frac{dp}{dt} = 0,$$

---

[3] See, for example, F. John, *Partial Differential Equations* (Springer, Berlin, 1978).

where t is a parameter. The first two equations imply the ordinary differential equations, which may be combined to the equation

$$\frac{dx_2}{x_2(1-x_2)} = -\frac{dy_2}{y_2(1-y_2)}. \tag{13.55}$$

This equation is easily integrated to yield the solution

$$C\frac{x_2}{1-x_2} = \left(\frac{y_2}{1-y_2}\right)^{-1}, \tag{13.56}$$

where C is the integration constant. Solving this equation for y_2 we find

$$y_2 = \frac{Cx_2}{1+(C-1)x_2}. \tag{13.57}$$

By integrating the third characteristic equation we trivially find

$$p = A, \tag{13.58}$$

where A is an integration constant. It may be regarded as a function of x_2 alone, because of the relation (Eq. (13.57)) between x_2 and y_2. Differentiating it with x_2 we find

$$\frac{dA}{dx_2} = \frac{dp}{dx_2} = -\frac{x_2-y_2}{x_1x_2}p, \tag{13.59}$$

for which we have made use of the first equation in Eq. (13.52). Eliminating y_2 on the right-hand side of this equation by using Eq. (13.57), we obtain

$$\frac{dp}{dx_2} = \frac{(C-1)p}{1+(C-1)x_2}.$$

Integrating this equation we finally obtain

$$p = B[1+(C-1)x_2], \tag{13.60}$$

where B is the integration constant. To fix the integration constants we consider the case of $x_2 = 0$. Then the pressure will be equal to p_1^0, since the system consists of component 1 alone in that case. We find

$$B = p_1^0. \tag{13.61}$$

If $x_2 = 1$, then $p = p_2^0$, which implies

$$C = \frac{p_2^0}{p_1^0}. \tag{13.62}$$

Eliminating B and C from Eq. (13.60) with Eqs. (13.61) and (13.62) we finally obtain the following formula for p:

$$p = p_1^0 \left[1 + \frac{(p_2^0 - p_1^0) x_2}{p_1^0} \right]. \tag{13.63}$$

Replacing C in Eq. (13.57) with Eq. (13.62) and inverting the relationship of y_2 and x_2 we obtain

$$x_2 = \frac{y_2 p_1^0}{p_2^0 + (p_1^0 - p_2^0) y_2}.$$

When this is substituted into Eq. (13.63), there follows the formula for $p(y_2)$:

$$p(y_2) = \frac{p_1^0 p_2^0}{p_2^0 + (p_1^0 - p_2^0) y_2}. \tag{13.64}$$

These are the results we are looking for. We see that Eqs. (13.63) and (13.64) are precisely what we have obtained in the previous section, namely, Eqs. (13.45) and (13.47), by using a more elementary method. To be sure, the present method is more complicated and circumspect, but its aim is to illustrate a method of solution for Eq. (8.72) or (8.71) which must be solved by one means or another for the pressure–concentration relations of real solutions. In the case of real solutions, the principle and method are the same, but mathematically more complicated because of the more complicated concentration dependence of chemical potentials. Such equations may be solved numerically on a computer.

13.5 Margules Expansions

The thermodynamics of real solutions requires activity coefficients expressed as functions of concentrations. Activity coefficients for binary solutions may be expanded in power series in mole fractions:

$$RT \ln f_1 = A_1 x_2 + B_1 x_2^2 + C_1 x_2^3 + \cdots,$$

$$\tag{13.65}$$

$$RT \ln f_2 = A_2 x_1 + B_2 x_1^2 + C_2 x_1^3 + \cdots.$$

These expansions are equivalent to the virial expansion for a gas and are called the Margules expansions. The expansion coefficients are not all independent since the chemical potentials must satisfy the Gibbs–Duhem equation. Therefore it is possible to eliminate dependent coefficients by making use of the Gibbs–Duhem equation. If T and p are kept constant, the Gibbs–Duhem equation for a binary solution may be written as

$$RTx_1 d\ln f_1 + RTx_2 d\ln f_2 = 0, \tag{13.66}$$

when Eq. (13.4) is made use of for the chemical potentials. By using the identity $x_1 + x_2 = 1$ and substituting the Margules expansions into the aforementioned equation we obtain the equation

$$A_2 + (2B_2 - A_2)x_1 + (3C_2 - 2B_2)x_1^2 - 3C_2 x_1^3 + \cdots \tag{13.67}$$
$$= (A_1 + 2B_1 + 3C_1)x_1 - (2B_1 + 6C_1)x_1^2 + 3C_1 x_1^3 + \cdots.$$

Comparison of the terms of like power on both sides of the equation yields the following set of algebraic equations from the first order through the third in x_1

$$A_2 = 0,$$
$$2B_2 - A_2 = A_1 + 2B_1 + 3C_1,$$
$$3C_2 - 2B_2 = -(2B_1 + 6C_1),$$
$$C_2 = -C_1,$$
$$\vdots$$

When these equations are solved, there follows the following relations between coefficients:

$$A_1 = A_2 = 0,$$
$$B_2 = B_1 + \tfrac{3}{2}C_1,$$
$$C_2 = -C_1, \tag{13.68}$$
$$\vdots$$

Therefore the Margules expansions now take the forms

$$RT \ln f_1 = B_1 x_2^2 + C_1 x_2^3 + \cdots,$$

$$RT \ln f_2 = \left(B_1 + \tfrac{3}{2}C_1\right) x_1^2 - C_1 x_1^3 + \cdots. \tag{13.69}$$

Notice that the expansions are consistent with Convention I which we have chosen for the reference states of chemical potentials; see Sec. 11.4. The coefficients can be determined by measuring various properties of the solution. They are generally dependent on temperature.

13.6 Regular Solutions

Statistical mechanical considerations show that if a mixture of two liquids were to behave ideally, the two liquids must consist of similar molecules. The forces acting on any molecule are then quite similar to the forces in the pure liquids. Under these conditions the partial vapor pressure or the fugacity of each component, which is a measure of its tendency to escape from the solution, is expected to obey Raoult's law. Such is actually the case for some solutions, since liquid solutions known to behave ideally consist of similar molecules, for example, ethylene bromide and propylene bromide, *n*-hexane and *n*-heptane, *n*-butyl chloride and bromide, ethyl bromide and iodide, benzene and toluene, and so on. If the constituents of a mixture differ appreciably in their nature, a deviation from ideal behavior is generally the rule. For example, the vapor pressures show either positive or negative deviations from Raoult's law for such solutions. Here we discuss a special class of nonideal solutions before we take up the subject of real solutions.

In 1929, J. H. Hildebrand observed that there is a great similarity in the thermodynamic behavior of a class of nonideal solutions which are characterized by the absence of hydrogen bonding and acid–base association. He called them regular solutions. Regular solutions differ from ideal solutions in that the intermolecular forces between different components in the solution are no longer similar and molecules may be unequal unlike in the case of ideal solutions. Nevertheless, the differences are sufficiently small so that the entropy of mixing is nearly that of ideal solutions. We shall therefore define regular solutions as follows: *a solution is called regular if the partial molar entropies of components in the solution are those of ideal solutions*

$$\bar{s}_i = s_i^l - R \ln x_i. \tag{13.70}$$

This implies that for regular solutions the entropy of mixing is that of ideal solutions, since

$$\Delta s = \sum_{i=1}^{c} x_i(\bar{s}_i - s_i^l) = -R \sum_{i=1}^{c} x_i \ln x_i. \tag{13.71}$$

This definition of regular solutions gives rise to a number of characteristics. We list them here.

(1) *The partial molar specific heat of component i is equal to the molar specific heat of pure liquid i:*

$$\bar{c}_{pi} = c_{pi}^l.$$

This follows from Eq. (13.70) since

$$\bar{c}_{pi} = T\left(\frac{\partial \bar{s}_i}{\partial T}\right)_{p,x} = T\left(\frac{\partial s_i^l}{\partial T}\right)_p = c_{pi}^l.$$

(2) *For regular solutions the activity coefficients are given in terms of enthalpies.*

This is in contrast to real solutions for which the chemical potentials are given in terms of activity coefficients. Since the chemical potentials for pure liquids may be written as

$$\mu_i^l = h_i^l - Ts_i^l,$$

we find

$$\mu_i = \bar{h}_i - T\bar{s}_i$$
$$= \mu_i^l + RT \ln x_i + \bar{h}_i - h_i^l. \tag{13.72}$$

When this is compared with Eq. (13.4) in an appropriate convention for the reference state, we identify

$$f_i = \exp\left(\frac{\bar{h}_i - h_i^l}{RT}\right). \tag{13.73}$$

This equation suggests that it is possible to calculate the activity coefficients from the calorimetric data on the solution, if the solution is regular. We will elaborate on this point a little later.

(3) *The partial molar volumes of regular solutions are given by the pressure derivatives of* $\bar{h}_i - h_i^l$.

From Eq. (13.72) we find

$$\bar{v}_i - v_i^l = \left(\frac{\partial \mu_i}{\partial p}\right)_{T,x} - \left(\frac{\partial \mu_i^l}{\partial p}\right)_T$$
$$= \left[\frac{\partial(\bar{h}_i - h_i^l)}{\partial p}\right]_{T,x}. \tag{13.74}$$

Therefore the volume change of mixing is given by the formula

$$\Delta v = \sum_{i=1}^c x_i(\bar{v}_i - v_i)$$
$$= \left(\frac{\partial \, \Delta h}{\partial p}\right)_T$$
$$= RT \sum_{i=1}^c x_i \left(\frac{\partial}{\partial p} \ln f_i\right)_{T,x}, \tag{13.75}$$

where Δh is the heat of mixing.

Differentiating Eq. (13.74) with respect to T we find

$$\left(\frac{\partial \bar{v}_i}{\partial T}\right)_{p,x_i} - \left(\frac{\partial v_i^l}{\partial T}\right)_p = \frac{\partial}{\partial T}\left[\frac{\partial}{\partial p}(\bar{h}_i - h_i^l)\right]_{T,x}$$

$$= \frac{\partial}{\partial p}(\bar{c}_{pi} - c_{pi}^l)$$

$$= 0, \tag{13.76}$$

the last equality being due to the property 1 listed above. Therefore we obtain

$$\left(\frac{\partial \, \Delta v}{\partial T}\right)_{p,x} = 0 \tag{13.77}$$

for regular solutions.

For regular solutions the Margules expansions become expansions for partial molar enthalpies. Because of the relation (Eq. (13.73)) between the activity coefficients and partial molar enthalpies, we obtain

$$\bar{h}_1 - h_1^l = B_1 x_2^2 + C_1 x_2^3 + \cdots, \tag{13.78}$$

$$\bar{h}_2 - h_2^l = (B_1 + \tfrac{3}{2}C_1)x_1^2 - C_1 x_1^3 + \cdots.$$

Therefore it is possible to determine the coefficients from calorimetric data of the solution in the case of a regular solution. If the solution is dilute then it is possible to neglect the cubic or higher terms in Eq. (13.78) in which case we may write

$$\bar{h}_1 - h_1^l = B_1 x_2^2, \qquad \bar{h}_2 - h_2^l = B_1 x_1^2. \tag{13.79}$$

If these approximations are used for calculating the heat of mixing for a binary regular solution, there follows a rather simple formula for the heat of mixing

$$\Delta h = \sum_{i=1}^{2} x_i(\bar{h}_i - h_i^l) = B_1 x_1 x_2. \tag{13.80}$$

To this approximation, B_1 therefore is seen as four times the equimolar heat of mixing:

$$B_1 = 4 \, \Delta h \left(x_1 = x_2 = \tfrac{1}{2}\right). \tag{13.81}$$

This clearly provides a way to measure B_1, if C_1 is negligible.

13.7   Real Solutions

Most solutions are neither ideal nor regular, and thus possess none of the simplifying features of ideal and regular solutions. Here we study real solutions. We shall pay particular attention to dilute real solutions because of its relative simplicity which, nevertheless, does not lose the essential features of real solutions.

If two or more constituents markedly different in their nature are mixed together, the solution shows noticeable deviations from the ideal behavior if the concentrations of the solutes are finite. For example, the vapor pressures show either positive or negative deviations. If the deviation is positive, the partial pressure or fugacity of each constituent is greater than it should be if Raoult's law were obeyed, and if the deviation is negative, it is just the opposite. It has been experimentally observed that as the mole fraction of a given constituent of a solution tends to unity, the fugacity of the constituent approaches the value for an ideal solution. That is, in a dilute solution the behavior of the solvent approaches that required by Raoult's law, although it may depart markedly from the ideal behavior at higher concentrations. Since the approach to the ideal behavior is asymptotic, it is not expected that Raoult's law would hold for the solute as well, unless the system as a whole exhibits no deviation from the ideal behavior over the whole range of composition. Although the solute in the dilute solution does not necessarily obey Raoult's law, as $x_1 \to 1$, it does conform to the simple relation

$$p_{f2} = kx_2, \tag{13.82}$$

where k is a constant. This constant becomes the fugacity p_{f2}^0 of the pure solute if the solution is ideal. The aforementioned expression embodies Henry's law for the solutes. We may thus define *a dilute solution as one in which the solvent obeys Raoult's law whereas the solutes obey Henry's law.* It is possible to see that Eq. (13.82) must hold for the solute. We consider it for a binary solution.

At constant T and p the Gibbs–Duhem equation for the solution is

$$x_1 \, d\mu_1 + x_2 \, d\mu_2 = 0, \tag{13.83}$$

where x_1 and x_2 are the mole fractions of the solvent and the solute. Since the chemical potentials of the species in the vapor phase are given by

$$\mu_1^v = \mu_1^*(T) + RT \ln p_{f1},$$

$$\tag{13.84}$$

$$\mu_2^v = \mu_2^*(T) + RT \ln p_{f2},$$

and by the equilibrium conditions

$$\mu_1 = \mu_1'' \quad \text{and} \quad \mu_2 = \mu_2^v,$$

the Gibbs–Duhem equation (Eq. (13.83)) may be written as

$$\frac{\partial \ln p_{f1}}{\partial \ln x_1} = \frac{\partial \ln p_{f2}}{\partial \ln x_2}. \tag{13.85}$$

If Raoult's law holds for the solvent, namely, component 1, then

$$\frac{\partial \ln p_{f1}}{\partial \ln x_1} = 1$$

and there follows from Eq. (13.85) the equation

$$d \ln p_{f2} = d \ln x_2. \tag{13.86}$$

Integration of this differential equation yields

$$p_{f2} = k x_2, \tag{13.87}$$

where k is a constant. This is Henry's law for the solute. Therefore we see that Henry's law is the necessary condition for the solvent to show the ideality.

For binary solutions the Margules expansions still can be applied. If the solution is dilute the expansions may be truncated; for example, they may be limited to the first term:

$$RT \ln f_1 = B x_2^2, \quad RT \ln f_2 = B x_1^2, \tag{13.88}$$

where we have dropped the subscript from the coefficient B. As an improvement of the treatment of binary solutions by van Laar, Scatchard proposed the following expressions for the activity coefficients:

$$RT \ln f_1 = A v_1 \phi_2^2, \quad RT \ln f_2 = A v_2 \phi_1^2, \tag{13.89}$$

where ϕ_i is the volume fraction of species i

$$\phi_i = \frac{n_i v_i}{n_1 v_1 + n_2 v_2} (i = 1, 2), \tag{13.90}$$

where v_i are the molar volumes of pure liquids and A is a constant parameter. The Scatchard equations correlate quite well with the experimental data for many binary solutions of normal liquids. Notice that the Scatchard equations are expansions of activity coefficients in volume fractions, whereas the Margules expansions are expansions in mole fractions. In any case, since the expansions give excess chemical potentials, thermodynamics of real solutions ultimately amounts to measuring activity coefficients as a function of temperature, pressure, and concentrations in appropriate units. Scatchard equations turn out to be appropriate to use for macromolecular solutions, because mole fractions are a misleading measure of concentration effects for thermodynamic properties of such solutions. The reason is that macromolecules have a large number of monomer units which contribute to thermodynamic properties as if they are independent molecules. Therefore, if there are molecules with large molecular weights it is more appropriate to use volume fractions instead of mole fractions to measure deviations from the ideal behavior (Figs. 13.2 and 13.3).

If one activity coefficient of a binary solution is known, the other activity coefficient can be obtained from it by making use of the Gibbs–Duhem equation. If T and p are kept constant there follows from Eqs. (13.84) and (13.85) the equation

$$x_1 d\ln f_1 = -x_2 d\ln f_2. \tag{13.91}$$

Integration of this equation gives either f_1 or f_2 in terms of the other. We first find that one of the Margules expansions implies the other. Since

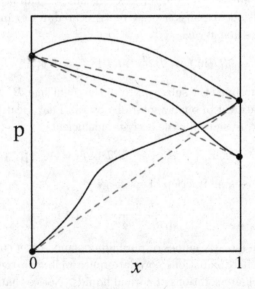

Fig. 13.2   Positive deviation from the ideal behavior.

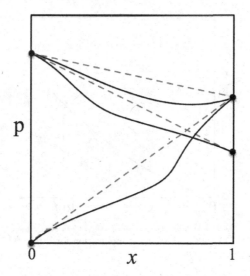

Fig. 13.3 Negative deviation from the ideal behavior.

$f_2 \to 1$ as $x_2 \to 1$ according to the convention adopted for activity coefficients (Convention I), integrating the aforementioned equation yields

$$\ln f_2 = - \int_1^{x_2} dx_2 \frac{x_1}{x_2} \frac{\partial \ln f_1}{\partial x_2} = \int_0^{1-x_2} dx_1 \frac{x_1}{x_2} \frac{\partial \ln f_1}{\partial x_2}. \qquad (13.92)$$

Substitution of the Margules expansion for $\ln f_1$ into the right-hand side and integration of the resulting series yield the series

$$RT \ln f_2 = \left(B_1 + \tfrac{3}{2} C_1 \right) x_1^2 - C_1 x_1^3 + \cdots. \qquad (13.93)$$

This is the other expansion of Eq. (13.69). The reverse is also true. If the information on the solute is more readily available, the procedure used here may be reversed and we find

$$\ln f_1 = - \int_0^{x_2} \frac{x_1}{x_2} d \ln f_2. \qquad (13.94)$$

If an analytic formula for $\ln f_2$ is not available, the right-hand side can be graphically integrated by plotting x_2/x_1 against $\ln f_2$ as in Fig. 13.4 and then measuring the area under the curve.

We now examine the entropy of mixing for real solutions with the example of a binary real solution. We will use the Margules expansions for the activity coefficients.

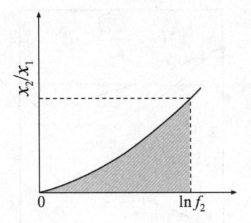

Fig. 13.4   Graphical integration for $\ln f_2$.

The chemical potentials are then given by the formulas

$$\mu_1 = \mu_1^l(T,p) + RT\ln x_1 + B_1 x_2^2 + C_1 x_2^3 + \cdots,$$
$$\mu_2 = \mu_2^l(T,p) + RT\ln x_2 + B_2 x_1^2 + C_2 x_1^3 + \cdots, \tag{13.95}$$

where B_2 and C_2 are related to B_1 and C_1 as in Eq. (13.68). The coefficients are functions of T and p in general. Here Convention I is used for the chemical potentials. The partial molar entropies are then given by

$$\bar{s}_1 = s_1^l - R\ln x_1 - \left(\frac{\partial B_1}{\partial T}\right)_p x_2^2 - \left(\frac{\partial C_1}{\partial T}\right)_p x_2^3 - \cdots,$$
$$\bar{s}_2 = s_2^l - R\ln x_2 - \left(\frac{\partial B_2}{\partial T}\right)_p x_1^2 - \left(\frac{\partial C_2}{\partial T}\right)_p x_1^3 - \cdots. \tag{13.96}$$

Therefore the entropy of mixing for the binary solution is

$$\Delta s = -R\sum_{i=1}^{2} x_i \ln x_i - \left(\frac{\partial B_1}{\partial T}\right)_p x_1 x_2$$
$$- \tfrac{1}{2}\left(\frac{\partial C_1}{\partial T}\right)_p^{\bullet} (1 + x_2)\, x_1 x_2 + \cdots \tag{13.97}$$

The first term on the right hand side is positive semidefinite, but the rest of the terms is not positive unless the derivatives of B_1 and C_1 are all negative. Therefore, unless the aforementioned derivatives are all negative or the first term on the right hand side is larger than the rest of the terms, the entropy of mixing for an isolated system cannot be concluded to be positive semidefinite, as is for the case of ideal solutions formed in the same condition. The terms involving the derivatives simply give a measure of the temperature effect. This analysis amplifies the statement made in Chapter 11 and Sec. 13.2 in connection with the entropy of mixing and the second law of thermodynamics which is often illustrated with the positive entropy of mixing.

13.8 Osmotic Coefficient of Bjerrum

Since Eq. (13.94) is not in a convenient form to integrate numerically because of the singular behavior of x_2^{-1}, it is useful to cast it into a more suitable form. This can be achieved by using the osmotic coefficient introduced by Bjerrum. It is defined by

$$\phi = -r^{-1}\ln a_1, \tag{13.98}$$

where

$$r = \frac{x_2}{x_1}. \tag{13.99}$$

Rearrangement of Eq. (13.98) and taking a differential yields

$$d\ln a_1 = -rd\phi - \phi dr = -rd\phi - r\phi d\ln r. \tag{13.100}$$

The Gibbs–Duhem equation for a two-component system at constant T and p is

$$x_2 d\ln a_2 = -x_1 d\ln a_1. \tag{13.101}$$

By combining Eqs. (13.100) and (13.101) we find a differential form for a_2:

$$d\ln a_2 = d\phi + \phi d\ln r. \tag{13.102}$$

Since

$$a_2 = x_2 f_2$$

and, as $x_2 \to 0$,

$$f_2 \to 1 \text{ and } x_1 \to 1,$$

we find

$$\frac{a_2}{r} = \frac{x_2 f_2}{x_2} x_1 = x_1 f_2 \to 1. \tag{13.103}$$

Therefore it is useful to recast Eq. (13.102) in the form

$$d\ln\left(\frac{a_2}{r}\right) = d\phi + (\phi - 1)d\ln r. \tag{13.104}$$

When this equation is integrated from $r = 0$ to an arbitrary value of r, we find

$$\ln\left(\frac{a_2}{r}\right) = \phi(r) - \phi(0) + \int_0^r dr \frac{\phi - 1}{r}, \tag{13.105}$$

for which we have made use of the equality

$$\ln\left(\frac{a_2}{r}\right)\bigg|_{r=0} = 0.$$

This holds since $a_2 \to x_2$ as $r \to 0$, namely, as $x_2 \to 0$. We now examine the behavior of ϕ as $r \to 0$. Since we have

$$RT\ln f_1 = Bx_2^2 + \cdots$$

and

$$\ln x_1 = \ln\left(\frac{x_1}{x_1 + x_2}\right) = -\ln(1 + r),$$

the osmotic coefficient may be put in the form

$$\phi = -r^{-1}\ln a_1$$
$$= r^{-1}\ln(1 + r) + k'x_1x_2 + \cdots, \tag{13.106}$$

where

$$k' = -\frac{B}{RT}.$$

It is also possible to write

$$x_1x_2 = rx_1^2$$
$$= r(1 + r)^{-2}$$
$$= r(1 - 2r + \cdots).$$

Upon substituting this result and expanding $\ln(1+r)$ into the Taylor series we obtain the following limiting form for ϕ:

$$\phi = 1 + \left(k' - \frac{1}{2}\right)r + O(r^2).$$

This implies that

$$\lim_{r \to 0} \frac{\phi - 1}{r} = k' - \frac{1}{2} \tag{13.107}$$

and
$$\phi(0) = 1. \tag{13.108}$$

Therefore the integral of Eq. (13.100) is convergent at $r = 0$.

The aforementioned result may be easily adapted for data in the units of molality. For a solvent of molecular weight M_1,

$$r = \frac{mM_1}{1000}$$

and

$$\phi = -\frac{1000}{mM_1} \ln a_1.$$

With these results it is then easy to find

$$\ln \gamma_2 = \ln\left(\frac{a_2}{m}\right)$$
$$= \phi - 1 + \int_0^m dm \, \frac{\phi - 1}{m}. \tag{13.109}$$

This equation forms a basis of measurements for activity coefficients which will be discussed in later sections.

As we have seen with regular solutions, studies of changes in thermodynamic functions can provide information on parameters appearing in the expansions such as the Margules expansions or the Scatchard equations for activity coefficients. Studies generally involve the free energy, enthalpy, and entropy of mixing which show marked deviations from ideal behavior. Here we calculate them in the case of binary solutions.

We define the Gibbs free energy of mixing by the equation

$$\Delta g = \sum_{i=1}^{2} x_i(\mu_i - \mu_i^l). \tag{13.110}$$

Substitution of chemical potentials (Eq. (13.84)) produces it in the form

$$\Delta g = RT \sum_{i=1}^{2} x_i \ln x_i + RT \sum_{i=1}^{2} x_i \ln f_i. \tag{13.111}$$

The first term on the right-hand side of Eq. (13.111) is the Gibbs free energy of mixing for the ideal solution and the second term represents the deviation from ideal behavior. We call it the excess Gibbs free energy of mixing:

$$\Delta g_{ex} = RT \sum_{i=1}^{2} x_i \ln f_i. \tag{13.112}$$

The excess free energy and other excess quantities are the quantities to which we will pay attention in the following. If the Margules expansions (Eq. (13.88)) are substituted into Eq. (13.112), we find the excess free energy in a relatively simple form:

$$\Delta g_{ex} = Bx_1x_2. \tag{13.113}$$

We observe that this is in the same form as the heat of mixing (Eq. (13.80)) for regular solutions. The coefficient B is generally dependent on temperature, although the dependence is weak. The excess entropy of mixing is then derived by taking the temperature derivative of Δg_{ex}:

$$\Delta s_{ex} = -x_1x_2\frac{dB}{dT}. \tag{13.114}$$

The entropy of mixing is of course

$$\Delta s = -R\sum_{i=1}^{2} x_i \ln x_i - x_1x_2\frac{dB}{dT}. \tag{13.115}$$

This is an approximation of the more general formula in Eq. (13.97). If $B(T)$ is independent of T, the entropy of mixing is then that of an ideal solution.

The heat of mixing can be derived by using Eqs. (13.111), (13.113), and (13.115) in the relation

$$\Delta h = \Delta g + T\,\Delta s.$$

It gives rise to the equation

$$\Delta h = x_1x_2\left(B - T\frac{dB}{dT}\right). \tag{13.116}$$

Since the heat of mixing is equal to zero for ideal solutions, the entire Δh is an excess quantity unlike other excess quantities. If Δh is differentiated with T, we obtain the change of heat capacity on mixing:

$$\Delta C_{pm} = -x_1x_2T\frac{d^2B}{dT^2}. \tag{13.117}$$

Similar quantities can be derived by using the Scatchard equations instead of the Margules expansions. We find, by using Eq. (13.89), the excess free energy in the form

$$\Delta g_{ex} = Av\phi_1\phi_2, \tag{13.118}$$

where v is the mean volume

$$v = x_1v_1 + x_2v_2.$$

Since the coefficient A is also a function of temperature, the entropy of mixing can be obtained from Eq. (13.118) and the procedure for other quantities is the same as before.

The Henry's law constant may be determined by measuring the solute fugacity in the limit of infinite dilution. This coefficient can be related to the coefficient B. In order to see this connection let us return to Eq. (13.84), which may be cast in the form

$$p_{f1} = x_1 f_1 p_{f1}^0, \qquad p_{f2} = x_2 f_2 p_{f2}^0, \qquad (13.119)$$

where p_{fi}^0 $(i = 1, 2)$ are the fugacities of pure liquids (see Eqs. (13.40) and (13.41)). Substituting the Margules expansions (Eq. (13.84)) into Eq. (13.119) we obtain

$$p_{f1} = x_1 p_{f1}^0 \exp\left(\frac{B x_2^2}{RT}\right),$$
$$p_{f2} = x_2 p_{f2}^0 \exp\left(\frac{B x_1^2}{RT}\right). \qquad (13.120)$$

Therefore in the limit of $x_1 \to 1$ Raoult's law

$$p_{f1} = x_1 p_{f1}^0 \qquad (13.121)$$

is recovered for the solvent, whereas Henry's law

$$p_{f2} = k x_2 \qquad (13.122)$$

is recovered for the solute, where

$$k = p_{f2}^0 \exp\left(\frac{B}{RT}\right). \qquad (13.123)$$

The Henry's law constant k thus can be determined from the coefficient B and vice versa.

13.9 Determination of Activity Coefficients

We have indicated some ways to determine activity coefficients by measuring excess thermodynamic quantities of mixing. There are other ways to determine them. These methods usually exploit phase equilibria and in this sense they are basically different from those based on excess quantities of mixing. We discuss them in the following.

13.9.1 *Vapor Fugacity*

Relation (13.119) between the vapor fugacity and the activity coefficient points to a way to determine the activity coefficient. To make the discussion general we cast Eq. (13.119) in a general form

$$p_{fi} = f_i x_i p_{fi}^0 \quad (i = 1, 2, \ldots, r). \tag{13.124}$$

This formula indicates that it is possible to determine the activity coefficient by measuring the fugacity at a given composition of the solution.

Let us assume that the vapor pressure is such that the Lewis–Randall rule is applicable to the fugacity. If the mole fraction of the component i in the vapor phase is y_i then

$$p_{fi} \simeq y_i p_{fi}^0 \tag{13.125}$$

and therefore Eq. (13.124) reduces to the form

$$f_i \simeq \frac{y_i}{x_i}. \tag{13.126}$$

In this approximation the activity coefficient is simply the ratio of the concentrations of the substance in the vapor phase and in the solution. It is then sufficient to measure the compositions of the solution and the vapor in order to determine the activity coefficients.

13.9.2 *Freezing Point Depression*

Two-phase equilibria between a binary solution and its solid solvent component can be used for measuring activity coefficients. For example, consider ice in equilibrium with an aqueous solution of glycerol. We designate the solvent as component 1 and the solute as component 2. If T and p are kept constant the equilibrium condition for the phase equilibrium is

$$\mu_1^s = \mu_1,$$

where μ_1^s is the chemical potential of pure solid of the component 1. Since

$$\mu_1 = \mu_1^l(T, p) + RT \ln(x_1 f_1),$$

the equilibrium condition may be written as

$$T^{-1}(\mu_1^s - \mu_1^l) = R \ln(x_1 f_1). \tag{13.127}$$

This expression does not clearly indicate how the activity coefficient depends on temperature. In order to make it a little more obvious it is

necessary to obtain an expression for the rate of change in the activity coefficient with regard to temperature. To obtain such an expression let us
differentiate Eq. (13.127) with respect to T at constant p. We then find

$$\frac{\partial}{\partial T} \ln(x_1 f_1) = \frac{\Delta h_m}{RT^2}, \tag{13.128}$$

where Δh_m is the molar heat of melting:

$$\Delta h_m = h_1^l - h_1^s.$$

We have used the relation

$$\left(\frac{\partial}{\partial T} \frac{\mu}{T}\right)_p = -\frac{h}{T^2}$$

to obtain Eq. (13.128) from Eq. (13.127). Let us denote by T_m the melting
temperature at which the solid and liquid phases of pure substance 1 are
in equilibrium in the absence of the solute. We now integrate Eq. (13.128)
from the melting point T_m to an arbitrary temperature T to obtain

$$\ln(x_1 f_1) = \ln(x_1 f_1)|_{T=T_m} + \int_{T_m}^{T} dT' \frac{\Delta h_m}{RT'^2}. \tag{13.129}$$

From Eq. (13.127)

$$\ln(x_1 f_1)|_{T=T_m} = (RT_m)^{-1}[\mu_1^s(T_m, p) - \mu_1^l(T_m, p)], \tag{13.130}$$

but since at $T = T_m$ the solid and liquid phases of pure substance 1 are in
equilibrium, there holds the equilibrium condition

$$\mu_1^s(T_m, p) = \mu_1^l(T_m, p).$$

Consequently, by the definition of T_m, $\ln(x_1 f_1)|_{T=T_m} = 0$ identically, and
Eq. (13.129) reduces to the form

$$\ln(x_1 f_1) = \int_{T_m}^{T} dT \frac{\Delta h_m}{RT^2}. \tag{13.131}$$

Since $x_1 f_1 \leq 1$, the integral on the right is negative. Since Δh_m is positive
by definition, the integrand is positive. Therefore the only way the integral
is negative is that the upper end T of the integral be less than the lower
end T_m which is the melting point of pure solid of component 1, namely,
the solvent:

$$T \leq T_m.$$

In other words, since T is the temperature at which the pure solid of component 1 (solvent) is in equilibrium with the solution, the presence of the

solute lowers the melting point of the solid consisting of the solvent species. This phenomenon is called the freezing point depression.

Equation (13.131) provides a procedure to determine the activity coefficient by measuring the lowering of the melting point from the value of the melting point that the pure solid would have if it were in equilibrium with the pure solvent. To see this we perform the integration by using the following expansion for the heat of melting:

$$\Delta h_m = \Delta h_m^0 + \Delta c_{pm}(T - T_m) + \cdots, \qquad (13.132)$$

where

$$\Delta c_{pm} = \left(\frac{\partial \, \Delta h_m}{\partial T}\right)_p.$$

We also expand T^{-2}:

$$T^{-2} = T_m^{-2}\left(1 + 2\frac{\Delta T}{T_m} + \cdots\right), \qquad (13.133)$$

where

$$\Delta T = T_m - T.$$

By substituting these expansions into Eq. (13.131) and performing the integration we obtain to second order in ΔT the equation

$$\ln(x_1 f_1) = -\frac{\Delta h_m^0}{RT_m^2}\Delta T - \left(\frac{\Delta h_m^0}{RT_m} - \frac{\Delta c_{pm}}{2R}\right)\left(\frac{\Delta T}{T_m}\right)^2. \qquad (13.134)$$

Here ΔT denotes the degree of deviation in the melting point from that of the pure solvent–solid equilibrium and is called the freezing point depression. The value of ΔT depends on the concentration of the solution in equilibrium with the solid and measuring ΔT can provide the concentration dependence of f_1. If $\Delta T/T_m \ll 1$ then it is possible to neglect the second order term in Eq. (13.134) and hence the equation is simplified:

$$\ln(x_1 f_1) = -\frac{\Delta h_m^0}{RT_m^2}\Delta T. \qquad (13.135)$$

This may be used to a good approximation if the solution is dilute. Since at infinite dilution $f_1 \to 1$, it is possible to put $f_1 = 1$ to an approximation in Eq. (13.135), if the solution is sufficiently dilute:

$$\ln x_1 = -\frac{\Delta h_m^0}{RT_m^2}\Delta T. \qquad (13.136)$$

Since the logarithmic function can be expanded in Taylor series

$$\ln x_1 - \ln(1 - x_2) = -x_2 - \tfrac{1}{2}x_2^2 + \cdots,$$

Eq. (13.136) may be further approximated to the form

$$x_2 \simeq \frac{\Delta h_m^0}{RT_m^2} \Delta T. \tag{13.137}$$

In this approximation the freezing point depression is directly proportional to the concentration of the solute, if the solution is sufficiently dilute. Of course, Eqs. (13.136) and (13.137) are not useful for determining the activity coefficient, but Eq. (13.137) may be employed for determining the molecular weight of the solute. If we denote by W_1 and W_2 the weights of the solute and solvent component in the solution, because the solution is dilute, the mole fraction x_2 then may be approximated by the formula

$$x_2 \simeq \frac{n_2}{n_1} = \frac{W_2 M_1}{W_1 M_2},$$

where M_1 and M_2 are the molecular weights of the solvent and the solute, respectively. We may then write Eq. (13.137) in the form

$$M_2 = \frac{RT_m^2 M_1 W_2}{\Delta h_m^0 W_1} \Delta T^{-1}, \tag{13.138}$$

when the solution is sufficiently dilute. Strictly speaking, this formula is applicable if the solution is very dilute so that the activity coefficient is equal to unity and $\ln x_2$ may be replaced with $-x_2$ and if Δh_m is constant with respect to T, and if ΔT is not too large compared with T_m.

To derive the formulas given above for freezing point depression we have made an important assumption that the solid phase consists of the pure solvent component only. This may not be generally the case, however. It is sometimes possible that the solid phase consists of a solid solution of a composition different from that of the liquid solution. Even in this case the equilibrium condition is still

$$\mu_1(s) = \mu_1(l),$$

where $\mu_1(s)$ and $\mu_1(l)$ are the chemical potential of component 1 in the solid and liquid phases, respectively. The former may be given in the form

$$\mu_1(s) = \mu_1^s(T, p) + RT \ln(x_1^s f_1^s), \tag{13.139}$$

where x_1^s is the mole fraction of component 1 in the solid solution, and f_1^s the corresponding activity coefficient. When Eq. (13.139) is used in the aforementioned equilibrium condition, Eq. (13.127) is now modified to the following form:

$$T^{-1}(\mu_1^s - \mu_1^l) = R \ln\left(\frac{x_1 f_1}{x_1^s f_1^s}\right). \tag{13.140}$$

By proceeding in the same manner as for Eq. (13.131) we obtain

$$\ln\left(\frac{x_1 f_1}{x_1^s f_1^s}\right) = \int_{T_m}^{T} dT \frac{\Delta h_m}{RT^2}. \tag{13.141}$$

If Eq. (13.132) is employed for Δh_m, we obtain from Eq. (13.141)

$$\ln\left(\frac{x_1 f_1}{x_1^s f_1^s}\right) = -\frac{\Delta h_m^0}{RT_m^2} \Delta T - \left(\frac{\Delta h_m^0}{RT_m} - \frac{\Delta c_{pm}}{2R}\right)\left(\frac{\Delta T}{T_m}\right)^2 \tag{13.142}$$

to second order in ΔT. The argument of the logarithmic function on the left-hand side represents a distribution of the solvent component between the two phases. We define the partition coefficient K_{part} for the two solutions by the ratio

$$K_{\text{part}} = \frac{x_1^s f_1^s}{x_1 f_1}. \tag{13.143}$$

Since in the case of very dilute solutions

$$f_1 \to 1 \quad \text{and} \quad f_1^s \to 1,$$

we find

$$\lim_{x_1 \to 1} K_{\text{part}} = \frac{x_1^s}{x_1} = K_{\text{part}}^0, \tag{13.144}$$

which shows that K_{part} indeed indicates the degree of partitioning of component 1 between the solid and liquid solutions. Therefore measurement of ΔT can supply the partitioning coefficient. If the solution is sufficiently dilute and $\Delta T/T_m \ll 1$, we may approximate Eq. (13.142) as follows:

$$x_2 - x_2^s = \frac{\Delta h_m^0}{RT_m^2} \Delta T \tag{13.145}$$

or

$$x_2\left(1 - \frac{x_2^s}{x_2}\right) = \frac{\Delta h_m^0}{RT_m^2} \Delta T. \tag{13.146}$$

If the liquid phase is richer in component 2 so that $x_2^s/x_2 < 1$ then the left-hand side is positive and hence $\Delta T > 0$; that is, the freezing point is lowered by the presence of a solute, but if the solid phase is richer in component 2 then $x_2^s/x_2 > 1$ and consequently $\Delta T < 0$, which means that the freezing point is elevated by the presence of the solute in both solutions. These phenomena are consequences of the second law of thermodynamics, inasmuch as the equilibrium conditions are consequences of it.

13.9.3 *Boiling Point Elevation*

Activity coefficients may be determined from boiling point measurements. The treatment for the method of determination is completely analogous to that of the method of freezing point depression. We consider a solution in equilibrium with the vapor of the solvent at a given pressure and temperature. It is assumed that the solute is nonvolatile so that the vapor consists of the solvent species only. The solvent will be designated as component 1. By the material equilibrium condition, the chemical potential of the solvent components in the vapor phase must be equal to its chemical potential in the solution. Thus we have the equation

$$\ln(x_1 f_1) = \frac{\mu_1(v) - \mu_1^l(T, p)}{RT}, \qquad (13.147)$$

where $\mu_1(v)$ is the chemical potential of the solvent (component 1) of the vapor in equilibrium with the solution phase. Differentiating this equation with T at constant p and x_1 yields the equation

$$\frac{\partial}{\partial T} \ln(x_1 f_1) = -\frac{\Delta \bar{h}_v}{RT^2}, \qquad (13.148)$$

where

$$\Delta \bar{h}_v = \bar{h}_1^v - h_1^l \qquad (13.149)$$

with $\bar{h}_1^v$ and h_1^l denoting the partial molar and molar enthalpy of the solvent in the vapor and the solution, respectively. Since the vapor consists of component 1 (i.e., the solvent) only, $\bar{h}_1^v = h_1^v$ and therefore

$$\Delta \bar{h}_v = \Delta h_v$$
$$= h_1^v - h_1^l. \qquad (13.150)$$

Replacing $\Delta \bar{h}_v$ with Δh_v in Eq. (13.148) and integrating the equation we obtain

$$\ln(x_1 f_1) = \ln(x_1 f_1)|_{T=T_b} - \int_{T_b}^{T} dT \, \frac{\Delta h_v}{RT^2}, \qquad (13.151)$$

where T_b is the boiling point of the pure solvent liquid. Then, by the choice of standard state we have made for the solvent chemical potential μ_1, we find

$$\ln(x_1 f_1)|_{T=T_b} = 0. \qquad (13.152)$$

Furthermore, the heat of vaporization may be expanded in temperature around T_b:

$$\Delta h_v = \Delta h_v^0(T_b) + \Delta c_{pv} \Delta T + \cdots,$$

where

$$\Delta T = T - T_b,$$

$$\Delta c_{pv} = \left(\frac{\partial \, \Delta h_v}{\partial T} \right)_{p,T=T_b}.$$

Substituting this expansion into Eq. (13.151) and carrying out the integration finally yields the equation

$$\ln(x_1 f_1) = -\frac{\Delta h_v^0}{RT_b^2} \, \Delta T + \left(\frac{\Delta h_v^0}{RT_b} - \frac{\Delta c_{pv}}{2R} \right) \left(\frac{\Delta T}{T_b} \right)^2 + \cdots \qquad (13.153)$$

for which we have also made use of the expansion

$$T^{-1} = T_b^{-1} \left(1 + \frac{\Delta T}{T_b} \right)^{-1} = T_b^{-1} \left[1 - \frac{\Delta T}{T_b} + \left(\frac{\Delta T}{T_b} \right)^2 - \cdots \right].$$

Equation (13.153) plays the same role for the boiling point elevation as Eq. (13.127) does for the freezing point depression. It can be used for determining not only the activity of the solvent, but also the molecular weight of the solute in the case of sufficiently dilute binary solutions. The approximation equivalent to Eq. (13.131) is

$$x_2 = \frac{\Delta h_v^0}{RT_b^2} \, \Delta T, \qquad (13.154)$$

which implies that $\Delta T > 0$, that is, the boiling point must be elevated owing to the presence of the solute.

Let us remove the assumption that the solute is nonvolatile. Then the vapor consists of a mixture of the same species as those comprising of the solution, and the chemical potential of the solvent species in the vapor phase is given by the form

$$\mu_1(v) = \mu_1^*(T) + RT \ln p_{f1}$$
$$= \mu_1^0(T,p) + RT \ln(y_1^v f_1^v), \qquad (13.155)$$

where y_1^v is the mole fraction and f_1^v is the fugacity coefficient of component 1 in the vapor. Since at equilibrium

$$\mu_1(v) = \mu_1(l)$$
$$= \mu_1^l(T,p) + RT \ln(x_1 f_1),$$

there follows the equation

$$\ln\left(\frac{x_1 f_1}{x_1^v f_1^v}\right) = \frac{\mu_1^0(T,p) - \mu_1^l(T,p)}{RT}.$$

Note that $\mu_1^0(T,p)$ is the chemical potential of the vapor consisting of component 1 only. The right-hand side may be written as before:

$$\frac{\mu_1^0(T,p) - \mu_1^l(T,p)}{RT} = -\int_{T_b}^{T} dT \frac{\Delta h_v}{RT^2}.$$

Therefore we finally obtain

$$\ln\left(\frac{x_1 f_1}{y_1^v f_1^v}\right) = -\int_{T_b}^{T} dT \frac{\Delta h_v}{RT^2}. \qquad (13.156)$$

The left-hand side is not necessarily negative since it is possible that

$$x_1 f_1 > y_1^v f_1^v.$$

In this case, the boiling point can be lowered, namely,

$$T < T_b.$$

This situation is the counterpart of the freezing point elevation which occurs when the solid phase consists of a solid solution richer than the liquid solution phase with regard to the solute species, as we have discussed in the previous section.

The effect of removing the nonvolatility assumption for the solutes may be examined in another way, if Eq. (13.148) is used. Since the vapor is a mixture, the partial molar enthalpy $\bar{h}_1^v$ is no longer equal to h_1^v. In this case the $\Delta \bar{h}_v$ may be decomposed into the heat of vaporization and the heat of mixing:

$$\Delta \bar{h}_v = \Delta h_v + \Delta h_{mix},$$
$$\Delta h_v = h_1^v - h_1^l, \qquad\qquad (13.157)$$
$$\Delta h_{mix} = \bar{h}_1^v - h_1^v,$$

where h_1^v denotes the molar enthalpy of the pure gaseous solvent component. Therefore Δh_v is obviously the heat change accompanying the vaporization of 1 mole of the pure liquid solvent and Δh_{mix} represents the change in enthalpy of the solvent component when a gaseous mixture is formed with the constituents of the solution. According to our previous calculations (see Eq. (11.44))

$$\Delta h_{mix} = \int_0^p dp \left[\Delta \bar{v}_1 - T \left(\frac{\partial \Delta \bar{v}_1}{\partial T} \right)_{T,x} \right].$$

On integration of Eq. (13.148) where $\Delta \bar{h}_1$ is given by Eq. (13.157) there follows the equation

$$\ln(x_1 f_1) = -\int_{T_b}^{T} dT' \frac{\Delta h_v + \Delta h_{mix}}{RT'^2},$$

for which Eq. (13.152) is made use of. The integral on the right-hand side may be evaluated in an approximation. For example, the Lewis-Randall rule is applicable at pressures less than about 100 atm. Under this rule

$$\Delta \bar{v}_1 = \bar{v}_1 - v_1 \simeq 0,$$

which means that

$$\Delta h_{mix} \simeq 0.$$

If the vapor pressure is not too high, the aforementioned approximation allows us to ignore Δh_{mix} and recover the result for the case of the vapor consisting of the solvent only:

$$\ln(x_1 f_1) = -\int_{T_b}^{T} dT' \frac{\Delta h_v}{RT'^2},$$

which is Eq. (13.148).

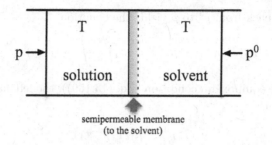

semipermeable membrane
(to the solvent)

Fig. 13.5　Osmosis.

13.9.4　*Osmotic Pressure*

Let us imagine a liquid solution separated from the pure solvent by a nondeformable, diathermal membrane which is permeable to the solvent molecules only (see Fig. 13.5). We assume that the two phases are in thermal equilibrium. Since one phase is a solution, the chemical potential of the solvent in the solution is generally different from that of the pure solvent, if the pressures and temperatures are the same in both phases. Since we are assuming thermal equilibrium between both phases, material equilibrium of the solvent component between the two phases can be possible only if there is a difference in pressure in the two phases. This pressure difference is called the *osmotic pressure*. We define it more precisely as follows: the osmotic pressure is the excess pressure that must be applied to a solution to prevent the passage of the solvent into the solution through a semipermeable membrane which separates the two liquids and is permeable to the solvent molecules only. Thus, if we denote the pressure on the pure solvent phase by p^0 and the osmotic pressure by π then the pressure on the solution necessary to maintain material equilibrium is

$$p = p^0 + \pi. \tag{13.158}$$

This is the equilibrium condition for pressure. Since there is thermal equilibrium, the only other equilibrium condition to consider is

$$\mu_1(T, p, x_1) = \mu_1^l(T, p^0). \tag{13.159}$$

Since

$$\left(\frac{\partial \mu_1}{\partial p} \right)_{T, x_1} = \bar{v},$$

integrating it back from p^0 to p yields the equation

$$\mu_1(T, p, x_1) - \mu_1(T, p^0, x_1) = \int_{p^0}^{p} dp\, \bar{v}_1(p). \qquad (13.160)$$

Because of the equilibrium condition (Eq. (13.159)), the left-hand side may be replaced with

$$\mu_1(T, p^0) - \mu_1(T, p^0, x_1).$$

Since the solvent chemical potential of the solution at p^0 is

$$\mu_1(T, p^0, x_1) = \mu_1^l(T, p^0) + RT \ln\left[x_1 f_1(p^0)\right] \qquad (13.161)$$

and $\mu_1(T, p^0) = \mu_1^l(T, p^0)$ by the convention on the chemical potential of the pure liquid (the solvent), Eq. (13.159) is given by the form

$$RT \ln\left[x_1 f_1(p^0)\right] = -\int_{p^0}^{p} dp\bar{v}_1(p). \qquad (13.162)$$

Since $x_1 f_1 \leq 1$, the left-hand side of Eq. (13.162) is negative and hence it is possible to deduce $p - p^0 = \pi \geq 0$. That is, the osmotic pressure must be positive. To proceed further it is necessary to have the pressure dependence of the partial molar volume of the solvent in the solution. We may expand $\bar{v}_1$ in the power series of $p - p^0 = \pi$:

$$\bar{v}_1 = \bar{v}_1^0 - \kappa_1^0 \bar{v}_1^0 \pi + \cdots, \qquad (13.163)$$

where

$$\bar{v}_1^0 = \bar{v}_1(p^0)$$

and κ_1^0 is the isothermal compressibility of the pure solvent defined by the formula

$$\kappa_1^0 = -\frac{1}{\bar{v}_1^0}\left(\frac{\partial \bar{v}_1^0}{\partial p^0}\right)_T.$$

Use of Eq. (13.163) for the integral in Eq. (13.162) yields

$$RT \ln\left[x_1 f_1(p^0)\right] = -\bar{v}_1^0 \pi + \tfrac{1}{2}\kappa_1^0 \bar{v}_1^0 \pi^2 + \cdots, \qquad (13.164)$$

for which κ_1^0 is assumed constant. This equation provides a way to determine the activity coefficient by measuring the osmotic pressure. Note that the activity coefficient is determined at the value of pressure p^0.

If the liquid is incompressible, then $\kappa_1^0 = 0$, and to first order in osmotic pressure

$$RT \ln x_1 f_1 = -\bar{v}_1^0 \pi. \tag{13.165}$$

If the solution is very dilute so that $f_1 \simeq 1$ then it is possible to approximate Eq. (13.165) as follows:

$$\ln x_1 = -\bar{v}_1^0 \pi. \tag{13.166}$$

If the solution is binary and dilute then

$$\ln x_1 \simeq -x_2$$

and consequently Eq. (13.166) reduces to the form

$$\pi \bar{v}_1^0 = RT x_2, \tag{13.167}$$

which is the van't Hoff equation for osmotic pressure.

If the Margules expansion is employed for $RT \ln f_1$ in the case of a binary solution, Eq. (13.165) becomes

$$\pi \bar{v}_1^0 = x_2 RT \left[1 + \left(\tfrac{1}{2} - \frac{B_1}{RT} \right) x_2 + \cdots \right], \tag{13.168}$$

for which we have expanded $\ln x_1$ in the Taylor series of x_2.

Measurement of the osmotic pressure can also provide the molecular weight of the solute molecule. Since $x_2 \simeq n_2/n_1$ if the solution is sufficiently dilute, we obtain from Eq. (13.167)

$$M_2 = \frac{RT W_2 M_1}{\pi \bar{v}_1^0 W_1}, \tag{13.169}$$

where we are using the same notation as in Eq. (13.138). In polymer physical chemistry, measurement of osmotic pressure is routinely made use of to determine polymer molecular weights.

13.9.5 *Solubility*

The activity coefficients of solutes can be obtained by measuring their solubility in the solvent. In order to make our discussion simple let us consider a binary solution which is in equilibrium with the solid solute. The solute will be designated as component 2, and the solvent as component 1. The temperature and pressure are kept constant. The remaining equilibrium condition is then

$$\mu_2^s(T, p) = \mu_2(T, p, x_2), \tag{13.170}$$

where $\mu_2^s(T, p)$ is the chemical potential of the pure solid solute. But the solute chemical potential in the solution is given by

$$\mu_2(T, p, x_2) = \mu_2^l(T, p) + RT \ln(x_2 f_2). \qquad (13.171)$$

Combining this equation with the equilibrium condition yields

$$RT \ln(x_2 f_2) = \mu_2^s(T, p) - \mu_2^l(T, p).$$

Differentiating it with T at constant p and x_2 we obtain the differential equation

$$\left(\frac{\partial \ln x_2 f_2}{\partial T} \right)_{p, x_2} = \frac{\Delta h_m}{RT^2}, \qquad (13.172)$$

where Δh_m is the molar heat of melting of the solute:

$$\Delta h_m = h_2^l - h_2^s$$

with h_2^l and h_2^s denoting the molar enthalpy of pure liquid 2 and pure solid 2, respectively. We may now choose a temperature T_0 such that the solubility of the solute is very low and consequently the solution behaves practically as an ideal solution. We will denote the solubility at that temperature by x_2^0. Then at $x_2 = x_2^0$ the activity coefficient f_2 may be put equal to unity. By integrating Eq. (13.172) from T_0 to an arbitrary temperature T we obtain the following equation for f_2:

$$\ln f_2 = \ln \left(\frac{x_2^0}{x_2} \right) + \int_{T_0}^{T} dT \frac{\Delta h_m}{RT^2}. \qquad (13.173)$$

The integral may be evaluated easily by applying the expansion (Eq. (13.132)) for Δh_m. Thus we see that the heat of melting data of the solute and the solubility data at two temperatures enable us to determine the activity coefficient.

Equation (13.173) may be cast into a more conventional and familiar form if the solubility constant K_{sol} is defined by

$$K_{sol} = x_2 f_2. \qquad (13.174)$$

By the definition of T_0 made earlier the solubility constant K_{sol}^0 at $T = T_0$ is given by

$$K_{sol}^0 = x_2^0. \qquad (13.175)$$

then Eq. (13.174) takes the form

$$\ln \left(\frac{K_{sol}}{K_{sol}^0} \right) = \int_{T_0}^{T} dT \frac{\Delta h_m}{RT^2}, \qquad (13.176)$$

which is reminiscent of the van't Hoff equation for a chemical equilibrium constant.

If the solid phase is a solid solution of, say, two components 1 and 2 then the chemical potential of component 2 of the solid may be written as

$$\mu_2(s) = \mu_2^s(T, p) + RT \ln(y_2 f_2^s), \tag{13.177}$$

where y_2 is the mole fraction of component 2, and f_2^s its activity coefficient. It is then easy to show that

$$\left[\frac{\partial}{\partial T} \ln\left(\frac{x_2 f_2}{y_2 f_2^s} \right) \right]_p = \frac{\Delta h_m}{RT^2}.$$

On integration this yields

$$\ln\left(\frac{x_2 f_2}{y_2 f_2^s} \right) = \ln\left(\frac{x_2 f_2}{y_2 f_2^s} \right)_{T=T_0} + \int_{T_0}^{T} dT \frac{\Delta h_m}{RT^2}. \tag{13.178}$$

Choose T_0 as the melting point of the pure solid solute into pure liquid. Since the activity coefficients f_2^s and f_2 in such states are equal to unity, we have

$$\lim_{x_2, y_2 \to 1} \ln\left(\frac{x_2 f_2}{y_2 f_2^s} \right)_{T=T_0} = 0.$$

Therefore there follows the equation

$$\ln\left(\frac{x_2 f_2}{y_2 f_2^s} \right) = \int_{T_0}^{T} dT \frac{\Delta h_m}{RT^2}. \tag{13.179}$$

This equation describes the temperature dependence of the solubility of the binary solid solution in solvent 1. Depending on whether the ratio $(x_2 f_2 / y_2 f_2^s)$ is larger than unity or not, $T > T_0$ or $T < T_0$.

Problems

(1) Give in full the derivation of the statement that heat is evolved upon mixing two liquids which form a system exhibiting negative deviations from ideal behavior. The deviation is said to be negative if the actual partial pressure of each constituent is less than it should be if Raoult's law were obeyed. The deviation is said to be positive in the opposite case.

(2) On the basis of the lowest order Margules expansions, show that for a liquid mixture exhibiting positive deviations from Raoult's law the

activity coefficient of each constituent must be greater than unity, whereas for negative deviations it must be less than unity.

(3) The vapor pressure of liquid ethylene is 40.6 atm at 273.15 K and 24.8 atm at 253.15 K. Estimate the ideal solubility of the gas in a liquid at 298.15 K and 1 atm pressure. How many grams of ethylene should dissolve in 1 kg of benzene under these conditions, if the gas and solution behave ideally.

(4) The freezing point of benzene is 278.55 K and its latent heat of melting is 126.36 J g^{-1}. A solution containing 6.054 g of triphenylmethane in 1 kg of benzene has a freezing point which is 273.2763 K below that of the pure solvent. Calculate the molecular weight of the solute.

(5) Mixtures of benzene and toluene behave almost ideally; at 303.15 K, the vapor pressure of pure benzene is 0.1182 m Hg and that of pure toluene is 0.0367 m Hg. Determine the partial pressures and weight composition of the vapor in equilibrium with a liquid mixture consisting of equal weights of the two constituents.

Chapter 14

Thermodynamics of Surfaces

In the previous chapter on thermodynamics of phase equilibria we have
assumed that the densities of matter as well as the energy and entropy
densities of phases in equilibrium are uniform up to a mathematical surface that separates the contiguous phases. This presumes, for example,
densities of two contiguous phases can be discontinuous across the surface. However, in reality this is not generally true evidently with respect to
density, energy, and entropy, but only an idealized representation. There
exist interfaces between contiguous phases which are different from the bulk
phases themselves. In this chapter, we shall discuss the thermodynamics of
interfaces separating chemically inert phases. The interfacial phenomena
manifest themselves the more significantly the smaller the size of a phase
becomes relative to the contiguous phases. The phenomena range over wide
ranging subjects. The surface effects cannot be ignored in numerous fields
like colloids, soil science, petroleum engineering, botany, etc. Here in this
chapter, we will only present the gist of the thermodynamics of interfaces
as an elementary introduction to the subject matter. For more thorough
treatments the reader is referred to numerous monographs[1] on the subjects
mentioned.

Microscopically, molecules close to the interface experience an environ-
ment different from the environments of molecules in the bulk phase. The
interface extends over a thin region of a few or a few tens of molecular
diameters in thickness between the contiguous phases, and the physical
properties of the system vary continuously across the interface from one
phase to the other, see Fig. 14.1. Hence, if the phases are internally in

---

[1]For example, see A. W. Adamson, *Physical Chemistry of Surfaces* (Wiley, New York, 1990).

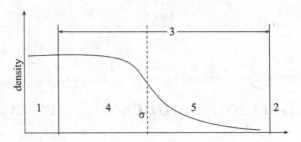

Fig. 14.1   Interfacial profile of density. The density, for example, changes continuously from one bulk phase to the other across the interface. The numerals refer to the parts of the volume enclosing the surface σ, see Fig. 14.2.

equilibrium, the physical properties of phases can be assumed uniform up to the thin interfacial region.

The foundations of thermodynamics of interface and surface were essentially laid by J. W. Gibbs in his monumental work on the thermodynamics of heterogeneous phases some 130 years ago. We follow his treatment in this chapter. For a more complete treatment we refer to his work.[2]

14.1   Dividing Surface

We consider a surface of discontinuity in the fluid which is in equilibrium under the assumption that gravity effects are absent. In the neighborhood of the physical surface of discontinuity, choose a point and a geometrical surface of equidensity with respect to a component that passes through the point. Thus all points on the surface correspond to that of the same local density with respect to the component chosen. This surface is called the dividing surface. It will be denoted by σ. Its position is arbitrary to some extent, but the normal directions are determined owing to the fact that all surfaces drawn in the manner prescribed have the parallel normal directions.

Imagine a closed surface that cuts through the dividing surface perpendicularly on each side and extends well into the homogeneous phases on each side of the dividing surface. This surface is parallel to the normal direction of σ. The space enclosed by the closed surface is then divided into three parts by placing two surfaces σ_{13} and σ_{23} placed sufficiently far from

---

[2]J. W. Gibbs, *The Scientific Papers of J. Willard Gibbs, Vol. 1, Thermodynamics*, pp. 219–328 (Dover, New York, 1961).

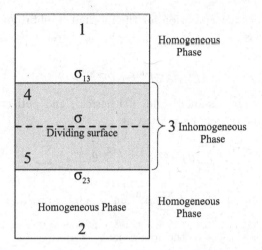

Fig. 14.2   Cross section of the interface. The lines represented by σ_{13} and σ_{23} on both sides of σ marks the thin film. The surfaces σ_{13} and σ_{23} are placed sufficiently far into 1 and 2.

σ so that the surfaces are outside the influence of the inhomogeneous film (bounded by σ_{13} and σ_{23} in the figure) surrounding σ, see Fig. 14.2. The homogeneous parts beyond σ_{13} and σ_{23} outside the film are designated 1 and 2, and that part containing the film is 3. The corresponding masses are denoted by M_1 and M_2, and that part enclosing the dividing surface σ by M_3. The area of σ is a. The part of 3 enclosed by σ and σ_{13} is designated 4, and that part enclosed by σ and σ_{23} is designated 5. Their masses are denoted by M_4 and M_5, respectively, which obviously make up mass M_3: $M_3 = M_4 + M_5$.

14.2   Gibbs Relation for Interface

To carry on study of thermodynamics of interfaces it is required to have the fundamental thermodynamic equation for interface. To this end, we now consider the interface to be diathermal and perfectly permeable to all the species in the two phases involved. *We also assume energy is assignable to each mass element and thus continuous as is the density.* It is then possible to assume that the total internal energy E of mass $(M_1 + M_2 + M_3)$ is given by

$$E = E^{(1)} + E^{(2)} + E^{(3)}, \tag{14.1}$$

where $E^{(i)}$ is the internal energy of mass M_i $(i = 1, 2, 3)$.

Now the necessary condition for the internal equilibrium of the system composed of masses M_1, M_2, and M_3 for all variations subject to fixed boundaries is

$$\delta E^{(1)} + \delta E^{(2)} + \delta E^{(3)} \geq 0. \tag{14.2}$$

The boundaries are assumed rigid, diathermal, and perfectly permeable. Then this inequality may be written as

$$T^{(1)} \, \delta S^{(1)} + \sum_{j=1}^{c} \mu_j^{(1)} \, \delta n_j^{(1)} + T^{(2)} \, \delta S^{(2)} + \sum_{j=1}^{c} \mu_j^{(2)} \, \delta n_j^{(2)}$$

$$+ T^{(3)} \, \delta S^{(3)} + \sum_{j=1}^{c} \mu_j^{(3)} \, \delta n_j^{(3)} \geq 0. \tag{14.3}$$

This variation is subject to the constraints

$$\delta S^{(1)} + \delta S^{(2)} + \delta S^{(3)} = 0,$$

$$\sum_{j=1}^{c} \delta n_j^{(\alpha)} = 0 \; (\alpha = 1, 2, 3). \tag{14.4}$$

The necessary and sufficient conditions for internal equilibrium of the system then are

$$T^{(1)} = T^{(2)} = T^{(3)} \equiv T, \tag{14.5}$$

$$\mu_j^{(1)} = \mu_j^{(2)} = \mu_j^{(3)} \equiv \mu_j \; (1 \leq j \leq c). \tag{14.6}$$

This conclusion follows upon applying the method used for the equilibrium conditions of heterogeneous phases in Chapter 8.

On account of the boundaries including σ dividing region 3 into regions 4 and 5 being rigid (fixed), the volumes of M_4 and M_5 remain constant. Therefore by virtue of the conditions (Eqs. (14.5) and (14.6)), while there holds for M_3 the equation

$$\delta E^{(3)} = T \, \delta S^{(3)} + \sum_{j=1}^{c} \mu_j \, \delta n_j^{(3)}, \tag{14.7}$$

there hold the equations for M_4 and M_5

$$\delta E^{(4)} = T \, \delta S^{(4)} + \sum_{j=1}^{c} \mu_j \, \delta n_j^{(4)}, \tag{14.8}$$

$$\delta E^{(5)} = T \, \delta S^{(5)} + \sum_{j=1}^{c} \mu_j \, \delta n_j^{(5)}. \tag{14.9}$$

On subtracting Eqs. (14.8) and (14.9) from Eq. (14.7), we obtain

$$\delta E^s = T\,\delta S^s + \sum_{j=1}^{c} \mu_j\,\delta n_j^s, \tag{14.10}$$

where

$$
\begin{aligned}
E^s &= E^{(3)} - E^{(1)} - E^{(2)}, \\
S^s &= S^{(3)} - S^{(1)} - S^{(2)}, \\
n_j^s &= n_j^{(3)} - n_j^{(1)} - n_j^{(2)} \ (1 \le j \le c).
\end{aligned}
\tag{14.11}
$$

It is emphasized that Eq. (14.10) holds, provided that the surfaces bounding M, including the surface σ, are fixed (rigid). The E^s, S^s, and n_j^s therefore define the excess energy, entropy, and densities of mass M_3 over the values of energy, entropy, and densities that mass M_3 would have if the quantities had the same uniform values up to the dividing surface σ as those of M_1 and M_2. It is convenient to define excess quantities per unit area a of σ:

$$E_\sigma = E^s/a, \quad S_\sigma = E^s/a, \quad \Gamma_j^\sigma = n_j^s/a, \tag{14.12}$$

which are, respectively, the surface energy, surface entropy, and surface concentration. Thus we obtain

$$\delta E_\sigma = T\,\delta S_\sigma + \sum_{j=1}^{c} \mu_j\,\delta \Gamma_j^\sigma \tag{14.13}$$

subject to the conditions that the bounding surfaces of M_4 and M_5 are fixed. The presence of E_σ, S_σ, and Γ_j^σ is the difference between the previous theory of heterogeneous equilibria (Chapter 8) and the present. Now we remove the condition that σ is fixed and examine the effects of variations in the position and form of the surface σ.

The quantities E^s, S^s, $n_1^s, \ldots, n_c^s$ and E_σ, S_σ, $\Gamma_1^\sigma, \ldots, \Gamma_c^\sigma$ are determined partly by the state of the system under consideration and partly by the form and position of the dividing surface σ. However, the positions of the surfaces drawn in the homogeneous regions cannot affect the values of E^s, etc. or E_σ, etc. The reason is: if the plane dividing the surface is moved a distance ϵ toward M_1, $n_i^{(1)}$ is unchanged while $n_i^{(4)}$ is increased by $\epsilon a \Gamma_i^{(4)}$ and $n_i^{(5)}$ is decreased by $-\epsilon a \Gamma_i^{(5)}$ where $\Gamma_i^{(4)}$ and $\Gamma_i^{(5)}$ are densities of the species i per a in M_4 and M_5, respectively. Therefore

$$\Delta\Gamma_i^\sigma = \epsilon(\Gamma_i^{(4)} - \Gamma_i^{(5)}) = \epsilon(\Gamma_i^{(1)} - \Gamma_i^{(2)}) = 0, \tag{14.14}$$

which means that the location of the dividing surface does not affect the values of Γ_i^{σ}. One can argue similarly for E^s and S^s.

However, Eq. (14.13) is modified if the position and shape of σ are varied, because it is expected that E^s should be a function of S^s and n_j^s as well as the characteristics of surface σ, namely, the area a and the principal curvatures c_1 and c_2 of σ:

$$E^s = E^s(S^s, n_1^s, \ldots, n_c^s, a, c_1, c_2). \qquad (14.15)$$

Therefore with the definitions

$$T = \left(\frac{\partial E^s}{\partial S^s}\right)_{\Gamma^\sigma a, c_1, c_2}, \quad \mu_j^s = \left(\frac{\partial E^s}{\partial n_j^s}\right)_{S^\sigma, \Gamma^{\sigma'}, a, c_1, c_2}, \qquad (14.16)$$

$$\gamma = \left(\frac{\partial E^s}{\partial a}\right)_{S^\sigma, \Gamma^\sigma, a, c_1, c_2}, \qquad (14.17)$$

$$C_1 = \left(\frac{\partial E^s}{\partial c_1}\right)_{S^\sigma, \Gamma^\sigma, a, c_2}, \qquad (14.18)$$

$$C_2 = \left(\frac{\partial E^s}{\partial c_2}\right)_{S^\sigma, \Gamma^\sigma, a, c_1}, \qquad (14.19)$$

we obtain the variational form for E^s

$$\delta E^s = T\,\delta S^s + \sum_{j=1}^{c} \mu_j\,\delta n_j^s + \gamma\,\delta a + C_1\,\delta c_1 + C_2\,\delta c_2. \qquad (14.20)$$

Here C_1 and C_2 are the constitutive properties of the interface, as are T, γ, and μ_j. This variational form may be rearranged to the form

$$\delta E^s = T\,\delta S^s + \sum_{j=1}^{c} \mu_j\,\delta n_j^s + \gamma\,\delta a$$

$$+ \frac{1}{2}(C_1 + C_2)\,\delta(c_1 + c_2) + \frac{1}{2}(C_1 - C_2)\,\delta(c_1 - c_2). \qquad (14.21)$$

For a planar surface, since $c_1 = c_2 = 0$ the fourth and fifth terms on the right of Eq. (14.21) vanish. For nearly planar surfaces it can be shown that

$$C_1 + C_2 = 0, \quad \delta(c_1 - c_2) = 0 \qquad (14.22)$$

and hence the second line on the right of Eq. (14.21) can be omitted. Equation (14.22) is shown to be true in the following.

We imagine a planar surface σ of area $(x + \delta x)(y + \delta y)$, see Fig. 14.3. In the neighborhood of this surface on the positive side of curvature, imagine

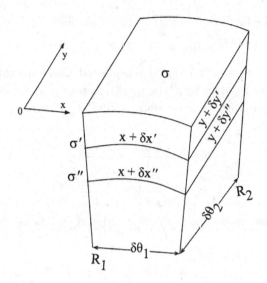

Fig. 14.3 Surface σ is planar and parallel to the dividing surface, and surfaces σ' and σ'' are spherical with the curvature vectors downward. R_1 and R_2 are the principal radii of curvature of surface σ'' and the principal radii of curvature of σ' are $R_1 + \epsilon$ and $R_2 + \epsilon$. $\delta\theta_1$ is the angle between R_1 and its displacement in the x direction whereas $\delta\theta_2$ is the corresponding angle for R_2 and its displacement in the y direction.

a surface σ' of area $(x + \delta x')(y + \delta y')$ and another surface σ'' of area $(x + \delta x'')(y + \delta y'')$ at distance λ from σ'. These surfaces are constructed such that their normal vectors of σ' and σ'' are parallel. The principal curvature vectors vary from zero to $1/R_1$ and $1/R_2$, respectively. That is,

$$\delta c_1'' = \frac{1}{R_1}, \quad \delta c_2'' = \frac{1}{R_2}. \tag{14.23}$$

Then it follows

$$\delta c_1' = \frac{1}{R_1 + \epsilon}, \quad \delta c_2' = \frac{1}{R_2 + \epsilon}. \tag{14.24}$$

On eliminating R_1 and R_2 by using Eq. (14.23), we obtain

$$\delta c_1' = \frac{\delta c_1''}{1 + \epsilon \delta c_1''}, \quad \delta c_2' = \frac{\delta c_2''}{1 + \epsilon \delta c_2''}. \tag{14.25}$$

Therefore to the first order in variations we obtain

$$\delta(c_1' + c_2') = \delta(c_1'' + c_2''), \tag{14.26}$$

where the term of $O(\epsilon\delta c_1''^2 + \epsilon\delta c_2''^2)$ is neglected. Since the varied areas are $(x + \delta x')(y + \delta y')$ and $(x + \delta x'')(y + \delta y'')$, respectively, the angles between the principal curvature vectors before and after the variation $\delta\theta_1$ and $\delta\theta_2$ are, respectively,

$$\delta\theta_1 = \frac{x + \delta x'}{R_1 + \epsilon} = \frac{x + \delta x''}{R_1}, \tag{14.27}$$

$$\delta\theta_2 = \frac{y + \delta y'}{R_2 + \epsilon} = \frac{y + \delta y''}{R_2}. \tag{14.28}$$

From these relations follow, to the first order in variations, the relations

$$\delta x' = \delta x'' + \epsilon x\, \delta c_1'', \tag{14.29}$$

$$\delta y' = \delta y'' + \epsilon y\, \delta c_2''. \tag{14.30}$$

With these relations we find the variation in area a' of σ':

$$\delta a' = (x + \delta x')(y + \delta y') - xy \tag{14.31}$$
$$= x\, \delta y' + y\, \delta x' + \text{ higher order,}$$

which, on use of Relations (14.29) and (14.30), gives rise to the relation

$$\delta a' = \delta a'' + a\epsilon\delta(c_1'' + c_2''). \tag{14.32}$$

For imaginary surfaces σ' and σ''

$$\delta E^\sigma - T\,\delta S^s - \sum_{j=1}^c \mu_j\, \delta n_j^\sigma$$

remains unchanged since M_3 remains unchanged. Therefore we obtain for planar surfaces for which $\delta(c_1 - c_2) = 0$

$$\gamma'\delta a' + \frac{1}{2}(C_1' + C_2')\,\delta(c_1' + c_2') = \gamma''\delta a'' + \frac{1}{2}(C_1'' + C_2'')\,\delta(c_1'' + c_2''). \tag{14.33}$$

Upon use of Eq. (14.26) and the fact that $\gamma'' = \gamma' \equiv \gamma$ for a plane surface (this is shown in Sec. 5 below) we obtain

$$C_1' + C_2' + 2\gamma a\epsilon = C_1'' + C_2''. \tag{14.34}$$

If σ' is held fixed in position, the quantities C_1', C_2', and γ have fixed values. Therefore ϵ can be chosen such that

$$C_1'' + C_2'' = 0. \tag{14.35}$$

For a nearly planar surface, evidently

$$\delta\,(c_1 - c_2) = O(\epsilon), \quad C_1 + C_2 = O(\epsilon) \tag{14.36}$$

according to Eqs. (14.34) and (14.35).

Therefore if the inhomogeneous film surrounding the surface σ is sufficiently thin compared with the principal radii of curvature then $\frac{1}{2}(C_1 + C_2)\,\delta(c_1 + c_2)$ is negligible. Therefore the dividing surface can be chosen such that this condition is met for a planar or nearly planar surface and it is possible to omit the term

$$\frac{1}{2}(C_1 + C_2)\,\delta(c_1 + c_2) + \frac{1}{2}(C_1 - C_2)\,\delta(c_1 - c_2)$$

in Eq. (14.21). For such surfaces Eq. (14.21) reduces to the differential form

$$\delta E^s = T\,\delta S^s + \sum_{j=1}^{c} \mu_j\,\delta n_j^s + \gamma\,\delta a. \tag{14.37}$$

This is the fundamental thermodynamic equation for planar surfaces σ. Equation (14.37) is called the Gibbs relation for interface. The meaning of γ is now clearly seen as the surface tension, and $\gamma\delta a$ as the surface work.

14.3 Nearly Planar Surface and Surface Tension

We now consider the effects caused by curvature, especially, on pressure. We shall assume that the surface tension is not affected by surface curvature. This assumption is justifiable as long as the radius of curvature is large compared with the thickness of the film.

For this purpose we now investigate the conditions for heterogeneous equilibrium on the pressures of the homogeneous phases on either side of σ. The criterion for equilibrium for M_3 (region 3) is

$$(\delta E^{(3)})_{S^{(3)}, n^{(3)}} \geq 0 \tag{14.38}$$

when the bounding surface is rigid. Since

$$\delta E^{(3)} = a\,\delta E_\sigma + \delta E^{(4)} + \delta E^{(5)}, \tag{14.39}$$

where

$$\delta E^{(\alpha)} = T\,\delta S^{(\alpha)} - p^{(\alpha)}\,\delta V^{(\alpha)} + \sum_{j=1}^{c} \mu_j\,\delta n_j^{(\alpha)} \quad (\alpha = 4, 5), \tag{14.40}$$

substitution of Eqs. (14.39) and (14.40) into Inequality (14.38) gives rise to the equilibrium conditions for pressure

$$p^{(4)} = p^{(1)}, \quad p^{(5)} = p^{(2)}. \tag{14.41}$$

Since the properties of M_1 and M_2 are the same as those of the bulk properties of phases on either side of σ by construction, if we denote the phases by α and β for 1 and 2, respectively, then we obtain the inequality

$$\gamma \delta a - p^{(\alpha)} \, \delta V^{(4)} - p^{(\beta)} \, \delta V^{(5)} \geq 0. \tag{14.42}$$

Since all variations are either positive or negative this gives rise to the conclusion

$$\gamma \delta a - p^{(\alpha)} \, \delta V^{(4)} - p^{(\beta)} \, \delta V^{(5)} = 0. \tag{14.43}$$

If $\delta a = 0$, then since

$$\delta V^{(4)} + \delta V^{(5)} = 0,$$

we conclude

$$p^{(\alpha)} = p^{(\beta)}. \tag{14.44}$$

Thus the pressure has the same value on both side of σ in the case of an undeformed plane surface.

In the case of a curved surface, if the dividing surface is uniformly moved by a distance $\delta \epsilon$ in the direction of its normals, in Eq. (14.43) the change in area is given by

$$\delta a = a(c_1 + c_2) \, \delta \epsilon, \tag{14.45}$$

while

$$\delta V^{(4)} = a \, \delta \epsilon, \quad \delta V^{(5)} = -a \, \delta \epsilon. \tag{14.46}$$

By Eq. (14.43) we obtain

$$[\gamma(c_1 + c_2) - (p^{(\alpha)} - p^{(\beta)})]a \, \delta \epsilon = 0.$$

Since $\delta \epsilon$ is arbitrary, we obtain

$$p^{(\alpha)} - p^{(\beta)} = \gamma \, (c_1 + c_2) \tag{14.47}$$

with the curvatures positive when their centers lie in the phase α. If the surface is part of a spherical surface of curvature $c = 1/R$, then

$$p^{(\alpha)} - p^{(\beta)} = 2\gamma c$$

$$= \frac{2\gamma}{R}. \tag{14.48}$$

According to this formula, the pressure inside a spherical drop of liquid of radius R is greater than the external pressure by $2\gamma/R$. Therefore the equilibrium pressure of a smaller sphere is greater than that of a larger sphere of the liquid at the same temperature and external pressure.

Relation (14.48) can be applied to the following situation of a thin spherical shell having two concentric surfaces, a typical example being a bubble of detergent. Let the outer surface has a radius of R_1 and the inner surface a radius of R_2. The pressure $p^{(film)}$ in the film is greater than the pressure $p^{(\beta)}$ in the exterior phase by $2\gamma/R_1$, namely,

$$p^{(film)} - p^{(\beta)} = \frac{2\gamma}{R_1},$$

whereas the pressure within the bubble $p^{(\beta)}$ is greater than $p^{(film)}$, namely,

$$p^{(\alpha)} - p^{(film)} = \frac{2\gamma}{R_2}.$$

Therefore

$$p^{(\alpha)} - p^{(\beta)} = 2\gamma\left(\frac{1}{R_1} + \frac{1}{R_2}\right). \tag{14.49}$$

Let

$$R = \frac{1}{2}(R_1 + R_2), \quad \varepsilon = R_1 - R_2.$$

Then there holds the Young–Laplace equation

$$p^{(\alpha)} - p^{(\beta)} = \frac{4\gamma}{R}\left[1 + O\left(\frac{\varepsilon^2}{R^2}\right)\right]. \tag{14.50}$$

The pressure inside the bubble is greater than the exterior pressure and the bubble is stabilized by the surface tension γ.

We consider another related example: the vapor pressure of a spherical drop of a liquid. When the vapor phase is in equilibrium with the liquid phase at T, there holds the Poynting equation

$$\frac{dp^{(v)}}{dp^{(l)}} = \frac{v_l}{v_v}, \tag{14.51}$$

where $p^{(v)}$ and $p^{(l)}$ are the vapor pressure and the pressure within the liquid, respectively, and v_l and v_v are molar volumes of the vapor and the liquid, respectively. For the sake of simplicity let us assume that the vapor is ideal. Then

$$\frac{dp^{(v)}}{dp^{(l)}} = v_l\frac{p^{(v)}}{RT}. \tag{14.52}$$

Integrating this equation, under the assumption of incompressibility of the liquid to an approximation, from the state of a planar liquid to a spherical liquid of radius r we obtain

$$\ln \frac{p^{(v)}}{p_0} = \frac{v_l}{RT}(p_l - p_0). \tag{14.53}$$

Note that initially $p_0 = p_l^{(initial)}$ in the planar state of the liquid at equilibrium with its vapor. Using Eq. (14.48) this can be written in the form

$$\ln \frac{p^{(v)}}{p_0} = \frac{v_l}{RT}\left[\frac{2\gamma}{r} + \left(p^{(v)} - p_0\right)\right], \tag{14.54}$$

where r is the radius of the liquid drop. Since r is assumed small, this may be approximated as follows:

$$\ln \frac{p^{(v)}}{p_0} = \frac{2\gamma v_l}{RTr}. \tag{14.55}$$

This shows that the smaller the drop the larger the vapor pressure becomes. This formula enables us to estimate the vapor pressure above a spherical drop from its properties.

14.4   Gibbs–Duhem Relation for Interface

The excess Helmholtz free energy A^s and the excess Gibbs free energy G^s of interface are, respectively, defined by

$$A^s = E^s - TS^s \tag{14.56}$$

and

$$G^s = E^s - TS^s - \gamma a. \tag{14.57}$$

Upon using Eq. (14.37) in these expressions, we obtain

$$\delta A^s = -S^s\,\delta T + \sum_{j=1}^{c} \mu_j\,\delta n_j^s + \gamma\delta a \tag{14.58}$$

and

$$\delta G^s = -S^s\,\delta T + \sum_{j=1}^{c} \mu_j\,\delta n_j^s - a\,\delta\gamma. \tag{14.59}$$

Since it is known experimentally that, as is the Gibbs free energy of the bulk phases, G^s is a first-order homogeneous function of densities

$\{n^s_j : j = 1, \ldots, c\}$, it is possible to express it as

$$G^{s} = \sum_{j=1}^{c} \mu_j n^s_j. \tag{14.60}$$

In view of Eq. (14.59) it follows

$$\sum_{j=1}^{c} n^s_j \, \delta\mu_j = -S^s \, \delta T - a \, \delta\gamma. \tag{14.61}$$

Alternatively, this may be written as

$$\sum_{j=1}^{c} \Gamma^\sigma_j \, \delta\mu_j = -S_\sigma \, \delta T - \delta\gamma. \tag{14.62}$$

This is the Gibbs–Duhem equation for interface, which is the integrability condition for the differential form (Eq. (14.59)). Just as does the Gibbs–Duhem equation for bulk phases, Eq. (14.61) plays an important role in the thermodynamics of interfaces.

14.5 Location of the Dividing Surface and Surface Tension

For the derivation of the interfacial equations and, in particular, Eq. (14.37) we have introduced a dividing surface at an arbitrary location in the vicinity of the physical surface of discontinuity. Therefore the surface tension and other interfacial quantities involved are expected to depend on the location of the dividing surface. Contrary to the expectation, this is not true; that is, they are independent of the location of the dividing surface. We have made use of this fact in the previous sections.

In this section, we show the statement is indeed true before proceeding to further discussions of interfacial thermodynamics. We will specialize our treatment to a plane interface. Let us now consider the difference between the values of various extensive properties of the surface as the plane dividing surface is moved a distance ϵ into, say, phase β from location 1 to location 2, keeping the area and state of the system fixed. The changes in volume of the phases α and β therefore are

$$\Delta V^{(\alpha)} = \epsilon a, \quad \Delta V^{(\beta)} = -\epsilon a. \tag{14.63}$$

The difference in E^s arising from the change in the location of the dividing surfaces is

$$\begin{aligned}
\Delta E^{s} &= E^s_2 - E^s_1 \\
&= \left(E^{(3)} - E^{(\alpha)}_2 - E^{(\beta)}_2 \right) - \left(E^{(3)} - E^{(\alpha)}_1 - E^{(\beta)}_1 \right) \\
&= -\left(E^{(\alpha)}_2 - E^{(\alpha)}_1 \right) - \left(E^{(\beta)}_2 - E^{(\beta)}_1 \right),
\end{aligned} \tag{14.64}$$

where $E_i^{(\alpha)}$ and $E_i^{(\beta)}$ $(i = 1, 2)$ are the internal energies of phases α and β with the dividing surface located at position i, respectively. Similarly,

$$\Delta S^s = -\left(S_2^{(\alpha)} - S_1^{(\alpha)}\right) - \left(S_2^{(\beta)} - S_1^{(\beta)}\right), \tag{14.65}$$

$$\Delta n_j^s = -\left(n_{j2}^{(\alpha)} - n_{j1}^{(\alpha)}\right) - \left(n_{j2}^{(\beta)} - n_{j1}^{(\beta)}\right). \tag{14.66}$$

By the integrability of the Gibbs equation—the Gibbs–Duhem equation, there holds Eq. (14.57), and we find

$$\Delta G^s = \Delta E^s - T\,\Delta S^s + \sum_{j=1}^{c} \mu_j\,\Delta n_j^s - a\,\Delta\gamma, \tag{14.67}$$

but since

$$\Delta G^s = G_2^s - G_1^s = -\left(G_2^{(\alpha)} - G_1^{(\alpha)}\right) - \left(G_2^{(\beta)} - G_1^{(\beta)}\right) \tag{14.68}$$

$$= -\left(E_2^{(\alpha)} - E_1^{(\alpha)}\right) + T\left(S_2^{(\alpha)} - S_1^{(\alpha)}\right) - \sum_{j=1}^{c} \mu_j\left(n_{j2}^{(\alpha)} - n_{j1}^{(\alpha)}\right)$$

$$- \left(p_2^{(\alpha)} - p_1^{(\alpha)}\right)a\epsilon - \left(E_2^{(\beta)} - E_1^{(\beta)}\right)$$

$$+ T\left(S_2^{(\beta)} - S_1^{(\beta)}\right) - \sum_{j=1}^{c} \mu_j\left(n_{j2}^{(\beta)} - n_{j1}^{(\beta)}\right) - \left(p_2^{(\beta)} - p_1^{(\beta)}\right)a\epsilon$$

$$= \Delta E^s - T\,\Delta S^s + \sum_{j=1}^{c} \mu_j\,\Delta n_j^s - \left(p_2^{(\alpha)} - p_1^{(\alpha)}\right)a\epsilon - \left(p_2^{(\beta)} - p_1^{(\beta)}\right)a\epsilon,$$

balancing Eqs. (14.67) and (14.68) yields

$$\left(p_1^{(\alpha)} - p_2^{(\alpha)}\right)\epsilon + \left(p_1^{(\beta)} - p_2^{(\beta)}\right)\epsilon = \Delta\gamma. \tag{14.69}$$

Since $p_2^{(\alpha)} - p_1^{(\alpha)} = 0$ and $p_2^{(\beta)} - p_1^{(\beta)} = 0$ on account of mechanical equilibrium condition we conclude

$$\Delta\gamma = \gamma_2 - \gamma_1 = 0.$$

That is, the surface tension is independent of the position of the dividing surface.

14.6 Gibbs Phase Rule Including Interface

The two phase system of c components with an interface obeys a phase rule which is modified from the Gibbs phase rule in the absence of interfaces. The intensive variables of the bulk phases α and β are $T^{(\alpha)}, p^{(\alpha)}, x_1^{(\alpha)}, \ldots, x_{c-1}^{(\alpha)}$ and $T^{(\beta)}, p^{(\beta)}, x_1^{(\beta)}, \ldots, x_{c-1}^{(\beta)}$ whereas the intensive variables for the interface are $T^\sigma, \gamma, \Gamma_1^\sigma, \ldots, \Gamma_{c-1}^\sigma$. Therefore there are $3(c+1)$ variables in total. The number of constraining equations—the equilibrium conditions—are $2(c+1)$ according to Eqs. (14.5) and (14.6). Therefore the number of degree of freedom is

$$f = 3(c+1) - 2(c+1) = c+1. \tag{14.70}$$

This is the Gibbs phase rule for two phase equilibrium for a c-component system with an interface. It is in contrast to the two phase equilibrium of the same system without an interface for which $f = c$.

14.7 Thermodynamics of Interface

If variations are limited to those in which the varied state is one of equilibrium then the variation sign δ can be replaced with the differential sign d. In this case, in place of Eq. (14.37) we have the differential form

$$dE^s = T \, dS^s + \sum_{j=1}^{c} \mu_j \, dn_j^s + \gamma \, da \tag{14.71}$$

and the Gibbs–Duhem equation is written as

$$S^s \, dT + \sum_{j=1}^{c} n_j^s \, d\mu_j = -a \, d\gamma \tag{14.72}$$

or

$$S_\sigma \, dT + \sum_{j=1}^{c} \Gamma_j^\sigma \, d\mu_j = -d\gamma. \tag{14.73}$$

This equation is in fact called the Gibbs adsorption equation.

14.7.1 *Invariance of Derivatives to the Position of the Dividing Surface*

It was shown that the surface tension is invariant to the position of the dividing surface. The derivatives of the surface tension is also invariant to the position of the dividing surface. It is shown below: It is possible to choose the dividing surface such that one of Γ_j^σ is equal to zero except when the corresponding quantities in the two phases are identical. For example, if the dividing surface is chosen such that Γ_1^σ vanishes, the Gibbs adsorption equation becomes

$$d\gamma = -S_\sigma \, dT - \sum_{j=2}^{c} \Gamma_{j(1)}^\sigma \, d\mu_j, \qquad (14.74)$$

where the subscript 1 in $\Gamma_{j(1)}^\sigma$ refers to the densities when the dividing surface is chosen such that $\Gamma_1^\sigma = 0$. It follows from Eq. (14.74)

$$\left(\frac{\partial \gamma}{\partial \mu_j} \right)_{T,\mu'} = -\Gamma_{j(1)}^\sigma, \qquad (14.75)$$

where $\mu' = \{\mu_k; \mu_k \neq \mu_j\}$. We would like to show this derivative is invariant to the position of the dividing surface. We consider a two-component system for simplicity. At $T = $ constant

$$d\gamma = -\Gamma_1^\sigma \, d\mu_1 - \Gamma_2^\sigma \, d\mu_2, \qquad (14.76)$$

but

$$dp^{(\omega)} = \rho_1^{(\omega)} \, d\mu_1 + \rho_2^{(\omega)} \, d\mu_2 \quad (\omega = \alpha, \beta), \qquad (14.77)$$

where $\rho_i^{(\omega)} = n_i^{(\omega)}/V^{(\omega)}$. Since $p^{(\alpha)} = p^{(\beta)}$ at two-phase equilibrium it follows

$$d\mu_1 = \frac{\rho_2^{(\beta)} - \rho_2^{(\alpha)}}{\rho_1^{(\alpha)} - \rho_1^{(\beta)}} d\mu_2. \qquad (14.78)$$

Substitution of this into Eq. (14.76) yields

$$\left(\frac{\partial \gamma}{\partial \mu_2} \right)_T = -\left[\Gamma_2^\sigma - \frac{\rho_2^{(\alpha)} - \rho_2^{(\beta)}}{\rho_1^{(\alpha)} - \rho_1^{(\beta)}} \Gamma_1^\sigma \right]. \qquad (14.79)$$

We note that Γ_2^σ and Γ_1^σ are the excess quantities of components per unit area of the surface when the dividing surface is located at the surface of tension. Since $\epsilon(\rho_1^{(\alpha)} - \rho_1^{(\beta)})$ is the amount by which Γ_1^σ is increased when the dividing surface is moved normally a distance ϵ toward the inside α and Γ_1^σ is the excess quantity of component 1 when the dividing surface is given the position of the surface of tension, the distance that the dividing surface must be moved toward α to make the excess quantity of 1 at the surface equal to zero is

$$\epsilon = -\frac{\Gamma_1^\sigma}{\rho_1^{(\alpha)} - \rho_1^{(\beta)}}. \qquad (14.80)$$

The amount that must be added to Γ_2^σ is then

$$-\frac{\left(\rho_2^{(\alpha)} - \rho_2^{(\beta)}\right)\Gamma_1^\sigma}{\rho_1^{(\alpha)} - \rho_1^{(\beta)}}. \tag{14.81}$$

Therefore we conclude that

$$\Gamma_{2(1)}^\sigma = \Gamma_2^\sigma - \frac{\left(\rho_2^{(\alpha)} - \rho_2^{(\beta)}\right)}{\rho_1^{(\alpha)} - \rho_1^{(\beta)}}\Gamma_1^\sigma \tag{14.82}$$

and Eq. (14.79) is identical with Eq. (14.75) in the case of a two-component system. This shows the invariance of the derivatives to the position of the dividing surface.

14.7.2 *Various Thermodynamic Relations for Interface*

Earlier, we have introduced the Helmholtz and Gibbs free energies for interface. It is also convenient to define the enthalpy for the interface. Formally and in analogy to the enthalpy of a bulk phase, it may be defined by

$$H^s = E^s + p^s V^s,$$

but since the excess volume V^s of a surface of discontinuity is equal to zero by definitions of the excess quantities, we find

$$H^s = E^s. \tag{14.83}$$

Since

$$dE^s = T\ dS^s + \sum_{j=1}^c \mu_j\ dn_j^s + \gamma\ da, \tag{14.84}$$

it follows

$$dH^s = T\ dS^s + \sum_{j=1}^c \mu_j\ dn_j^s + \gamma\ da. \tag{14.85}$$

From the definitions of A^s and G^s follow the differential forms—the fundamental equations

$$dA^s = -S^s\ dT + \sum_{j=1}^c \mu\ dn_j^s + \gamma\ da \tag{14.86}$$

and

$$dG^s = -S^s\ dT + \sum_{j=1}^c \mu_j\ dn_j^s - a\ d\gamma. \tag{14.87}$$

Various thermodynamic relations for interface can be derived from these fundamental equations—the Gibbs relations for interface. We will examine some of thermodynamic relations in the case of a binary mixture for simplicity.

14.7.3 *Liquid–Vapor Equilibrium*

14.7.3.1 *Density Dependence of* γ

Earlier, we have calculated the relation of surface tension to the interfacial density; see Eq. (14.75). Here we consider the case of a binary solution at equilibrium with its vapor phase. At constant temperature there hold the equations

$$d\gamma = -\Gamma^\sigma_{2(1)} \, d\mu_2, \tag{14.88}$$

$$d\gamma = -\Gamma^\sigma_{1(2)} \, d\mu_1, \tag{14.89}$$

$$-d\gamma = \Gamma^\sigma_1 \, d\mu_1 + \Gamma^\sigma_2 \, d\mu_2. \tag{14.90}$$

We designate component 2 for the solute and component 1 for the solvent. To make further progress in thermodynamics it is necessary to know the chemical potentials—the constitutive relations. We shall assume the chemical potentials are given by the form

$$\mu_i(p, T, x) = \mu_i^0(p, T) + RT \ln a_i \quad (i = 1, 2), \tag{14.91}$$

$$a_i = x_i f_i, \tag{14.92}$$

where x_i is the mole fraction and f_i the activity coefficient of component i. Note that this is the same as the chemical potential of component i in the bulk owing to the equilibrium conditions. An appropriate convention should be used for the activity coefficients; see Chapter 11. Then the excess densities $\Gamma^\sigma_{2(1)}$ and $\Gamma^\sigma_{1(2)}$ can be calculated from the information on the surface tension and the bulk properties:

$$\Gamma^\sigma_{1(2)} = -\frac{1}{RT} \left(\frac{\partial \gamma}{\partial \ln x_1 f_1} \right)_T, \tag{14.93}$$

$$\Gamma^\sigma_{2(1)} = -\frac{1}{RT} \left(\frac{\partial \gamma}{\partial \ln x_2 f_2} \right)_T. \tag{14.94}$$

Equation (14.94) shows how the excess interfacial density of the solute—that is, the adsorption of component 2—can be determined by measuring the variation of the surface tension with respect to the solute concentration in the bulk phase. Note that for dilute solutions the activity coefficients may be taken equal to unity. These equations show that if the solute is positively adsorbed in the surface, namely, $\Gamma^\sigma_{2(1)}$ is positive, the surface tension decreases with increasing concentration of solute in the dilute solution limit. However, in the case of $\Gamma^\sigma_{2(1)}$ negative, since Γ^σ_2 cannot be negative there holds the bound in solute density

$$\frac{\left(\rho_2^{(\alpha)} - \rho_2^{(\beta)}\right)\Gamma^\sigma_1}{\rho_1^{(\alpha)} - \rho_1^{(\beta)}} \geq 0, \tag{14.95}$$

which puts a bound on the increase in surface tension.

It is possible to make use of vapor pressure to express Eqs. (14.93) and (14.94). If the vapor may be regarded as ideal then since at constant T

$$d\mu_i = RTd\ln p_i, \tag{14.96}$$

where p_i is the vapor pressure, it follows that

$$\Gamma^\sigma_{1(2)} = -\frac{1}{RT}\left(\frac{\partial\gamma}{\partial\ln p_1}\right)_T = -\frac{p_1}{RT}\left(\frac{\partial\gamma}{\partial p_1}\right)_T, \tag{14.97}$$

$$\Gamma^\sigma_{2(1)} = -\frac{1}{RT}\left(\frac{\partial\gamma}{\partial\ln p_2}\right)_T = -\frac{p_2}{RT}\left(\frac{\partial\gamma}{\partial p_2}\right)_T. \tag{14.98}$$

From the Gibbs–Duhem equation for the vapor phase at constant temperature and pressure

$$x_1 d\ln p_1 + x_2 d\ln p_2 = 0, \tag{14.99}$$

where x_1 and x_2 are the mole fractions of vapor species, and from Eqs. (14.97) and (14.98) follows the relation

$$x_2\Gamma^\sigma_{1(2)} = -x_1\Gamma^\sigma_{2(1)}. \tag{14.100}$$

Since Eq. (14.90) can be written for the situation under consideration

$$-d\gamma = \Gamma^\sigma_1 RTd\ln p_1 + \Gamma^\sigma_2 RTd\ln p_2, \tag{14.101}$$

this may be written as

$$-d\gamma = \left(\Gamma^\sigma_2 - \frac{x_2}{x_1}\Gamma^\sigma_1\right) RTd\ln p_2 = \left(\Gamma^\sigma_1 - \frac{x_1}{x_2}\Gamma^\sigma_2\right) RTd\ln p_1, \tag{14.102}$$

which implies

$$\Gamma^\sigma_{1(2)} = \Gamma^\sigma_1 - \frac{x_1}{x_2}\Gamma^\sigma_2, \tag{14.103}$$

$$\Gamma^\sigma_{2(1)} = \Gamma^\sigma_2 - \frac{x_2}{x_1}\Gamma^\sigma_1. \tag{14.104}$$

This shows $\Gamma^\sigma_{1(2)}$ and $\Gamma^\sigma_{2(1)}$ can be measured from the variation in γ with respect to the vapor pressure. The results obtained for binary mixtures can be easily generalized to multicomponent mixtures.

14.7.3.2 *Temperature Dependence of γ*

Since by the equilibrium conditions

$$\mu_j = \mu_j^{(\alpha)} = \mu_j^{(\beta)} \qquad (14.105)$$

and

$$\left(\frac{\partial \mu_j}{\partial T}\right)_{x^{(\alpha)}} = -s_j^{(\alpha)} + \left(\frac{\partial p}{\partial T}\right)_{x^{(\alpha)}} \bar{v}_j^{(\alpha)}, \qquad (14.106)$$

differentiating the Gibbs–Duhem equation for interface at constant $x^{(\alpha)}$ yields

$$\left(\frac{\partial \gamma}{\partial T}\right)_{x^{(\alpha)}} = -S_\sigma + \sum_{j=1}^{c} \Gamma_j^\sigma s_j^{(\alpha)} - \sum_{j=1}^{c} \Gamma_j^\sigma \bar{v}_j^{(\alpha)} \left(\frac{\partial p}{\partial T}\right)_{x^{(\alpha)}}. \qquad (14.107)$$

Furthermore, since

$$d\mu_j^{(\alpha)} = -s_j^{(\alpha)} dT + \bar{v}_j^{(\alpha)} dp + \sum_{i=1}^{c-1} \left(\frac{\partial \mu_j^{(\alpha)}}{\partial x_i^{(\beta)}}\right)_{T,p,x_{j\neq i}^{(\alpha)}} dx_i^{(\beta)}, \qquad (14.108)$$

from the second equality of Eq. (14.105) follows

$$\sum_{j=1}^{c} x_j^{(\beta)} \Delta \bar{s}_j \, dT - \sum_{j=1}^{c} x_j^{(\beta)} \Delta \bar{v}_j \, dp + \sum_{i=1}^{c} \mu_i^{(\alpha\beta)} dx_i^{(\beta)} = 0, \qquad (14.109)$$

where

$$\Delta \bar{s}_j = \bar{s}_j^{(\beta)} - \bar{s}_j^{(\alpha)}, \qquad (14.110)$$

$$\Delta \bar{v}_j = \bar{v}_j^{(\beta)} - \bar{v}_j^{(\alpha)}, \qquad (14.111)$$

$$\mu_i^{(\alpha\beta)} = \sum_{j=1}^{c} x_j^{(\beta)} \left(\frac{\partial \mu_j^{(\alpha)}}{\partial x_i^{(\beta)}}\right)_{T,p,x_{j\neq i}^{(\alpha)}}. \qquad (14.112)$$

Therefore we obtain from Eq. (14.109) the Clapeyron equation for the two-phase mixture

$$\left(\frac{\partial p}{\partial T}\right)_{x^{(\alpha)}} = \frac{\sum_{j=1}^{c} x_j^{(\beta)} \Delta \bar{s}_j}{\sum_{j=1}^{c} x_j^{(\beta)} \Delta \bar{v}_j} = \frac{\Delta s}{\Delta v}, \qquad (14.113)$$

where Δs and Δv are the entropy and volume change per mole of the mixture:

$$\Delta s = \sum_{j=1}^{c} x_j^{(\beta)} \Delta \bar{s}_j, \quad \Delta v = \sum_{j=1}^{c} x_j^{(\beta)} \Delta \bar{v}_j.$$

Upon substitution of this derivative into Eq. (14.107) we obtain

$$\left(\frac{\partial \gamma}{\partial T}\right)_{x^{(\alpha)}} = \sum_{j=1}^{c} \Gamma_j^{\sigma} s_j^{(\alpha)} - S_{\sigma} - \sum_{i=1}^{c} \Gamma_i^{\sigma} \bar{v}_i^{(\alpha)} \frac{\Delta s}{\Delta v}. \tag{14.114}$$

In the case of one-component system this equation reduces to

$$\frac{d\gamma}{dT} = \frac{\left(s^{(\alpha)} v^{(\beta)} - v^{(\alpha)} s^{(\beta)}\right)}{v^{(\beta)} - v^{(\alpha)}} \Gamma^{\sigma} - S_{\sigma}. \tag{14.115}$$

If the surface is chosen such that $\Gamma^{\sigma} = 0$ then

$$\frac{d\gamma}{dT} = -S_{\sigma}. \tag{14.116}$$

This relation makes it possible to evaluate the surface entropy.

Experimentally, it is known that in the vicinity of the critical temperature T_c the surface tension varies with T as[3]

$$\gamma = \gamma_0 (1 - T/T_c)^{11/9}. \tag{14.117}$$

That is, the surface tension vanishes at $T = T_c$. According to Guggenheim, this form was originally suggested by J. van der Waals who took 1.234 for the exponent. Thus in the neighborhood of T_c the surface entropy is deduced to behave as

$$S_{\sigma} = \frac{11\gamma_0}{9T_c}(1 - T/T_c)^{2/9}. \tag{14.118}$$

Since $H_{\sigma} = E_{\sigma}$ as shown earlier, we find

$$E_{\sigma} = G_{\sigma} + TS_{\sigma} \tag{14.119}$$

or it may be written as

$$E_{\sigma} = \gamma - T\frac{d\gamma}{dT}. \tag{14.120}$$

Therefore in the neighborhood of $T_c > T$

$$E_{\sigma} = \gamma_0 (1 - T/T_c)^{2/9}\left(1 + \frac{2T}{9T_c}\right) \tag{14.121}$$

according to the empirical Formula (14.117). Therefore the surface excess energy decreases with increasing T and eventually vanishes at T_c.

---

[3]See A. Ferguson, *Trans. Faraday Soc.* **19**, 408 (1923); E. A. Guggenheim, *J. Chem. Phys.* **13**, 253 (1945); L. Riedel, *Chem.-Ing. Tech.* **27**, 209 (1955); M. Yu. Gorbachev, *Phys. Chem. Liq.* **39**, 315 (2001).

Since the surface specific heat C_σ may be defined by

$$C_\sigma = \frac{dE_\sigma}{dT}, \tag{14.122}$$

we note that it is possible to deduce from Formula (14.122) that in the neighborhood of T_c the surface specific heat is singular and negative:

$$C_\sigma = -\frac{22\gamma_0}{81T_c^2} \frac{T}{(1 - T/T_c)^{7/9}} \quad (T < T_c). \tag{14.123}$$

This negative surface excess heat capacity near the critical temperature $(T_c > T)$ is in contrast to the bulk heat capacities, which must be positive; see Chapter 5, Section 5.3. Since the surface excess energy in essence is a relative energy of the surface over the bulk energy, the surface specific heat, C_σ, is the rate of change in the relative energy. Therefore C_σ given in Eq. (14.123) is not in violation of the second law of thermodynamics, which demands a positive heat capacity as we have shown in Chapter 5.

Chapter 15

Electrolyte Solutions

The treatment of activities and activity coefficients of nonelectrolytes given in Chapter 13 can be applied to electrolyte solutions. An electrolyte is called strong if it completely dissociates into constituent ions in solution. Otherwise, it is called weak, weak acids or bases being examples. We will study solutions of strong electrolytes in this chapter. Nonideal behavior of nonelectrolyte solutions arises from molecular interactions, which are generally finite ranged. Therefore, if the solution is dilute, then the interactions between solute molecules are rather weak because of large separations of solute molecules. However, the situation changes drastically in the case of ionic solutions, owing to the long range nature of Coulomb interactions between ions in the solution. For example, ionic solutions would not display Henry's law even at a concentration at which nonelectrolyte solutions would exhibit almost ideal behavior. We have seen that according to the Margules expansions for the activities of binary solutions the logarithm of an activity coefficient starts with a quadratic term in concentration, but in the case of ionic solutions it is known experimentally that the same quantity starts with the square root of the ionic strength of the solution or, roughly speaking, with $m^{1/2}$ where m is the molality of the solution. This behavior can be understood by means of the Debye–Hückel theory, which will be discussed in the next chapter. Our aim here is to discuss methods of phenomenologically determining activities and activity coefficients of ions. Therefore we shall develop some notions and definitions necessary for the purpose.

15.1   Mean Activity and Mean Activity Coefficient

For electrolyte solutions the standard state of each ionic species is chosen such that the activity of an ion becomes equal to the concentration at infinite dilution at 1 atm pressure and the temperature of the solution. For electrolyte solutions it is conventional to express the concentrations in terms of molalities.

Let us consider an electrolyte represented by the formula $C_{\nu_+}^{z+} A_{\nu_-}^{z-}$ which on dissociation produces ν_+ cations C^{z+} of charge number z_+ (positive), and ν_- anions A^{z-} of charge number z_- (negative). As an example, take $BaSO_4$ for which $\nu_+ = 1$, $\nu_- = 1$, $z_+ = 2$, and $z_- = -2$. The dissociation of an electrolyte may be looked upon as a chemical reaction; thus

$$C_{\nu_+}^{z+} A_{\nu_-}^{z-} \rightleftharpoons \nu_+ C^{z+} + \nu_- A^{z-}. \tag{15.1}$$

The electroneutrality demands

$$\nu_+ z_+ + \nu_- z_- = 0. \tag{15.2}$$

The chemical potential of each ionic species is looked for in the form

$$\mu_+ = \mu_+^0(T, p) + RT \ln a_+ \tag{15.3}$$

and

$$\mu_- = \mu_-^0(T, p) + RT \ln a_-, \tag{15.4}$$

where the subscripts $+$ and $-$ refer to the cation and anion, respectively, and a_+ and a_- are the activities. These activities may be expressed in terms of the activity coefficient γ_+ or γ_- and the molality m_+ or m_-:

$$a_+ = \gamma_+ m_+; \quad a_- = \gamma_- m_-. \tag{15.5}$$

The standard chemical potentials are then defined by the limits

$$\mu_+^0 = \lim_{m_+ \to 0} (\mu_+ - RT \ln m_+),$$
$$\mu_-^0 = \lim_{m_- \to 0} (\mu_- - RT \ln m_-), \tag{15.6}$$

which imply that

$$\lim_{m_+ \to 0} \gamma_+ = 1, \tag{15.7}$$

$$\lim_{m_- \to 0} \gamma_- = 1. \tag{15.8}$$

Since we are assuming complete dissociation of the electrolyte, the chemical potential $\mu_{\pm}$ of $C_{\nu_+}^{z_+} A_{\nu_-}^{z_-}$ must be the sum of ionic chemical potentials:

$$\mu_{\pm} = \nu_+ \mu_+ + \nu_- \mu_-. \tag{15.9}$$

Let us denote the molality of the electrolyte solution by m. Then, obviously

$$m_+ = \nu_+ m \quad \text{and} \quad m_- = \nu_- m. \tag{15.10}$$

Substitution of Eqs. (15.3) and (15.4) into Eq. (15.9) yields the chemical potential $\mu_{\pm}$ in the form

$$\mu_{\pm} = \mu_{\pm}^0 + RT \ln[(\gamma_+ m_+)^{\nu_+}(\gamma_- m_-)^{\nu_-}], \tag{15.11}$$

where

$$\mu_{\pm}^0 = \nu_+ \mu_+^0 + \nu_- \mu_-^0. \tag{15.12}$$

The ionic chemical potentials μ_+ and μ_- or the ionic activity coefficients γ_+ and γ_- are not separately measurable in the laboratory. They are in fact only collectively measured. It therefore is convenient to define the mean activity coefficient by

$$\gamma_{\pm} = (\gamma_+^{\nu_+} \gamma_-^{\nu_-})^{1/\nu}, \tag{15.13}$$

where

$$\nu = \nu_+ + \nu_-, \tag{15.14}$$

and the mean molality by

$$m_{\pm} = (m_+^{\nu_+} m_-^{\nu_-})^{1/\nu}. \tag{15.15}$$

With these definitions the chemical potential $\mu_{\pm}$ may be given in the form

$$\mu_{\pm} = \mu_{\pm}^0 + \nu RT \ln(\gamma_{\pm} m_{\pm}). \tag{15.16}$$

This form of chemical potential is operational and can be measured in the laboratory by using the methods discussed in Chapter 13. It can also be measured by electrochemical methods which will be discussed later. When Eq. (15.10) is made use of, the mean molality may be written in terms of the molality m of the electrolyte:

$$m_{\pm} = (\nu_+^{\nu_+} \nu_-^{\nu_-})^{1/\nu} m. \tag{15.17}$$

The mean quantities defined above suggest that the mean activity may be defined similarly:

$$a_{\pm} = (a_+^{\nu_+} a_-^{\nu_-})^{1/\nu} = \gamma_{\pm} m_{\pm}. \tag{15.18}$$

In order to better understand the meaning of the mean activity let us return to Eq. (15.9) and interpret it somewhat differently. This equation may be regarded as the chemical equilibrium condition for the chemical reaction (15.1), if we write the chemical potential for the electrolyte in the form

$$\mu_{\pm} = \mu_2^0 + RT \ln a_2, \tag{15.19}$$

where a_2 is the activity of the electrolyte and μ_2^0 is the standard chemical potential. Lewis and Randall proposed to choose this such that

$$\mu_2^0 = \mu_{\pm}^0. \tag{15.20}$$

This choice is tantamount to choosing the standard state such that the equilibrium constant of chemical reaction (15.1) becomes unity. This is easily seen as follows. Since the equilibrium constant is

$$K = \frac{a_+^{\nu_+} a_-^{\nu_-}}{a_2} \tag{15.21}$$

and it is also given in terms of standard chemical potentials by the formula

$$K = \exp\left(\frac{\mu_{\pm}^0 - \mu_2^0}{RT}\right), \tag{15.22}$$

Eq. (15.20) implies that $K = 1$ and thus

$$a_2 = a_+^{\nu_+} a_-^{\nu_-} = a_{\pm}^{\nu}. \tag{15.23}$$

The second line follows from Eq. (15.18). It clearly shows that $a_{\pm}$ is the geometric mean of a_2.

Mean activities and mean activity coefficients can be determined by applying the methods discussed in Chapter 13 in addition to the electrochemical methods we are going to discuss in the next chapter. Since electrolytes are not generally volatile, vapor pressure measurements cannot be employed for direct measurement of activity coefficients. However, indirect measurements may be carried out for the solvent by using various methods already mentioned for nonelectrolytes and then employing the Gibbs–Duhem equation to calculate the activity coefficient of the solute.

15.2   Isopiestic Method

Here we shall discuss a method of vapor pressure measurement which is called the isopiestic method. This method enables us to determine the activity of an electrolyte solution whose solvent vapor is in equilibrium with that of a solution for which the activity is accurately known over a concentration range. To make the discussion simple we consider two binary electrolytic solutions consisting of different electrolytes. As usual, the solvent will be designated as 1 and the two electrolytes as 2 and 3, respectively. Thus one solution consists of components 1 and 2, and the other consists of components 1 and 3. We shall assume that the 1–3 solution is the reference solution for which the activities are already known. We will denote this solution by solution 2. The other will be denoted by solution 1. The two solutions are maintained at a given temperature and are enclosed in a single cover such that they acquire a common vapor pressure, that is, an equilibrium vapor pressure (see Fig. 15.1). When an equilibrium is reached, the solutions are analyzed for their compositions. The equilibrium conditions then are

$$\mu_1(v) = \mu_1(sol.\,1), \quad \mu_1(v) = \mu_1(sol.\,2). \tag{15.24}$$

Let us denote the activities of the solvent in solutions 1 and 2 by $a_1^{(1)}$ and $a_1^{(2)}$, respectively. Then the chemical potentials may be written as

$$\mu_1(sol.\,1) = \mu_1^l(T,p) + RT \ln a_1^{(1)},$$

$$\mu_1(sol.\,2) = \mu_1^l(T,p) + RT \ln a_1^{(2)}, \tag{15.25}$$

with obvious meaning for other symbols. The equilibrium conditions therefore imply the equation

$$a_1^{(1)} = a_1^{(2)}, \tag{15.26}$$

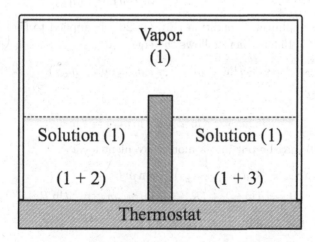

Fig. 15.1   Isopiestic method of determining activity coefficients.

which is another form of the equilibrium condition for the system. This may be rewritten in the form

$$x_1^{(1)} f_1^{(1)} = x_1^{(2)} f_1^{(2)}, \tag{15.27}$$

where the superscripts refer to the solutions. Therefore the activity coefficient $f_1^{(1)}$ at the concentration $x_1^{(1)}$ can be determined from the known value of $f_1^{(2)}$ at $x_1^{(2)}$. To determine the solute activity coefficient we now define osmotic coefficients ϕ_1 and ϕ_2 for the two solutions:

$$
\begin{aligned}
\phi_1 &= -\left(\frac{1000}{\nu_2 m_2 M_1}\right) \ln a_1^{(1)}, \\
\phi_2 &= -\left(\frac{1000}{\nu_3 m_3 M_1}\right) \ln a_1^{(2)},
\end{aligned}
\tag{15.28}
$$

where ν_2 is the number of ions produced by 1 mole of electrolyte 2, ν_3 the number of ions produced by 1 mole of electrolyte 3, and m_2 and m_3 are the molalities of electrolytes 2 and 3. Then the Gibbs–Duhem equations for the two solutions are

$$
\begin{aligned}
\left(\frac{1000}{M_1}\right) d\ln a_1^{(1)} &= -m_2 d\ln a_2^{(1)} \\
&= -\nu_2 m_2 d\ln(\gamma_{\pm 2} m_{\pm 2}), \\
\left(\frac{1000}{M_1}\right) d\ln a_1^{(2)} &= -m_3 d\ln a_2^{(2)} \\
&= -\nu_3 m_3 d\ln(\gamma_{\pm 3} m_{\pm 3}).
\end{aligned}
\tag{15.29}
$$

When the equilibrium condition (Eq. (15.26)) is applied to the two equations in Eq. (15.29), there follows the equation

$$\nu_2 m_2 d\ln(\gamma_{\pm 2} m_{\pm 2}) = \nu_3 m_3 d\ln(\gamma_{\pm 3} m_{\pm 3}). \tag{15.30}$$

Since

$$d\ln m_{\pm 2} = d\ln m_2, \quad d\ln m_{\pm 3} = d\ln m_3,$$

Eq. (15.30) may be put into a more convenient form

$$
\begin{aligned}
d\ln \gamma_{\pm 2} &= \mathcal{R}\, d\ln(\gamma_{\pm 3} m_{\pm 3}) - d\ln m_2 \\
&= d\ln \gamma_{\pm 3} + (\mathcal{R} - 1) d\ln(\gamma_{\pm 3} m_{\pm 3}) + d\ln \mathcal{R},
\end{aligned}
\tag{15.31}
$$

where $\mathcal{R}$ is the isopiestic ratio defined by

$$\mathcal{R} = \frac{\nu_3 m_3}{\nu_2 m_2}. \tag{15.32}$$

By integrating Eq. (15.31) we find the activity coefficient by the isopiestic method:

$$\ln \gamma_{\pm 2} = \ln \gamma_{\pm 3} + \ln \mathcal{R} + \int_0^{m_2} d(\gamma_{\pm 3} m_{\pm 3}) \frac{(\mathcal{R} - 1)}{\gamma_{\pm 3} m_{\pm 3}}. \tag{15.33}$$

Therefore a plot of $[(\mathcal{R} - 1)/\gamma_{\pm 3} m_{\pm 3}]$ vs $\gamma_{\pm 3} m_{\pm 3}$ and a graphical integration of the curve will yield the integral and thus the mean activity coefficient of electrolyte 2. It is sometimes necessary to combine the isopiestic method with other methods such as osmotic pressure and freezing point measurements to cover the required range of concentration, since it is not possible to obtain accurate values for $\mathcal{R}$ below $0.1m$ in concentration.

15.3 Activity Coefficient from Freezing Point Measurement

Since freezing point measurements of electrolytes give highly accurate activity coefficients, they are an important source of information. If the expansions (13.132) and (13.133) are used in Eq. (13.131), it can be recast into the form more suitable for discussion:

$$d\ln a_1 = -\left(RT_m^2\right)^{-1}\left[\Delta h_m^0 + b\theta + O(\theta^2)\right] d\theta, \tag{15.34}$$

where a_1 is the activity of the solvent,

$$\theta = T_m - T,$$
$$b = 2\Delta h_m^0 / T_m - \Delta C_{pm},$$

and we have retained only two terms in the expansion. Since the Gibbs–Duhem equation may be written at constant T and p as

$$d\ln a_2 = -\frac{1000}{mM_1} d\ln a_1,$$

the equation for a_2 is

$$d\ln a_2 = \frac{1000}{RT_m^2 M_1}\left(\Delta h_m^0 + b\theta\right)\frac{d\theta}{m}. \tag{15.35}$$

With the definitions

$$\lambda = \frac{RT_m^2 M_1}{1000 \, \Delta h_m^0},$$
$$\eta = \frac{1000b}{RT_m^2 M_1},$$

and with the aid of Eqs. (15.23) and (15.18), we obtain the following equation for the activity coefficient of the electrolyte

$$d \ln \gamma_\pm = (\nu\lambda m)^{-1} \, d\theta + \eta(\nu m)^{-1}\theta \, d\theta - d\ln m. \tag{15.36}$$

Integration of this equation will supply the mean activity coefficient. Since the freezing point depression θ is a function of m, the integration requires measurements of θ over a range of m.

Since $\theta = \nu\lambda m$ at very low concentrations as can easily be inferred from Eq. (13.136), Eq. (15.36) is not in a suitable form for integration because of the logarithmic behavior of the first and last terms. The following transformation proposed by Lewis is found useful for such a purpose:

$$\Omega = 1 - \frac{\theta}{\nu\lambda m}.$$

Then

$$d\Omega = \theta(\nu\lambda m^2)^{-1} \, dm - (\nu\lambda m)^{-1} \, d\theta$$

so that

$$d\ln \gamma_\pm = -\Omega d\ln m - d\Omega + \eta\theta(\nu m)^{-1} \, d\theta. \tag{15.37}$$

By integrating this equation we obtain

$$\ln \gamma_\pm = -\Omega - \int_0^m \Omega d\ln m + \int_0^m \nu^{-1}\eta\theta\left(\frac{d\theta}{dm}\right) d\ln m. \tag{15.38}$$

This form of equation has no logarithmic singularity and a graphical integration becomes quite straightforward and not prone to inaccuracy.

15.4 Activity Coefficient from Osmotic Pressure Measurement

The osmotic pressure expression (Eq. (13.160)) for nonelectrolytes can be also directly made use of for determining the activity coefficients of electrolytes. We consider a binary solution. Changing the variable from p to the osmotic pressure π, we write Eq. (13.160) in the form

$$RT \ln a_1 = -\int_0^\pi dz \, \bar{v}_1, \tag{15.39}$$

where $\bar{v}_1$ is a function of z: $\bar{v}_1 = \bar{v}_1(p^0 + z)$. We introduce the osmotic coefficient of the electrolyte defined by the expression

$$\phi = -\left(\frac{1000}{\nu m M_1}\right) \ln a_1. \tag{15.40}$$

Note that there appears the total number of ions produced by the electrolyte in the definition of osmotic coefficient, since the molality is increased by the factor ν in the case of a strong electrolyte. Replacing with ϕ the activity term on the left-hand side of Eq. (15.39) we obtain

$$\left(\frac{\nu M_1 RT}{1000}\right) m\phi = \int_0^\pi dz\, \bar{v}_1. \tag{15.41}$$

Since the compressibility of the solvent is generally negligible, the partial molar volume on the right in Eq. (15.41) may be regarded as a constant and the integration may be easily performed. We then obtain

$$\pi = \left(\frac{\nu M_1 RT}{1000\bar{v}_1^0}\right) m\phi, \tag{15.42}$$

where $\bar{v}_1^0 = \bar{v}_1(p^0)$. Compare this expression for the equivalent formula for nonelectrolytes. The term osmotic coefficient originated from this equation. This equation suggests that measurements of the osmotic pressure yield the values of the osmotic coefficient which can be used to find the activity coefficient of the electrolyte by applying the method leading to Eq. (13.105):

$$\ln \gamma_\pm = \phi - 1 + \int_0^m dm \frac{\phi - 1}{m}, \tag{15.43}$$

where ϕ is given by Eq. (15.40).

The osmotic-pressure effect can be significantly noticeable even for dilute solutions, but its measurement requires a reliable semipermeable membrane. Recent developments in polymeric membranes and zeolites could make this effect achieve its potential for many systems of interest and usefulness in electrochemistry.

Chapter 16

Debye–Hückel Theory of Strong Electrolyte Solutions

Solutions of strong electrolytes exhibit nonideal behavior even at very low concentrations at which nonelectrolyte solutions would normally obey Henry's law. We have seen that in the case of binary solutions the Margules expansions begin with the square of a mole fraction. That is, for example, $\ln f_1$ tends to zero quadratically with respect to the concentration of the solute. This indicates that the activity coefficients tend fast to unity for neutral molecules. The situation drastically changes for electrolytes. While studying thermodynamics of strong electrolyte solutions, G. N. Lewis in 1913 observed that in contrast to nonelectrolyte solutions the mean excess free energies of electrolytes in solutions tend to zero as $m^{1/2}$. This means that the nonideality of the solutions persists down to much lower concentrations than for nonelectrolyte solutions. In order to make this statement more precise let us introduce the concept of ionic strength which was defined by Lewis as

$$\Gamma = \frac{1}{2} \sum_i m_i z_i^2,$$ (16.1)

here m_i is the molality of ionic species i and z_i is its charge number. Lewis showed experimentally that

$$\ln \gamma_\pm = -B\sqrt{\Gamma},$$ (16.2)

where B is a constant dependent on the temperature and other properties such as the dielectric constant of the solution. This was a vexing phenomenon at that time until Debye and Hückel came up with a theory[1] in 1923. Their theory is a statistical mechanical theory. Such a theory

---

[1]P. Debye and E. Hückel, *Phys. Z.* **185**, 305 (1923).

would not normally have a place in a textbook that treats thermodynamics phenomenologically. However, their theory has far reaching implications for many physical phenomena related to ionic solutions and plasmas. In fact, it was the first theory which showed how one might treat interactions in many particle systems in an average way (mean field theory) and consequently has had a considerable influence on the thinking in physical chemistry and physics. For this reason we make a break from our basic philosophy of describing phenomenology in this textbook and discuss their theory in an elementary manner.

One of the most important basic assumptions in the Debye–Hückel theory is that all deviations from ideality in ionic solutions are caused by the Coulomb interactions between ions produced by the electrolyte(s) in the solution. In the second assumption, it is postulated that the solvent is a structureless dielectric continuum of dielectric constant D. In the third assumption, ions are assumed to be hard spheres of radius r_i with the charge distributed uniformly on the surface. Therefore we imagine that an ionic solution is a continuous medium in which charged ions are immersed, while interacting with each other through long-ranged Coulomb potentials.

16.1 Ionic Atmosphere

Ions in the solution do not remain in one position, but continuously move around (Brownian motion), while mutually interacting, owing to the thermal agitation exerted on them by the medium. Since ions of opposite charges tend to attract each other, if we take an ion and count the distribution of ions around it, there will be a tendency for it to attract more ions of the opposite charge around it, but as we move farther out from the ion, the clustering of oppositely charged ions will gradually diminish and then the distribution of positively and negatively charged ions will be evened out and become that of the bulk. This region of space of an uneven charge distribution around the ion in question is called the ionic atmosphere. Since the ionic atmosphere is, on the average, charged oppositely to the ion in question, there is a Coulomb interaction between the ion and its ionic atmosphere, and the first task is then to calculate this interaction potential Φ. Since the charges in the ionic atmosphere are distributed and these distributions depend on the concentration of ions, the potential is expected to depend on the concentration of the solution. The potential on an ion is determined by the charge distribution around it through the Poisson equation

$$-\nabla^2 \Phi(\mathbf{R}) = 4\pi\rho/D, \tag{16.3}$$

where ρ is the charge distribution and D is the dielectric constant of the solvent. To solve Eq. (16.3) for Φ it is necessary to know the charge distribution ρ in the ionic atmosphere. Debye and Hückel assumed that the distribution is given by the Boltzmann distribution function. If the potential on a charge e_i at position $\mathbf{R}$ is $\Phi(\mathbf{R})$, then the potential energy is

$$V(\mathbf{R}) = e_i\Phi(\mathbf{R}). \tag{16.4}$$

According to the assumption just made, the number n_i' of ion i of charge e_i at position $\mathbf{R}$ is

$$n_i' = n_i \exp\left[-\frac{V(\mathbf{R})}{k_B T}\right] = n_i \exp\left[-\frac{e_i\Phi(\mathbf{R})}{k_B T}\right], \tag{16.5}$$

where n_i is the bulk number density of ion i defined by

$$n_i = \frac{\text{total number of ionic species } i \text{ in the solution}}{\text{volume of the solution}}. \tag{16.6}$$

The charge distribution ρ is then

$$\rho = \sum_i n_i e_i \exp\left[-\frac{e_i\Phi(\mathbf{R})}{k_B T}\right], \tag{16.7}$$

where the summation is over all ionic species. For example, if the solution is made up by a uni–univalent electrolyte only, then

$$\rho = ne\left\{\exp\left[\frac{e\Phi(\mathbf{R})}{k_B T}\right] - \exp\left[-\frac{e\Phi(\mathbf{R})}{k_B T}\right]\right\}, \tag{16.8}$$

where n is the number density of the electrolyte in the solution and e is the charge of an electron (the absolute value).

16.2 Mean Potential and the Excess Free Energy

When Eqs. (16.3) and (16.7) are combined, there follows the Poisson–Boltzmann equation:

$$\nabla^2 \Phi(\mathbf{R}) = -\frac{4\pi}{D} \sum_i n_i e_i \exp\left[-\frac{e_i\Phi(\mathbf{R})}{k_B T}\right]. \tag{16.9}$$

This equation unfortunately is nonlinear[2] with respect to Φ and not solvable in an analytic form. However, at low concentrations the Boltzmann factor may be expanded and the nonlinear terms in the expansion may be neglected to a good approximation. There then follows a linearized equation

$$\nabla^2 \Phi(\mathbf{R}) = -\frac{4\pi}{D} \sum_i n_i e_i \left[1 - \frac{e_i \Phi(\mathbf{R})}{k_B T} \right]. \tag{16.10}$$

Since by the electroneutrality

$$\sum_i n_i e_i = 0,$$

we finally obtain

$$\nabla^2 \Phi(\mathbf{R}) = \frac{4\pi}{D k_B T} \sum_i n_i e_i^2 \Phi(\mathbf{R}). \tag{16.11}$$

It is convenient to define

$$\kappa = \left(\frac{4\pi}{D k_B T} \sum_i n_i e_i^2 \right)^{1/2}, \tag{16.12}$$

which is called the Debye screening constant or the inverse Debye length. It has the dimension of reciprocal length and gives a measure of the size of the ionic atmosphere in the solution as we will see later.

Since ions are assumed to be hard spheres, the ionic atmospheres around them must be spherically symmetric and consequently the potential $\Phi(\mathbf{R})$ must be spherically symmetric. This means that Φ does not depend on the orientation of position vector $\mathbf{R}$. In that case, it is convenient to use the spherical coordinate system, and we cast the linearized Poisson–Boltzmann equation in the form

$$\frac{1}{R^2} \frac{d}{dR} R^2 \frac{d}{dR} \Phi(R) = \kappa^2 \Phi(R). \tag{16.13}$$

This ordinary differential equation is easily solved. Its general solution is

$$\Phi(R) = A \frac{e^{-\kappa R}}{R} + A' \frac{e^{\kappa R}}{R}, \tag{16.14}$$

---

[2]If the Poisson–Boltzmann (PB) equation is regarded as a differential equation that by itself defines $\Phi(\mathbf{R})$, then it is indeed nonlinear. However, from the more complete statistical mechanical viewpoint of electrolytes Φ or, more generally, the charge density in Eq. (16.7) obeys, for example, Fokker–Planck equations for pair correlation functions to which the PB equation is coupled. In this extended viewpoint, the PB equation is no longer nonlinear. L. Onsager solved such coupled differential equations in the low density limit in his work on conductance of electrolytes. See L. Onsager and R. M. Fuoss, *J. Phys. Chem.* **34**, 2689 (1932).

where A and A' are the integration constants. They are determined as follows. First let us observe that the potential must vanish as $R \to \infty$, since the ionic distribution becomes that of the bulk solution, namely, uniform with respect to the positive and negative charges, as R is increased from the ion in question. This boundary condition can be satisfied by the solution (Eq. (16.14)), only if $A' = 0$. Therefore the acceptable solution is

$$\Phi(R) = \frac{A}{R} \exp(-\kappa R). \tag{16.15}$$

The remaining constant A will be determined by making use of the electroneutrality of the ionic atmosphere. The potential obtained here has an exponential factor which decreases with increasing R and its rate of decrease is determined by the Debye screening factor κ. If κ is equal to zero, the potential is like a Coulomb potential. This potential is called the Debye potential.

The charge e_i on ion i must be exactly balanced by other ions in the ionic atmosphere. Since the charge density in the volume element $4\pi R^2\, dR$ around the central ion i in the ionic atmosphere is

$$4\pi \rho R^2\, dR,$$

the electroneutrality condition is

$$e_i = -\int_{r_i}^{\infty} dR\ R^2 4\pi\rho. \tag{16.16}$$

Since to the linear approximation in Φ

$$\rho = -\sum_i \frac{n_i e_i^2}{k_B T} \Phi(R), \tag{16.17}$$

substitution into Eq. (16.16) yields

$$e_i = AD\kappa^2 \int_{r_i}^{\infty} dR\ R\ \exp(-\kappa R), \tag{16.18}$$

which on integration gives rise to the equation for A:

$$A = \frac{e_i \exp(\kappa r_i)}{D(1 + \kappa r_i)}.$$

On substitution of this result into Eq. (16.15) we finally obtain the potential in the form

$$\Phi(R) = \frac{e_i \exp[-\kappa(R - r_i)]}{D(1 + \kappa r_i)R}. \tag{16.19}$$

At $R = r_i$ the potential has the form

$$\Phi(r_i) = \frac{e_i}{Dr_i} - \frac{e_i \kappa}{D(1 + \kappa r_i)}$$

$$\equiv \Phi_s + \Phi_c. \tag{16.20}$$

The first term on the right Φ_s is the potential created by the charge of ion i itself which is uniformly distributed on the surface of the sphere, and this part is obviously independent of other ions. The second part Φ_c is the contribution from the ionic atmosphere and depends on the concentrations of other ions present in the solution. This is that part of the potential contributing to the excess chemical potential.

Since by the first assumption the excess chemical potential arises from the Coulomb interaction, this excess free energy must be equal to the work done to bring an ion against the potential Φ_c. This work may be calculated by the following device. The work is divided up into two parts: one part is that of bringing a neutralized ion to the point $R = r_i$, and the other is that of charging the neutralized ion brought to $R = r_i$ to the full ionic charge e_i. The first part is small compared to the second in magnitude and therefore can be neglected. Then the total work is that of charging the ion and is given by

$$W_{electrical} = \int_0^{e_i} de\, \Phi_c(r_i)$$

$$= -\frac{e_i^2 \kappa}{2D(1 + \kappa r_i)}. \tag{16.21}$$

This electrical work must be equal to the excess free energy of the ion:

$$k_B T \ln \gamma_i = -\frac{z_i^2 e^2 \kappa}{2D(1 + \kappa r_i)}, \tag{16.22}$$

where γ_i is the activity coefficient of ion i and z_i is its charge number: $e_i = z_i e$ where z_i is positive for cations and negative for anions. The mean activity coefficient of the electrolyte $C_{\nu_+}^{z_+} A_{\nu_-}^{z_-}$ is now easy to calculate with Eq. (16.22):

$$\nu k_B T \ln \gamma_\pm = \nu_+ k_B T \ln \gamma_+ + \nu_- k_B T \ln \gamma_-$$

$$= -\frac{(\nu_+ z_+^2 + \nu_- z_-^2) e^2 \kappa}{2D(1 + \kappa r)},$$

for which we have replaced the ionic radii r_i with their mean value $r = (r_+ + r_-)/2$ in the denominator.

Since $\nu_+ z_+ + \nu_- z_- = 0$ by the electroneutrality of the electrolyte, we find

$$\nu_+ z_+^2 + \nu_- z_-^2 = \nu_+ z_+ (z_+ - z_-),$$

and

$$\frac{\nu_+ z_+^2 + \nu_- z_-^2}{\nu_+ + \nu_-} = \frac{\nu_+ z_+ (z_+ - z_-)}{\frac{\nu_+ z_+}{z_+} + \frac{\nu_- z_-}{z_-}} = -z_+ z_- = |z_+ z_-|.$$

Using this relation, we rewrite the mean activity coefficient in the form

$$\ln \gamma_\pm = -\frac{|z_+ z_-| e^2 \kappa}{2 D k_B T (1 + \kappa r)}. \tag{16.23}$$

This is the Debye–Hückel expression for the mean excess chemical potential of an electrolyte. To put Eq. (16.23) into a more useful form we express the Debye screening factor κ in molality:

$$\kappa = \left(\frac{8\pi e^2 N^2 \rho_0}{1000 D R T} \Gamma \right)^{1/2}, \tag{16.24}$$

where ρ_0 is the density of the solvent and N is the Avogadro number. It must be stressed that κ is proportional to the square root of ionic strength. The radius of the ionic atmosphere therefore decreases with increasing ionic strength, and the excess chemical potential is approximately proportional to $\Gamma^{1/2}$; see Eq. (16.2) for the definition of ionic strength given in terms of molality. If the concentration is so small that $\kappa r \ll 1$ then the κr term in the denominator can be neglected, and we obtain the excess chemical potential in a form similar to that by Lewis: at 298.15 K

$$\ln \gamma_\pm = -B' |z_+ z_-| \sqrt{\Gamma}$$
$$= -0.509 |z_+ z_-| \sqrt{\Gamma}, \tag{16.25}$$

where B' is given by

$$B' = \frac{e^2}{2 D k_B T} \left(\frac{8\pi e^2 \rho_0 N^2}{1000 D R T} \right)^{1/2}. \tag{16.26}$$

This gives the molecular theoretic meaning of the empirical constant B in the Lewis formula for the excess free energy of an electrolyte.

The Debye–Hückel theory provides the limiting law for the concentration dependence of mean activity coefficients. As can be expected on the basis of the assumptions and approximations made and evident from Fig. 14.1, the theory is limited to the low concentration regime and, more specifically, below about $0.01m$ in concentration. Above this value of molality the theory

Chemical Thermodynamics

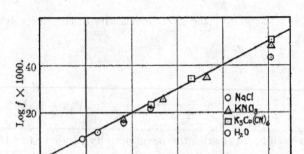

Fig. 16.1   Mean activity vs ionic strength (denoted μ). The solid line is the Debye–Hückel theory and the symbols are experiment. Reproduced with permission from N. J. Brönsted and V. K. La Mer, *J. Amer. Chem. Soc.* **46**, 557 (1924) © 1924 American Chemical Society.

begins to show considerable deviations. This limitation should be expected from the linearization approximation made to Poisson–Boltzmann equation, giving rise to the linear differential equation (Eq. (16.11)). Improving the theory beyond the limitation is not trivial and poses theoretical challenge, which stems from the fact that since Coulomb potentials are infinitely ranged and the integrals involved in the statistical mechanical treatment are divergent and thus require unconventional theoretical stratagems. This subject is beyond the scope of this chapter.

Chapter 17

Galvanic Cells and Electromotive Forces

17.1 Reversible Galvanic Cells and Reversible Electrodes

A galvanic cell is an electrochemical device consisting of more than two phases of electrolytes and two electrodes in which a chemical reaction occurs and the free energy of the chemical reaction is transformed into a current which is conducted through the two electrodes of the cell. In reversible cells, chemical reactions are made to occur infinitesimally slowly by imposing an opposing potential. See Fig. 17.1 for an example of galvanic cells.

There are several types of electrodes used in constructing galvanic cells. Regardless of the types of electrodes employed, they are based on the same fundamental principle in that an oxidized and reduced state of an element are always involved. Here oxidation refers to the loss of electrons and reduction to the gain of electrons by the element or the ion. The first type of electrode consists of an element in contact with a solution of its own ion as in the case of a metallic electrode which is immersed in a solution of its soluble salt. Another example is the hydrogen electrode which may be in contact with a solution of, say, HCl. Since hydrogen is nonconducting, an inert metallic conductor, for example, Pt or Au, is used to maintain electrical contact. The second type of electrode consists of a metal and its salt of low solubility which is in contact with a solution of the anion of the salt. For example, Ag, AgCl(s), MCl(solution), where M is an element which forms a halide. The third type of electrode consists of an inert metal such as gold or platinum immersed in a solution which contains both oxidized and reduced states of an oxidation–reduction system, for example, Fe^{2+} and Fe^{3+}.

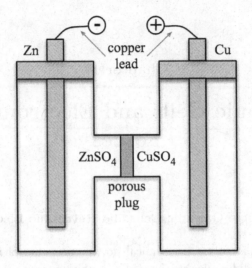

Fig. 17.1   A galvanic cell without a liquid junction.

Galvanic cells are assemblies of two electrodes of the types just described and in each electrode the following chemical reaction occurs:

$$M \rightleftharpoons M^{z\pm} \pm ze.$$

17.2   Electrochemical Potentials

The thermodynamic formalism for reversible galvanic cells and electrochemical systems in general is the same as for nonelectrolytes except for the fact that the electroneutrality condition

$$\sum_i n_i e_i = 0 \qquad (17.1)$$

must be imposed, where n_i is the mole number and e_i is the charge on ionic species i. Therefore $z_i = e_i/e$ is the charge number of the ion, e being the unit of charge. In order to appreciate the significance and magnitude of the effect in the event that it is locally violated because of local fluctuations in concentrations let us consider a sphere of radius 1 cm which contains an excess charge[1] of 10^{-12} moles. The excess charge will be distributed on the

---

[1]The absolute value of the charge on an electron is $e = 1.6021 \times 10^{-19}$ Coulombs and the charge of ionic species of charge number 1 is 1 Faraday defined as $F = 0.9687 \times 10^5$ Coulombs mol^{-1} for 1 mole of ions.

surface of the sphere and the potential of this charge distribution on the surface is

$$\psi = \frac{Q}{4\pi\epsilon_0 R},$$

where Q is the charge on the sphere, R is the radius, and ϵ_0 is the vacuum permittivity. Since $4\pi\epsilon_0 = 1.11 \times 10^{-10}\,\mathrm{CV^{-1}\,m^{-1}}$, for $Q = 10^{-12} \times 0.96 \times 10^5$ C and $R = 10^{-2}$ m, we find

$$\psi = \frac{0.96 \times 10^{-7}}{1.11 \times 10^{-10}} \times 10^2\,\mathrm{V} = 0.86 \times 10^5\,\mathrm{V}.$$

This simple calculation indicates that a fluctuation in ionic concentration in the order of pico moles results in a potential of an order of 10^5 V, but the concentration change of 10^{-12} moles is not accessible to a laboratory detection. Therefore two phases of practically the same chemical composition may have different electrical potentials because of undetectable charge fluctuations. For this reason the chemical potential may be written in two parts; one depending on the composition and the other depending on the electrical work and thus on the electrical potential:

$$\mu_{ei} = \mu_i + z_i F \psi. \tag{17.2}$$

Here μ_i is the composition part of the chemical potential, ψ is the electrical potential, and F denotes Faraday for a mole of charge ($1F = 9.64931 \times 10^4$ C). The μ_{ei} is called the electrochemical potential of ionic species i. This division of μ_{ei} into two parts, however, has no operational meaning, since it is not possible to measure the two parts separately. The difference in the electrochemical potentials of two phases α and β of an identical chemical composition is then

$$\mu_{ei}^\alpha - \mu_{ei}^\beta = z_i F(\psi^\alpha - \psi^\beta), \tag{17.3}$$

where the quantities superscripted with α and β refer to those in the phases α and β, respectively. Phases α and β are the conducting wires attached to the two electrodes of the cell. The potential difference $\psi^\alpha - \psi^\beta$ is called the electromotive force (EMF) of the cell. It must be stressed that *the potential differences and thus the electromotive forces are definable, only if the two phases are identical in chemical composition.* The two phases are in equilibrium with respect to ionic species i, only if

$$\mu_{ei}^\alpha = \mu_{ei}^\beta, \tag{17.4}$$

and therefore the potential difference vanishes at equilibrium.

If the potential is variable then the Gibbs relation for a phase is

$$dE = T \, dS - p \, dV + q \, d\psi + \sum_i \mu_{ei} \, dn_i, \qquad (17.5)$$

where

$$q = \sum_i n_i z_i F. \qquad (17.6)$$

This is a generalized form of the Gibbs relation used for nonelectrolytes. If the potential is constant then it takes the form

$$dE = T \, dS - p \, dV + \sum_i \mu_{ei} \, dn_i. \qquad (17.7)$$

The internal energy is generally dependent on the electric energy, and an electrochemical energy may be defined so as to include the electrical contribution. This is of course suggested by the fact that the electrochemical potential may be given as a derivative of the internal energy,

$$\mu_{ei} = \left(\frac{\partial E}{\partial n_i} \right)_{S,V,n',\psi}. \qquad (17.8)$$

It is customary to define the electrochemical energy by the relation

$$E_e = E + \sum_i n_i z_i F \psi, \qquad (17.9)$$

where E denotes the internal energy in the absence of potential ψ and the second term on the right arises from the electrical work on the ionized species. This division is as arbitrary as is for the electrochemical potential partitioned as in Eq. (17.2). With the electrochemical energy so defined, the fundamental equation may be written as

$$dE_e = T \, dS - p \, dV + \psi \, dq + \sum_i \mu_{ei} \, dn_i. \qquad (17.10)$$

It is possible to develop a thermodynamic theory of electrochemical cells by using Eq. (17.5) or Eq. (17.10). However, we will not have an occasion to use them in this chapter.

17.3 Galvanic Cells Without Liquid Junction

Galvanic cells are classified into two major categories of cells with and without liquid junctions. Liquid junctions occur when cells consist of electrodes in contact with a liquid solution of the salt of the element constituting the electrode; they are a source of irreversibility. This kind of irreversibility can be minimized if a salt bridge is employed. They generally give rise to a noticeable junction potential which complicates the matter. Here we consider cells without liquid junctions first.

17.3.1 *Cell Diagrams and the Sign Convention*

We have defined the electromotive force of a cell by the potential difference between two wires of the same metal connected to the two electrodes of the cell. However, there is still an ambiguity of sign in the definition and it is necessary to fix the sign by adopting a convention. To illustrate it we will take an example. We denote a galvanic cell with a diagram which indicates various components comprising the cell. We call it the cell diagram (see Fig. 17.2). Let us consider a cell denoted by the following cell diagram:

$$M(s)|MCl(m)|AgCl(s)|Ag(s), \qquad (17.11)$$

where M is either a metal or an element which can form a compound with chlorine, for example, H_2. Different phases are separated by a vertical bar in-between and the letter in the parentheses indicates the state or concentration of the substance in question. In constructing a cell diagram, the cell is arranged such that the following convention applies:

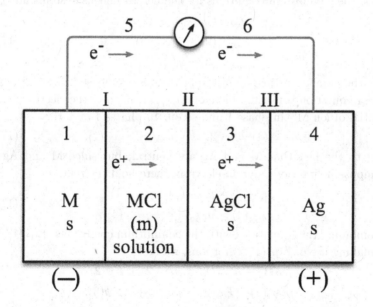

Fig. 17.2   Ag–metal galvanic cell.

Convention. *The oxidizing reaction occurs in the electrode on the left, the anode and the reducing reaction occurs in the electrode on the right, the cathode.*

In this convention the chemical reaction occurring in the cell is written such that the direction of the reaction corresponds to that of positive electricity flowing through the cell from left to right (or that of negative electricity flowing from right to left). Therefore, the chemical reaction for the cell diagram written is

$$M(s) + AgCl(s) = MCl(s) + Ag(s), \qquad (17.12)$$

where M is oxidized whereas Ag^+ is reduced. When put in terms of the direction of electron movement through the wire connecting the two electrodes, the convention is equivalent to saying that electrons are transferred from the left electrode to the right *through the wire connecting the electrodes.*

To facilitate discussions on electromotive force we schematically represent the cell (Eq. (17.11)) as a system of heterogeneous phases as in Fig. 17.2. Boundary I is assumed to be permeable to ion M^+ only, boundary II to ion Cl^- only, and boundary III to ion Ag^+ only. That is, the boundaries are semipermeable. If the temperature and pressure are the same for all phases, the equilibrium conditions for the heterogeneous systems are

$$\mu_{eM^+}(1) = \mu_{eM^+}(2),$$
$$\mu_{eCl^-}(2) = \mu_{eCl^-}(3), \qquad (17.13)$$
$$\mu_{eAg^+}(3) = \mu_{eAg^+}(4),$$

where the number in the parentheses indicates the phase in which the substance in question is found. Therefore $\mu_{eM^+}(1)$ denotes the electrochemical potential of ion M^+ in phase 1 and so on. In phases 1 and 4

$$\mu_{eAg}(4) = \mu_{Ag}(4), \quad \mu_{eM}(1) = \mu_M(1), \qquad (17.14)$$

owing to the fact that M and Ag are neutral, but since M and Ag are decomposable into ions and an electron, there hold the relations

$$\mu_{eM}(1) = \mu_{eM^+}(1) + \mu_{ee^-}(1),$$
$$\mu_{eAg}(4) = \mu_{eAg^+}(4) + \mu_{ee^-}(4). \qquad (17.15)$$

By combining these relations with the equilibrium conditions (Eq. (17.13)) and making use of Eq. (17.2) for ions, we find

$$\mu_M(1) - \mu_{ee^-}(1) = \mu_{M^+}(2) + F\psi(2),$$
$$\mu_{Cl^-}(2) - F\psi(2) = \mu_{Cl^-}(3) - F\psi(3),$$
$$\mu_{Ag}(4) - \mu_{ee^-}(4) = \mu_{Ag^+}(3) + F\psi(3).$$

Subtraction of the last equations from the first and rearranging the terms yield

$$\mu_{ee^-}(4) - \mu_{ee^-}(1) = [\mu_{Ag}(4) + \mu_{M^+}(2) + \mu_{Cl^-}(2)]$$
$$- [\mu_M(1) + \mu_{Ag^+}(3) + \mu_{Cl^-}(3)]$$
$$= [\mu_{Ag}(4) + \mu_{MCl}(2)] - [\mu_M(1) + \mu_{AgCl}(3)]. \quad (17.16)$$

Note that the right-hand side of the second equality of Eq. (17.16) is simply the affinity of the chemical reaction (Eq. (17.12)). We also observe that if the wire connecting the two electrodes is of a composition different from those of the electrodes then there hold the following equilibrium conditions for electrons:

$$\mu_{ee^-}(1) = \mu_{ee^-}(5), \quad \mu_{ee^-}(4) = \mu_{ee^-}(6),$$

where $\mu_{ee^-}(5)$ and $\mu_{ee^-}(6)$ are the electrochemical potentials of the wire attached to the electrodes in phases 1 and 4, respectively. Of course, the assumption here is that the boundaries between the electrodes and wire are permeable to the electrons only. Then the left-hand side of Eq. (17.16) may be written as

$$\mu_{ee^-}(4) - \mu_{ee^-}(1) = [\mu_{ee^-}(4) - \mu_{ee^-}(6)] + \mu_{ee^-}(6)$$
$$- [\mu_{ee^-}(1) - \mu_{ee^-}(5)] - \mu_{ee^-}(5)$$
$$= \mu_{ee^-}(6) - \mu_{ee^-}(5)$$
$$= -F[\psi(6) - \psi(5)].$$

The electromotive force of the cell is defined by

$$\epsilon = \psi(6) - \psi(5) = \frac{\mu_{ee^-}(4) - \mu_{ee^-}(1)}{-F}. \quad (17.17)$$

This is the American convention for EMF. It is opposite to the European convention. To put it simply as a summary, we adopt the sign of the EMF of a cell as the electrochemical potential $\mu_{ee^-}(4)$ of the right electrode minus the electrochemical potential $\mu_{ee^-}(1)$ of the left electrode divided by $-F$:

$$\epsilon = \frac{\mu_{ee^-}(right) - \mu_{ee^-}(left)}{-F}. \quad (17.18)$$

When this equation is combined with Eq. (17.16) and the chemical potentials are written as

$$\mu_i = \mu_i^0 + RT \ln a_i, \quad (17.19)$$

where a_i is the activity of species i and μ_i^0 is its appropriately chosen standard chemical potential, there follows the equation for EMF

$$\epsilon = \epsilon^0 - \frac{RT}{F} \ln\left(\frac{a_{MCl} a_{Ag}}{a_M a_{AgCl}}\right). \tag{17.20}$$

Here ϵ^0 is the standard electromotive force defined by

$$\epsilon^0 = -\frac{1}{F}\left[\mu_{Ag}^0 + \mu_{MCl}^0 - (\mu_M^0 + \mu_{AgCl}^0)\right]. \tag{17.21}$$

Since we may put

$$a_{Ag}(s) = a_M(s) = a_{AgCl}(s) = 1$$

according to the standard states chosen for activity, Eq. (17.20) may be written in the form

$$\epsilon = \epsilon^0 - \frac{RT}{F} \ln a_{MCl}. \tag{17.22}$$

It is called Nernst's equation. Since we may write the activity as

$$a_{MCl} = (\gamma_\pm m_\pm)^2,$$

the Nernst equation may be given in terms of the mean activity coefficient and molality of electrolyte $MCl(m)$:

$$\epsilon = \epsilon^0 - \frac{2RT}{F} \ln(\gamma_\pm m_\pm). \tag{17.23}$$

In the galvanic cell we have just considered, there is only 1 mole of electrons transferred in the course of the chemical reaction, since the metals are univalent. In order to generalize the result obtained above to a case where more than 1 mole of electrons are transferred by the reaction, we consider a galvanic cell of the following type:

$$L(s)|LX_n(m)|RX_n(s)|R(s), \tag{17.24}$$

where L and R are elements constituting the electrodes and LX_n and RX_n are the electrolytic compounds of L and R with univalent ion X. The cell reaction then is

$$L(s) + RX_n(s) = LX_n(m) + R(s), \tag{17.25}$$

and there are n moles of electrons transferred from L to R by the reaction. That is, L is oxidized by giving up n moles of electrons, whereas R is reduced by gaining n moles of electrons. The cell may be represented by the same figure as for Eq. (17.11) (see Fig. 17.2).

Now, by carrying out an analysis similar to that for the galvanic cell (Eq. (17.11)), we find

$$n[\mu_{ee^-}(4) - \mu_{ee^-}(1)] = [\mu_{LX_n}(2) + \mu_R(4)] - [\mu_L(1) + \mu_{RX_n}(3)]. \quad (17.26)$$

The electromotive force ϵ is still defined by Eq. (17.17). Therefore, substitution of Eqs. (17.17) and (17.19) into Eq. (17.26) gives rise to Nernst's equation in the form

$$\epsilon = \epsilon^0 - \frac{RT}{nF} \ln a_{LX_n} \quad (17.27)$$

for which we have put the activities of pure solid substances equal to 1 in accordance with the standard states chosen and

$$-nF\epsilon^0 = (\mu_R^0 + \mu_{LX_n}^0) - (\mu_L^0 + \mu_{RX_n}^0).$$

In terms of the mean activity coefficient and molality of LX_n,

$$a_{LX_n} = (\gamma_\pm m_\pm)^\nu, \quad (17.28)$$

where

$$\nu = \nu_+ + \nu_- = n + 1. \quad (17.29)$$

Therefore, the Nernst equation for the cell (Eq. (17.24)) is

$$\epsilon = \epsilon^0 - \frac{\nu RT}{nF} \ln(\gamma_\pm m_\pm), \quad (17.30)$$

where ν is given by Eq. (17.29). This is the desired generalization.

17.4   Fuel Cells

In this section, we discuss the basic thermodynamic aspect of fuel cells. Fuel cells, being open systems, operate differently from the galvanic cells, closed systems, discussed earlier. Nevertheless, the thermodynamic principles are the same as for the latter. For this reason and its relevance at present time we discuss the gist of it in this section, especially its thermodynamic aspect despite the appearance of this topic being somewhat misplaced in this chapter for galvanic cells.

A fuel cell is an electrochemical device generating electrical current from fuel supplied on the anode side and an oxidant supplied on the cathode side, which reacts in the presence of an electrolyte and metal serving also as a catalyst. The reaction products are continuously drawn out of the system while the electrolytes remain within the cell. For this reason the fuel cell is an open system, running as long as the fuel is supplied, whereas the

galvanic cells are closed and stop functioning unless recharged. We take
the example of a hydrogen–oxygen fuel cell for the purpose of illustration.
The hydrogen–oxygen fuel cell consists of two electrodes between which a
polymer electrolyte (semipermeable) membrane is enclosed that physically
separates the anode and cathode. The reaction product, water, is formed on
the cathode side and continuously drawn out of it. The electrons generated
by the reaction at the anode side is impermeable to the membrane, but
drawn out of the cell on the anode side through a metallic wire (circuit)
leading to the cathode. Thus the circuit is completed and electric current
is generated. Therefore the mode of operation of a fuel cell is seen to
be evidently different from those of galvanic cells discussed earlier in this
chapter. Nevertheless, the same thermodynamic principles apply as for
galvanic cells.

In this section, we employ a simplified model representing the essential
point of the hydrogen–oxygen fuel cell to illustrate the thermodynamics
involved. Expressed schematically according to the mode of cell diagram
adopted in this chapter, the hydrogen–oxygen fuel cell may be written as

$$|(Pt)H_2 - \text{electrolyte} \,|M|\, O_2 - \text{electrolyte}(Pt)|, \qquad (17.31)$$

where M is a semipermeable polymer electrolyte membrane that passes
charged masses (protons) but not electrons which get conducted through
the metallic wires connected to the platinum (electrodes) completing the
electrical circuit. See Fig. 17.3 for the schematic diagram of the system.

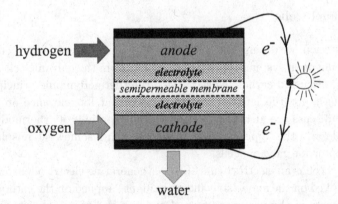

Fig. 17.3   Schematic diagram for a hydrogen–oxygen fuel cell. The electrolyte is impermeable to electrons which are conducted through the wires attached to the electrodes.
The broken lines represent the boundaries of the semipermeable membrane permeable to
hydrogen ions (protons). This completes the circuit with the external conducting wires.

At the cathode–electrolyte interface, catalyzed by electrolytes, the half reaction

$$\tfrac{1}{2}O_2 + 2e^- \rightleftharpoons O^{2-} \tag{17.32}$$

occurs, whereas at the anode–electrolyte interface, also catalyzed by the electrolytes, the half reaction

$$H_2 + O^{2-} \rightleftharpoons H_2O + 2e^- \tag{17.33}$$

occurs. In this particular case, there are 2 equivalents of electrons involved. Therefore the overall chemical reaction is

$$H_2 + \tfrac{1}{2}O_2 + \text{electrolyte} \rightleftharpoons H_2O + \text{electrolyte}. \tag{17.34}$$

This shows that in principle any exothermic chemical reaction can be made use of to construct a fuel cell, although there may be a number of practical aspects involved to make it practically feasible.

Expressed generally, a fuel cell involves a chemical reaction

$$\sum_i \nu_i M_i = 0 \tag{17.35}$$

and electrochemical potentials of charged species

$$\mu_{ei} = \mu_i + z_i F \psi. \tag{17.36}$$

If the charge transfer involved is denoted by q and the potential difference by ϵ, then the work done when charge dq is transported under ε is

$$dW = \epsilon\, dq \tag{17.37}$$

and the internal energy is given by

$$dE = dQ - dW. \tag{17.38}$$

If the process is reversible, we have already established that

$$dQ = T\, dS, \tag{17.39}$$

and the fundamental thermodynamic relation is given by

$$T\, dS = dE + \epsilon\, dq, \tag{17.40}$$

if there is no pressure–volume work. It should be noted that fuel cells operate at constant pressure and volume. Since dE is a total differential in (T, q), Eq. (17.40) can be written as

$$T\, dS = \left(\frac{\partial E}{\partial T}\right)_q dT + \left[\epsilon + \left(\frac{\partial E}{\partial q}\right)_T\right] dq. \tag{17.41}$$

Since S is also an exact differential, there follows the condition

$$\left[\frac{\partial}{\partial q}\frac{1}{T}\left(\frac{\partial E}{\partial T}\right)_q\right]_T = \left\{\frac{\partial}{\partial T}\frac{1}{T}\left[\epsilon + \left(\frac{\partial E}{\partial q}\right)_T\right]\right\}_q, \qquad (17.42)$$

which yields

$$\left(\frac{\partial \epsilon}{\partial T}\right)_q = \frac{1}{T}\left[\epsilon + \left(\frac{\partial E}{\partial q}\right)_T\right]. \qquad (17.43)$$

This means

$$T\,dS = C_q\,dT + T\left(\frac{\partial \epsilon}{\partial T}\right)_q dq. \qquad (17.44)$$

Note that the specific heat C_q is given by

$$C_q = \left(\frac{\partial E}{\partial T}\right)_q = T\left(\frac{\partial S}{\partial T}\right)_q. \qquad (17.45)$$

With the definition of enthalpy as usual

$$H = E + pV, \qquad (17.46)$$

we find at $d(pV) = 0$

$$dH = dQ - \epsilon\,dq = C_q\,dT + \left[T\left(\frac{\partial \epsilon}{\partial T}\right)_q - \epsilon\right]dq, \qquad (17.47)$$

and for an isothermal process the enthalpy change is given by

$$\Delta H = \left[T\left(\frac{\partial \epsilon}{\partial T}\right)_q - \epsilon\right]\Delta q. \qquad (17.48)$$

If n equivalents per mole of chemical is consumed by the fuel cell then the charge transfer involved is

$$\Delta q = nF. \qquad (17.49)$$

Hence we find

$$\Delta H = nF\left[T\left(\frac{\partial \epsilon}{\partial T}\right)_q - \epsilon\right] = nFT^2\left[\frac{\partial}{\partial T}\left(\frac{\epsilon}{T}\right)\right]_q. \qquad (17.50)$$

Since $\Delta H < 0$ for the process under consideration, it follows

$$T\left(\frac{\partial \epsilon}{\partial T}\right)_q < \epsilon \qquad (17.51)$$

or, on integration for a range of T,

$$\frac{\epsilon}{T} - \left(\frac{\epsilon}{T}\right)_{T_0} = \int_{T_0}^{T} dT' \frac{\Delta H}{nFT'^2} < 0. \tag{17.52}$$

The free energy change is given by

$$\Delta G = \Delta(H - TS), \tag{17.53}$$

which is the available work. The efficiency of "engine" — energy conversion device — is the work obtained per input of heat. Therefore the efficiency of the fuel cell can be defined by

$$\eta_f = \frac{\Delta G}{(\Delta H)_{max}}, \tag{17.54}$$

where $(\Delta H)_{max}$ is the maximum available heat input into the fuel cell. For an isothermal process we obtain for ΔG accompanying the fuel cell:

$$\begin{aligned}
\Delta G &= \Delta H - T \, \Delta S \\
&= nF\left[T\left(\frac{\partial \epsilon}{\partial T}\right)_q - \epsilon\right] - T\left(\frac{\partial \epsilon}{\partial T}\right)_q nF \\
&= -nF\epsilon.
\end{aligned} \tag{17.55}$$

Hence the efficiency is expressible as

$$\eta_f = \frac{\epsilon}{\left[\epsilon - T\left(\frac{\partial \epsilon}{\partial T}\right)_q\right]_{max}} \tag{17.56}$$

in terms of measurables. In this regard, it is useful to recognize that fuel cells do not involve a Carnot cycle like internal combustion engines. Nevertheless, it is subject to the Carnot theorem and the thermodynamic laws.

The Nernst equation can be also calculated by using the electrochemical potentials in the same form as for a galvanic cell by following a similar procedure as for the latter. For example, for the hydrogen–oxygen fuel cell considered above we find

$$\epsilon = \epsilon^0 - \frac{RT}{2F} \ln\left(\frac{a_{H_2O}}{a_{H_2} a_{O_2}^{1/2}}\right), \tag{17.57}$$

where a_i is the activity of species i and ϵ^0 is the standard potential difference (emf). It is seen to be in the usual form for the Nernst equation, indicating that the same thermodynamic principles are in operation for the fuel cell considered. The same principles apply to other types of fuel cells.

17.5   Donnan Membrane Equilibrium

Membrane equilibria in non-electrolyte solutions have been considered in connection with equilibrium conditions in the previous chapters. The same theory can be applied to membrane equilibria in electrolyte solutions, especially when there are species present in the solution, which are not capable of passing through the membrane. An example would be a solution of charged polymers and NaCl. Imagine two solutions separated by a semipermeable membrane. One solution consists of species R^-, C^+ and A^- in aqueous solution whereas the other solution consists of C^+ and A^- in the same solvent. The membrane is impermeable to the species R^- whereas C^+, A^-, and the solvent are permeable through the membrane. The solution containing the species R^- will be denoted by α, and the other solution by β. We will assume that the temperature is uniform throughout the system and that all the components are incompressible. Because of the impermeability of the species R^-, there will be a pressure difference—an osmotic pressure—and the equilibrium conditions are

$$p_\alpha v_1 + \mu_1^0 + RT \ln a_1^{(\alpha)} = p_\beta v_1 + \mu_1^0 + RT \ln a_1^{(\beta)}, \tag{17.58}$$

$$p_\alpha (v_c + v_a) + \mu_\pm^0 + RT \ln(a_+^{(\alpha)} a_-^{(\alpha)}) = p_\beta (v_c + v_a) + \mu_\pm^0 + RT \ln(a_+^{(\beta)} a_-^{(\beta)}), \tag{17.59}$$

where the subscript 1 denotes the solvent; p_α and p_β are the pressures of solutions α and β; v_1, v_c, and v_a are the molar volumes of the solvent, cation, and anion, respectively; and the rest of notation is standard. Rearranging the equilibrium conditions we obtain

$$p_\alpha - p_\beta = \frac{RT}{v_1} \ln \left(\frac{a_1^{(\beta)}}{a_1^{(\alpha)}} \right), \tag{17.60}$$

$$p_\alpha - p_\beta = \frac{RT}{v_c + v_a} \ln \left(\frac{a_+^{(\beta)} a_-^{(\beta)}}{a_+^{(\alpha)} a_-^{(\alpha)}} \right). \tag{17.61}$$

Elimination of the osmotic pressure $\Pi = p_\alpha - p_\beta$ from these equations yields the condition of equilibrium

$$\left(\frac{a_1^{(\beta)}}{a_1^{(\alpha)}} \right)^{v_c + v_a} = \left(\frac{a_+^{(\beta)} a_-^{(\beta)}}{a_+^{(\alpha)} a_-^{(\alpha)}} \right)^{v_1}. \tag{17.62}$$

If the solutions are sufficiently dilute the activities $a_1^{(\alpha)}$ and $a_1^{(\beta)}$ are approximately equal to unity. Since the activity coefficients of the ions may be set equal to unity in this case, the equilibrium condition is reduced to the form

$$n_+^{(\alpha)} n_-^{(\alpha)} = n_+^{(\beta)} n_-^{(\beta)}, \tag{17.63}$$

where $n_+^{(\alpha)}$ and so on are the mole numbers of the ions. This is the Donnan membrane equilibrium condition. The difference in concentrations of ions in solutions α and β gives rise to a potential difference, which is given by the formula

$$\epsilon = \frac{RT}{z_1 F} \ln \left(\frac{a_+^{(\beta)}}{a_+^{(\alpha)}} \right), \tag{17.64}$$

where we have put z_+ for generality, but $z_+ = 1$ for the case considered here. Equation (17.64) provides a method of determining the concentration imbalance by measuring the potential difference produced by the presence of an impermeable species in a solution. For example, one can study the thermodynamic properties of the species R^- by measuring the potential difference.

Chapter 18

Thermodynamics of Electric and Magnetic Fields

In the chapters on electrolytes and galvanic cells, electric fields entered in the thermodynamic formalism in the discussion of motions of strong electrolytes and their thermodynamic properties. Polarizable materials or dielectrics, although electrically neutral and non-conducting, can be altered in their properties by the presence of an electric field, static or time-dependent (i.e., dynamic). Similarly, a magnetic field can affect the thermodynamic properties of matter. In this chapter we consider the thermodynamics of dielectrics subjected to a static electric field and also the thermodynamics of magnetically polarizable materials in a static magnetic field. We will use the Gaussian units for electromagnetic fields and related quantities in the discussion. Since in the case of static fields the electric and magnetic fields are independent of each other, their effects can be discussed separately. We exclude conductors from the discussion.

18.1 Dielectrics in Electrostatic Field

We consider a capacitor of parallel plates α and β of area a separated by distance L. The volume of the capacitor is then $V = aL$. The surface charge density of the negative plate α is $-\sigma$, whereas the surface charge density of the positive plate β is σ, see Fig. 18.1. The potential difference between the plates is $\Delta\phi = \phi^{(\beta)} - \phi^{(\alpha)}$. The electric field strength $\mathbf{E}$ is then given by[1]

$$\mathbf{E} = \frac{\Delta\phi}{L}. \tag{18.1}$$

---

[1]It should be noted that the boldface letters $\mathbf{E}$, $\mathbf{P}$, and $\mathbf{p}$ used below are not vectors which they generally are. The boldface letters are used to distinguish them from lightface letters used for other quantities such as internal energy, pressure, etc.

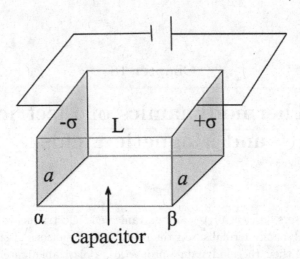

Fig. 18.1 A dielectric in a capacitor. Parallel plates α and β of area a have charge densities $-\sigma$ and $+\sigma$, respectively. They are separated at distance L and connected to a battery. The electric field strength is $\mathbf{E} = \Delta\phi/L$, where $\Delta\phi$ is the potential difference $\Delta\phi = \phi^{(\beta)} - \phi^{(\alpha)}$.

Recall that in the case of a continuous potential this may be written as

$$\mathbf{E} = -\nabla\phi.$$

It is well known in electrostatics that if $\mathbf{P}$ is the total polarization of the dielectric and $\mathbf{p}$ is the polarization density

$$\mathbf{p} = \frac{\mathbf{P}}{V}, \tag{18.2}$$

then the dielectric displacement $\mathbf{D}$ is given by

$$\mathbf{D} = \mathbf{E} + 4\pi\mathbf{p}. \tag{18.3}$$

For the capacitor under consideration

$$\mathbf{D} = 4\pi\sigma, \tag{18.4}$$

and if we write

$$\mathbf{D} = \varepsilon\mathbf{E}, \tag{18.5}$$

where ε is the dielectric constant of the non-conducting matter in the capacitor, then it follows from Eqs. (18.3)–(18.5)

$$\mathbf{p} = \chi\mathbf{E} \tag{18.6}$$

with χ defined by

$$\chi = \frac{\varepsilon - 1}{4\pi}. \tag{18.7}$$

Here χ is called the electric susceptibility per unit volume of the dielectric. It should be noted that Relation (18.6) holds for substances that are not ferroelectric. For ferroelectric materials there appears a term nonvanishing at $\mathbf{E} = 0$, that is,

$$\mathbf{p} = \mathbf{p}_0 + \chi\mathbf{E}, \tag{18.8}$$

where $\mathbf{p}_0$ represents the polarization density at $\mathbf{E} = 0$. In this chapter, we will not consider ferroelectrics.

If the charge density is uniform over the plates, the work to bring an infinitesimal charge $d\sigma$ from infinity onto the plate is then given by

$$dW_e = a(\phi^{(\beta)} - \phi^{(\alpha)})\, d\sigma = V\mathbf{E}\, d\sigma, \tag{18.9}$$

where $V = aL$. Therefore upon adding this work to the internal energy change of the dielectric in the absence of the field the overall internal energy change in the dielectric is given by

$$dE = dE_0 + dW_e = dE_0 + V\mathbf{E}\, d\sigma, \tag{18.10}$$

where dE_0 is the internal energy change in the absence of the electric field. For systems in equilibrium or for a reversible process this change may be written as

$$dE_0 = T\, dS - p\, dV + \sum_{i=1}^{c} \mu_i\, dn_i. \tag{18.11}$$

If the field is constant, from Eqs. (18.3)–(18.5) follows

$$V\, d\sigma = d\mathbf{P}. \tag{18.12}$$

Combining Eqs. (18.11) and (18.12) into Eq. (18.10) we obtain the fundamental equation—the Gibbs relation

$$dE = T\, dS - p\, dV + \sum_{i=1}^{c} \mu_i\, dn_i + \mathbf{E}\, d\mathbf{P}. \tag{18.13}$$

From this differential form follows the following constitutive derivatives for dielectrics in the presence of an electric field:

$$T^{-1} = \left(\frac{\partial S}{\partial E}\right)_{V,n,\mathbf{P}}, \tag{18.14}$$

$$p = -\left(\frac{\partial E}{\partial V}\right)_{S,n,\mathbf{P}}, \tag{18.15}$$

$$\mu_i = \left(\frac{\partial E}{\partial n_i}\right)_{S,V,n',\mathbf{P}}, \tag{18.16}$$

$$\mathbf{E} = \left(\frac{\partial E}{\partial \mathbf{P}}\right)_{S,V,n}. \tag{18.17}$$

Here the prime on n' in the subscript denotes excluding n_i. The enthalpy, Helmholtz free energy, and Gibbs free energy are defined as usual:

$$H = E + pV, \tag{18.18}$$
$$A = E - TS, \tag{18.19}$$
$$G = H - TS. \tag{18.20}$$

Using these definitions, we obtain from Eq. (18.13) the differential forms— the fundamental equations

$$dH = T\,dS + V\,dp + \sum_{i=1}^{c} \mu_i\,dn_i + \mathbf{E}\,d\mathbf{P}, \tag{18.21}$$

$$dA = -S\,dT - p\,dV + \sum_{i=1}^{c} \mu_i\,dn_i + \mathbf{E}\,d\mathbf{P}, \tag{18.22}$$

$$dG = -S\,dT + V\,dp + \sum_{i=1}^{c} \mu_i\,dn_i + \mathbf{E}\,d\mathbf{P}. \tag{18.23}$$

Since it is convenient to treat T, p, $n_1, \ldots, n_c$, and $\mathbf{E}$ as independent variables it is necessary to use Legendre transformation

$$\mathcal{F} = G - \mathbf{E} \cdot \mathbf{P} \tag{18.24}$$

to change the characteristic variables. Consequently, $d\mathcal{F}$ is given by the differential form

$$d\mathcal{F} = -S\,dT + V\,dp + \sum_{i=1}^{c} \mu_i\,dn_i - \mathbf{P}\,d\mathbf{E}, \tag{18.25}$$

in which T, p, $n_1, \ldots, n_c$, and $\mathbf{E}$ are characteristic variables instead of T, p, $n_1, \ldots, n_c$, and $\mathbf{P}$. Thermodynamic observables M are then regarded as functions of the new characteristic variables, and the partial molal quantities of M are defined by

$$\overline{M}_i = \left(\frac{\partial M}{\partial n_i}\right)_{T,p,n',\mathbf{E}}. \tag{18.26}$$

According to this definition, the chemical potentials are the partial molal quantities of free energy $\mathcal{F}$;

$$\mu_i = \left(\frac{\partial \mathcal{F}}{\partial n_i}\right)_{T,p,n',\mathbf{E}}. \tag{18.27}$$

The Maxwell relations between derivatives can be generalized to the case of electric fields. In particular, by using the differential form (18.25) we obtain

$$\left(\frac{\partial S}{\partial p}\right)_{T,n,\mathbf{E}} = -\left(\frac{\partial V}{\partial T}\right)_{p,n,\mathbf{E}}, \tag{18.28}$$

$$\left(\frac{\partial V}{\partial \mathbf{E}}\right)_{T,p,n} = -\left(\frac{\partial \mathbf{P}}{\partial p}\right)_{T,V,n,\mathbf{E}}, \tag{18.29}$$

$$\left(\frac{\partial \mu_i}{\partial \mathbf{E}}\right)_{T,p,n} = -\left(\frac{\partial \mathbf{P}}{\partial n_i}\right)_{T,p,n'\mathbf{E}} = -\overline{\mathbf{p}}_i, \tag{18.30}$$

where $\overline{\mathbf{p}}_i$ is the partial molar polarization. Since μ_i depends on T, p, $n_1, \ldots, n_{c-1}$, and $\mathbf{E}$, there follows from Eqs. (18.25) and (18.27) the differential form

$$d\mu_i = -\overline{s}_i\, dT + \overline{v}_i\, dp + \sum_{j=1}^{c-1} \mu_{ij}\, dn_j - \overline{\mathbf{p}}_i\, d\mathbf{E}, \tag{18.31}$$

where $\overline{s}_i$ and $\overline{v}_i$ are, respectively, the partial molar entropy and volume of component i

$$\overline{s}_i = -\left(\frac{\partial \mu_i}{\partial T}\right)_{p,n,\mathbf{E}}, \tag{18.32}$$

$$\overline{v}_i = \left(\frac{\partial \mu_i}{\partial p}\right)_{T,n,\mathbf{E}}, \tag{18.33}$$

and

$$\mu_{ij} = \left(\frac{\partial \mu_i}{\partial n_j}\right)_{T,p,n',\mathbf{E}}. \tag{18.34}$$

Of particular interest is the field dependence of $\overline{s}_i$ and $\overline{v}_i$. They can be examined by using the derivative relations

$$\left(\frac{\partial \overline{s}_i}{\partial \mathbf{E}}\right)_{T,p,n} = \left(\frac{\partial \overline{\mathbf{p}}_i}{\partial T}\right)_{p,n,\mathbf{E}}, \tag{18.35}$$

$$\left(\frac{\partial \overline{v}_i}{\partial \mathbf{E}}\right)_{T,p,n} = -\left(\frac{\partial \overline{\mathbf{p}}_i}{\partial p}\right)_{T,n,\mathbf{E}}. \tag{18.36}$$

These derivatives may be employed to examine the field dependence of thermodynamic functions of dielectrics.

18.2   Field Dependence of Thermodynamic Quantities

The molecular theory of dielectrics suggests that the electric susceptibility is the statistical average of dipole moments of the material. Consequently, it is generally a function of $\mathbf{E}$ and may be expressed in a power series of electric field $\mathbf{E}$. Since our discussion will be limited to a small field strength, we may consider χ constant with respect to $\mathbf{E}$.

In the case of weak fields, thermodynamic functions are generally quadratic with respect to $\mathbf{E}$. We can verify this is indeed true. We begin with the chemical potential. Integrating Eq. (18.30) over $\mathbf{E}$ while keeping other variables fixed, we obtain

$$\mu_i\,(T,p,n,\mathbf{E}) = \mu_i(T,p,n,0) - \int_0^{\mathbf{E}} \overline{\mathbf{p}}_i\, d\mathbf{E}$$

$$= \mu_i(T,p,n,0) - \frac{1}{2}\overline{\chi}_i\mathbf{E}^2, \tag{18.37}$$

for which we have assumed χ is independent of $\mathbf{E}$ and

$$\overline{\chi}_i = \left(\frac{\partial \chi_t}{\partial n_i}\right)_{T,p,n'} \quad (\chi_t = V\chi), \tag{18.38}$$

the partial molal electric susceptibility. The $\mu_i(T,p,n,0)$ is the chemical potential of component i in the absence of $\mathbf{E}$. Therefore the total electric susceptibility is given by

$$\chi_t = \sum_{i=1}^{c} \overline{\chi}_i n_i. \tag{18.39}$$

This shows that the free energy $\mathcal{F}$ is quadratic in $\mathbf{E}$ in the low field regime considered, as are the chemical potentials.

18.2.1   *Electrostriction*

In the same manner as for the chemical potentials, we find from Eq. (18.35)

$$\overline{s}_i(\mathbf{E}) = \overline{s}_i^0 + \frac{1}{2}\left(\frac{\partial \overline{\chi}_i}{\partial T}\right)_{p,n} \mathbf{E}^2, \tag{18.40}$$

where $\overline{s}_i^0 = \overline{s}_i(T,p,n,\mathbf{E}=0)$. Therefore the total entropy is given by

$$S(\mathbf{E}) = S^0 + \frac{\mathbf{E}^2}{2}\sum_{i=1}^{c} n_i\left(\frac{\partial \overline{\chi}_i}{\partial T}\right)_{p,n}. \tag{18.41}$$

Similarly, from Eq. (18.36) follows

$$\overline{v}_i(\mathbf{E}) = \overline{v}_i^0 - \frac{1}{2}\left(\frac{\partial \overline{\chi}_i}{\partial p}\right)_{T,n} \mathbf{E}^2 \qquad (18.42)$$

and

$$V = V^0 - \frac{\mathbf{E}^2}{2}\sum_{i=1}^{c} n_i\left(\frac{\partial \overline{\chi}_i}{\partial p}\right)_{T,n} = V^0 - \frac{\mathbf{E}^2}{2}\left(\frac{\partial \chi_t}{\partial p}\right)_{T,n}. \qquad (18.43)$$

Equation (18.43) is the formula for electrostriction given as a function of T, p, n, and $\mathbf{E}$. The sign of $(\partial \chi_t/\partial p)_{T,n}$ can be positive or negative, but the volume change is not reversible because it is proportional to $\mathbf{E}^2$, and hence the volume change is not affected by reversing the direction of the field.

18.2.2 *Electrocaloric Effects*

It is interesting to calculate the effect of $\mathbf{E}$ on temperature. To this end, we consider the partial molar enthalpy. Since from the definition of H the partial molar enthalpy is given by

$$\overline{h}_i = \mu_i + \overline{s}_i T + \mathbf{E}\overline{\mathbf{P}}_i,$$

upon using Eqs. (18.37), (18.40), and (18.42) we obtain

$$\overline{h}_i = \overline{h}_i^0 + \frac{1}{2}\left[\overline{\chi}_i + T\left(\frac{\partial \overline{\chi}_i}{\partial T}\right)_{p,n}\right]\mathbf{E}^2 \qquad (18.44)$$

and hence the enthalpy change arising from turning on $\mathbf{E}$ is given by

$$\Delta H = \frac{1}{2}\left[\chi_t + T\left(\frac{\partial \chi_t}{\partial T}\right)_{p,n}\right]\mathbf{E}^2. \qquad (18.45)$$

Let us now examine the effect on temperature by $\mathbf{E}$. Since it is convenient to take S, p, $n_1, \ldots, n_c$, and $\mathbf{E}$ for variables, if we wish to study the field dependence of temperature, Eq. (18.21) is rearranged to the form

$$d(H - \mathbf{E}\mathbf{P}) = T\,dS + V\,dp + \sum_{i=1}^{c}\mu_i\,dn_i - \mathbf{P}\,d\mathbf{E}, \qquad (18.46)$$

which gives the Maxwell relation

$$\left(\frac{\partial T}{\partial \mathbf{E}}\right)_{S,p,n} = -\left(\frac{\partial \mathbf{P}}{\partial S}\right)_{p,n,\mathbf{E}}. \qquad (18.47)$$

Since

$$\left(\frac{\partial \mathbf{P}}{\partial S}\right)_{p,n,\mathbf{E}} = \left(\frac{\partial \chi_t}{\partial S}\right)_{p,n,\mathbf{E}} \mathbf{E} \tag{18.48}$$

$$= \left(\frac{\partial v \overline{\chi}}{\partial T}\right)_{p,n,\mathbf{E}} \left(\frac{\partial T}{\partial s}\right)_{p,n,\mathbf{E}} \mathbf{E}$$

$$= \frac{T}{c_{p\mathbf{E}}} \left(\frac{\partial v \overline{\chi}}{\partial T}\right)_{p,n,\mathbf{E}} \mathbf{E},$$

where $\overline{\chi} = \chi/V$ is the electric susceptibility per volume, it follows

$$\left(\frac{\partial T}{\partial \mathbf{E}}\right)_{S,p,n} = -\frac{T}{c_{p\mathbf{E}}} \left(\frac{\partial v \overline{\chi}}{\partial T}\right)_{p,n,\mathbf{E}} \mathbf{E}, \tag{18.49}$$

where $c_{p\mathbf{E}}$ is the molar isobaric specific heat at constant $\mathbf{E}$

$$c_{p\mathbf{E}} = T \left(\frac{\partial s}{\partial T}\right)_{p,n,\mathbf{E}} \tag{18.50}$$

and v is the molar volume. We note that from Eq. (18.41) it follows that $c_{p\mathbf{E}}$ takes the form

$$c_{p\mathbf{E}} = c_p^0 + c_p^{(1)} \mathbf{E}^2, \tag{18.51}$$

where c_p^0 is the specific heat at $\mathbf{E} = 0$ and

$$c_p^{(1)} = \frac{1}{2} T \frac{\partial}{\partial T} \left(\frac{\partial v \overline{\chi}}{\partial T}\right)_{p,n,\mathbf{E}}. \tag{18.52}$$

Integrating Eq. (18.49) we find

$$T = T_0 - \int_0^{\mathbf{E}} \frac{\mathbf{E}}{\left(c_p^0 + c_p^{(1)} \mathbf{E}^2\right)} \left(\frac{\partial v \overline{\chi}}{\partial T}\right)_{p,n,\mathbf{E}} d\mathbf{E}. \tag{18.53}$$

Here the derivative $(\partial v \overline{\chi}/\partial T)_{p,n,\mathbf{E}}$ is generally negative. Therefore the effect of the field is to raise the temperature of the dielectric above the temperature at $\mathbf{E} = 0$. Reversely, if the magnitude of $\mathbf{E}$ is reduced reversibly and adiabatically from $\mathbf{E}$, the dielectric will be cooled by decreasing the field strength.

Before closing the discussion on electric fields, we remark that the equilibrium conditions for heterogeneous phases consisting of dielectrics are the same as for the case of external electric fields absent. The differences in various thermodynamic quantities from those of field-free systems originate from the field dependence of constitutive relations, such as the chemical potentials, which we have found depend on $\mathbf{E}$ quadratically in the leading order. Working out such $\mathbf{E}$ dependences for thermodynamic quantities is left to the reader as an exercise.

18.3 Static Magnetic Fields

If a paramagnetic substance is placed in a static magnetic field the magnetic work is done. It is given by

$$dW_m = \mathbf{H} \, d\mathbf{M}, \tag{18.54}$$

where $\mathbf{H}$ is the magnetic field and $\mathbf{M}$ is the total magnetization of the substance. Therefore the fundamental equation for the internal energy is given by the differential form

$$dE = T \, dS - p \, dV + \sum_{i=1}^{c} \mu_i \, dn_i + \mathbf{H} \, d\mathbf{M}. \tag{18.55}$$

We may define an analog of the enthalpy

$$\mathcal{H} = E + pV - \mathbf{HM}. \tag{18.56}$$

Similarly, an analog of the Helmholtz free energy may be defined by

$$\mathcal{A} = E - TS - \mathbf{HM}, \tag{18.57}$$

and the magnetic field equivalent of $\mathcal{F}$ defined earlier in the case of the electric field by

$$\mathcal{F} = E + pV - TS - \mathbf{HM} = \mathcal{H} - TS. \tag{18.58}$$

Combined with the fundamental equation for the internal energy E Eq. (18.55), these definitions give rise to the fundamental equations for the respective quantities:

$$d\mathcal{H} = T \, dS + V \, dp + \sum_{i=1}^{c} \mu_i \, dn_i - \mathbf{M} \, d\mathbf{H}, \tag{18.59}$$

$$d\mathcal{A} = -S \, dT - p \, dV + \sum_{i=1}^{c} \mu_i \, dn_i - \mathbf{M} \, d\mathbf{H}, \tag{18.60}$$

$$d\mathcal{F} = -S \, dT + V \, dp + \sum_{i=1}^{c} \mu_i \, dn_i - \mathbf{M} \, d\mathbf{H}. \tag{18.61}$$

Thus $\mathcal{H}$, $\mathcal{A}$, and $\mathcal{F}$ are characteristic functions of $(S, p, n_i, \mathbf{H})$, $(T, V, n_i, \mathbf{H})$, and $(T, p, n_i, \mathbf{H})$, respectively.

From Eq. (18.61) the chemical potential is seen to be the partial molar free energy of species i:

$$\mu_i = \left(\frac{\partial \mathcal{F}}{\partial n_i}\right)_{T,p,n',\mathbf{H}}. \tag{18.62}$$

Thus $\mathcal{F}$ is a first degree homogeneous function of n_i:

$$\mathcal{F} = \sum_{i=1}^{c} n_i \mu_i. \tag{18.63}$$

Furthermore, we find the Gibbs–Duhem equation for paramagnetic substances is given by

$$\sum_{i=1}^{c} n_i \, d\mu_i = -S \, dT + V \, dp - \mathbf{M} \, d\mathbf{H}. \tag{18.64}$$

Since the chemical potentials may be regarded as functions of T, p, $n_1, \ldots, n_{c-1}$, $\mathbf{H}$ we obtain

$$d\mu_i = -\bar{s}_i \, dT + \bar{v}_i \, dp + \sum_{j=1}^{c-1} \mu_{ij} \, dn_j - \overline{\mathbf{M}}_i \, d\mathbf{H}, \tag{18.65}$$

where $\overline{\mathbf{M}}_i$ is the partial molal magnetization of component i defined by $\overline{\mathbf{M}}_i = \left(\frac{\partial \mathbf{M}}{\partial n_i}\right)_{T,P,H,n'}$ and

$$\mu_{ij} = \left(\frac{\partial \mu_i}{\partial n_j}\right)_{T,p,n',\mathbf{H}}. \tag{18.66}$$

Various Maxwell relations can be derived from the differential forms (Eqs. (18.55) and (18.59)–(18.61)). For example, we obtain from Eq. (18.59) one of the Maxwell relations

$$\left(\frac{\partial T}{\partial \mathbf{H}}\right)_{S,p,n} = -\left(\frac{\partial \mathbf{M}}{\partial S}\right)_{p,n,\mathbf{H}}, \tag{18.67}$$

which will be made use of for studying the magnetocaloric effect. From Eq. (18.61) follow the relations

$$\left(\frac{\partial V}{\partial \mathbf{H}}\right)_{T,p,n} = -\left(\frac{\partial \mathbf{M}}{\partial p}\right)_{T,n,\mathbf{H}}, \tag{18.68}$$

$$\left(\frac{\partial S}{\partial \mathbf{H}}\right)_{T,p,n} = -\left(\frac{\partial \mathbf{M}}{\partial T}\right)_{p,n,\mathbf{H}}, \tag{18.69}$$

which may be used to examine the effects on volume and entropy of the magnetic field. From electrodynamics, the magnetic induction $\mathbf{B}$ may be written as

$$\mathbf{B} = \mu_{\mathrm{mp}}\mathbf{H}, \tag{18.70}$$

where μ_{mp} is called the magnetic permeability. Some authors call **B** the magnetic field strength. Since **M** is related to **B** and **H** by the relation

$$\mathbf{B} = \mathbf{H} + 4\pi\mathbf{m}, \tag{18.71}$$

where **m** is the magnetization density and $\mathbf{M} = V\mathbf{m}$, we may write

$$\mathbf{m} = \chi_m \mathbf{H}, \tag{18.72}$$

where

$$\chi_m = \frac{\mu_{mp} - 1}{4\pi}; \tag{18.73}$$

it is called the magnetic susceptibility. Recall that analogous relations and quantities were defined for the case of an electric field. As in the case of electric field, the thermodynamic functions are quadratic with respect to **H** if the field strength is weak. In the present case, μ_{mp} and χ_m are independent of the magnetic field strength. We will henceforth limit our discussion to such cases only.

18.3.1 *Magnetostriction*

Similarly to the case of the electric effect on the volume of a dielectric, the volume of a paramagnetic substance can be affected by a magnetic field. Integrating Eq. (18.68) from $\mathbf{H} = 0$ to $\mathbf{H}$ we find

$$V = V_0 - \frac{1}{2}\left(\frac{\partial V \chi_m}{\partial p}\right)_{T,n} \mathbf{H}^2, \tag{18.74}$$

where V_0 is the volume in the absence of the magnetic field. The derivative $(\partial V \chi_m / \partial p)_{T,n}$ is not large, but the volume can in principle change when the paramagnetic substance is subject to a magnetic field. This is the magnetic field analog of electrostriction.

18.3.2 *Magnetocaloric Effects*

In Chapter 6 on the third law of thermodynamics, we briefly discussed how low temperature could be achieved by applying the diamagnetic cooling method—the Debye–Giauque method, but its thermodynamic basis was not elucidated. Here we will discuss the thermodynamic foundation of the method. We begin the discussion by calculating the field effect on entropy.

The field effect on entropy can be calculated from Eq. (18.69). By integrating Eq. (18.69) over $\mathbf{H}$, we obtain

$$S(T, p, \mathbf{H}) = S(T, p, 0) - \int_0^{\mathbf{H}} \left(\frac{\partial V \chi_m}{\partial T} \right)_{p,n,\mathbf{H}} \mathbf{H} \, d\mathbf{H}. \tag{18.75}$$

This can be further evaluated. Since V and $\overline{\chi}_m$ are constant in the linear order in $\mathbf{H}$ we obtain

$$s(T, p, \mathbf{H}) = s(T, p, 0) - \frac{1}{2} v \left[\alpha \chi_m + \left(\frac{\partial \chi_m}{\partial T} \right)_{p,n} \right] \mathbf{H}^2, \tag{18.76}$$

where s is the molar entropy and α is the isobaric expansion coefficient

$$\alpha = V^{-1} \left(\frac{\partial V}{\partial T} \right)_{p,n,\mathbf{H}}. \tag{18.77}$$

Therefore the isobaric specific heat at constant $\mathbf{H}$ is given by

$$c_{p\mathbf{B}} = T \left(\frac{\partial s}{\partial T} \right)_{p,n,\mathbf{H}} = c_p + c_p^{(1)} \mathbf{H}^2, \tag{18.78}$$

where c_p is the isobaric specific heat in the absence of $\mathbf{H}$ and

$$c_p^{(1)} = \frac{1}{2} T \frac{\partial}{\partial T} \left\{ v \left[\alpha \overline{\chi}_m + \left(\frac{\partial \overline{\chi}_m}{\partial T} \right)_{p,n} \right] \right\}. \tag{18.79}$$

We may make use of Eq. (18.67). On integrating this equation over $\mathbf{H}$, we obtain

$$T = T_0 - \int_0^{\mathbf{H}} \left(\frac{\partial \chi_m}{\partial S} \right)_{p,n,\mathbf{B}} \mathbf{H} \, d\mathbf{H}. \tag{18.80}$$

Since

$$\left(\frac{\partial \chi_m}{\partial S} \right)_{p,n,\mathbf{H}} = \left(\frac{\partial \overline{\chi}_m}{\partial T} \right)_{p,n} \left(\frac{\partial T}{\partial s} \right)_{p,n,\mathbf{H}} = \frac{1}{c_{p\mathbf{H}}} \left(\frac{\partial \overline{\chi}_m}{\partial T} \right)_{p,n}, \tag{18.81}$$

where $\overline{\chi}_m$ is the magnetic susceptibility per mole

$$\overline{\chi}_m = \frac{\chi_m}{n}, \tag{18.82}$$

we find

$$T = T_0 - \left(\frac{\partial \overline{\chi}_m}{\partial T} \right)_{p,n} \int_0^{\mathbf{B}} \frac{\mathbf{H}}{c_{p\mathbf{H}}} d\mathbf{H} \tag{18.83}$$

$$\simeq T_0 - \frac{1}{2c_p} \left(\frac{\partial \overline{\chi}_m}{\partial T} \right)_{p,n} \mathbf{H}^2, \tag{18.84}$$

if $(\partial \overline{\chi}_m / \partial T)_{p,n}$ is assumed to be constant with respect to $\mathbf{H}$. If not, it should be kept under the integral sign. Since $(\partial \overline{\chi}_m / \partial T)_{p,n}$ is generally negative, $\Delta T = T(\mathbf{H}') - T(\mathbf{H})$ is negative if $\mathbf{H}' < \mathbf{H}$. That is, lowering the magnetic field strength the temperature of the system is lowered. This is the thermodynamic basis of the diamagnetic cooling method of Giauque and Debye discussed in Chapter 6.

Chapter 19

Thermodynamics of Nonequilibrium Processes

Although our principal interest lies in equilibrium (reversible) processes in matter in this work, we have treated the first and second laws of thermodynamics from the general viewpoint that encompasses reversible as well as irreversible processes occurring in matter. This represents a departure from the conventional treatment in classical thermodynamics. The main reason for the departure is that the thermodynamic laws, at least the first two laws, comprehensively apply to all natural macroscopic phenomena. As a matter of fact, reversible processes are idealized limits of irreversible phenomena occurring in reality. Therefore it is preferable to formulate the thermodynamics laws more comprehensively than usually done in the classical equilibrium thermodynamics practiced in the literature until now. Having formulated the thermodynamic laws in such a context, it is incumbent upon us to show, at least, one or two examples of how to use the general form for the thermodynamics laws presented in order to solve for irreversible phenomena. This subject area is still developing. For a fuller treatment of the thermodynamics of irreversible processes discussed here the reader is referred to monographs available in the literature.[1] In this chapter, we discuss thermodynamics of nonequilibrium processes with some examples of non-Newtonian flow in a Lennard–Jones fluid and chemical oscillations and pattern formation in a reacting fluid.

19.1 Extended Gibbs Relation for Calortropy

Recall that the extended notion of entropy enunciated for reversible processes by R. Clausius was given the terminology calortropy (*heat evolution*)

---

[1]For example, see B. C. Eu, *Generalized Thermodynamics: The Thermodynamics of Irreversible Processes and Generalized Hydrodynamics* (Kluwer, Dordrecht, 2002).

in Chapter 4. In irreversible thermodynamics the calortropy is regarded as dependent on conserved variables (e.g., density, momentum,[2] internal energy, and angular momentum) as well as non-conserved variables which are representative of irreversible processes occurring in the system in nonequilibrium, such as stress or strain, heat flow, mass flow, and so on. Since irreversible processes are generally time-dependent and spatially non-homogeneous, they are described in space-time. Therefore the calortropy Ψ of a system of volume V is expressed in terms of local density. We thus define the calortropy density $\widehat{\Psi}$ as follows:

$$\Psi(t) = \int_V dr \, \rho\widehat{\Psi}(\mathbf{r}, t), \qquad (19.1)$$

where ρ denotes the mass density at position $\mathbf{r}$ and time t. In irreversible thermodynamics this local calortropy density is subject to the local form of the second law of thermodynamics. The manifold spanned by the collection of conserved and non-conserved variables is called the thermodynamic space (manifold) $\mathfrak{M}$. We denote the aforementioned conserved variables by ρ, $\rho\mathbf{u}$ (momentum density), $\rho\mathcal{E}$ (internal energy density or just energy density), c_a (mass fraction; ratio of mass density ρ_a of species a to ρ) in the case of a mixture, and the non-conserved variables[3] collectively denoted by $\boldsymbol{\Phi}_{qa} = \rho\widehat{\boldsymbol{\Phi}}_{qa}$ $(q \geq 1)$ for $a = 1, \ldots, c$ for a c-component mixture. We will denote the pressure (stress) tensor by $\mathbf{P}$, the heat flux by $\mathbf{Q}$, mass fluxes by $\mathbf{J}_a$, etc. The $\mathbf{P}$ and $\mathbf{Q}$ may be decomposed into their species components:

$$\mathbf{P} = \sum_{a=1}^{c} \mathbf{P}_a, \quad \mathbf{Q} = \sum_{a=1}^{c} \mathbf{Q}_a. \qquad (19.2)$$

These non-conserved variables are suitably ordered in the set. More specifically, they are ordered as in Table 19.1. In the table, p_a denotes the local equilibrium pressure, $\widehat{h}_a$ the local enthalpy density, and $\boldsymbol{\delta}$ the unit second rank tensor which may be represented by the unit matrix

$$\boldsymbol{\delta} = \begin{pmatrix} 1 & 0 & 0 \\ 0 & 1 & 0 \\ 0 & 0 & 1 \end{pmatrix}. \qquad (19.3)$$

---

[2]Momentum, however, does not explicitly appear in the non-relativistic classical thermodynamic formalism.

[3]In the present discussion, we will not take volume transport phenomena into account. Therefore, the molar volume and its flux will not be included in $\mathfrak{M}$ for simplicity of formalism. Since their effects are generally of the second order with regard to the density gradient, their neglect is tolerable. For a theory of volume transport phenomena, see B. C. Eu, *J. Chem. Phys.* **129**, 094502, 134509 (2008).

Table 19.1 Non-conserved variables.

Name	Φ_{qa}	Hydrodynamic Symbols
Shear stress	Φ_{1a}	$[\mathbf{P}_a]^{(2)} = \frac{1}{2}\left(\mathbf{P}_a + \mathbf{P}_a^t\right) - \frac{1}{3}\delta \mathrm{Tr}\mathbf{P}_a$
Excess normal stress	Φ_{2a}	$\Delta_a = \frac{1}{3}\mathrm{Tr}\mathbf{P}_a - p_a$
Heat flux	Φ_{3a}	$\mathbf{Q}_a' = \mathbf{Q}_a - \hat{h}_a \mathbf{J}_a$
Mass flux	Φ_{4a}	$\mathbf{J}_a = \rho_a \hat{\mathbf{J}}_a$
etc.	$\vdots$	$\vdots$

Therefore $[\mathbf{P}_a]^{(2)}$ is the traceless symmetric part of the stress (pressure) tensor, and $\mathbf{Q}_a'$ is the apparent heat flux in excess of heat carried by mass.

19.1.1 *Differential Form for Calortropy*

The differential form for the calortropy density $\widehat{\Psi}$ is exact (total) in the thermodynamic manifold conjugate to the tangent manifold $\mathfrak{T}$ spanned by (intensive) variables $T, p, \widehat{\mu}_a$, and X_{qa}:

$$d\widehat{\Psi} = T^{-1}\left(d\mathcal{E} + p\,dv - \sum_{a=1}^{c} \widehat{\mu}_a\,dc_a + \sum_{q\geq 1}\sum_{a=1}^{c} X_{qa}\,d\widehat{\Phi}_{qa} \right). \qquad (19.4)$$

Here T, p, $\widehat{\mu}_a$, and X_{qa} are the (local) temperature, pressure, chemical potentials, and generalized potentials, respectively. The first three of $\mathfrak{T}$ have the classical thermodynamic counterparts whereas X_{qa} are new to the present formalism. The generalized potentials vanish as the system approaches equilibrium. It should be emphasized that the quantities involved in the differential one-form (Eq. (19.4)) are local and time-dependent. In this formalism the evolution of $\widehat{\Psi}$ is described by the evolution of variables in $\mathfrak{M}$, which obey their own evolution equations — the constitutive equations characteristic of nonequilibrium matter in question. It can be shown that the conserved variables $\mathcal{E}$, $v = 1/\rho$, c_a, and $\widehat{\mu}_a$ obey the conservation laws

$$\rho d_t v = \nabla \cdot \mathbf{u}, \qquad (19.5)$$

$$\rho d_t c_a = -\nabla \cdot \mathbf{J}_a \quad (1 \leq a \leq c), \qquad (19.6)$$

$$\rho d_t \mathbf{u} = -\nabla \cdot \mathbf{P} + \rho \mathbf{F}^{(ex)}, \qquad (19.7)$$

$$\rho d_t \mathcal{E} = -\nabla \cdot \mathbf{Q} - \mathbf{P} : \nabla \mathbf{u} + \sum_{a=1}^{c} \mathbf{J}_a \cdot \mathbf{F}_a^{(ex)}, \qquad (19.8)$$

where

$$d_t = \partial_t + \mathbf{u} \cdot \nabla$$

denotes the substantial time derivative in the coordinate system moving with the fluid velocity $\mathbf{u}$; $\mathbf{F}_a^{(ex)}$ is the external force density on species a; and $\mathbf{F}^{(ex)} = \sum_{a=1}^{c} c_a \mathbf{F}_a^{(ex)}$ is the total external force density. Equation (19.5) is the mass conservation law (balance equation); Eq. (19.6) is the mass fraction conservation law for which we have assumed no chemical reaction in the system; Eq. (19.7) is the momentum conservation law; and Eq. (19.8) is the internal energy conservation law for which we have excluded the effect of radiation. In nonrelativistic irreversible thermodynamics the momentum conservation law does not directly enter the differential form for $\widehat{\Psi}$, although the fluid velocity $\mathbf{u}$ is one of the conserved variables, obeying Eq. (19.7). For this reason we have not included $\mathbf{u}$ in $\mathfrak{M}$ explicitly.

The evolution equations of the non-conserved variables in $\mathfrak{M}$ may be compactly expressed in the form

$$\rho d_t \widehat{\Phi}_{qa} = -\nabla \cdot \psi_{qa} + \mathcal{Z}_{qa} + \Lambda_{qa} \ (1 \le a \le c; q \ge 1), \tag{19.9}$$

where ψ_{qa} is the flux of $\mathbf{\Phi}_{qa}$, $\mathcal{Z}_{qa}$ is the kinematic term which is generally nonlinear with respect to the variables in $\mathfrak{M}$ and $\mathbf{u}$ or variables spanning $\mathfrak{T}$, and Λ_{qa} is the dissipation term which is the seat of energy dissipation arising from process $\mathbf{\Phi}_{qa}$; it is closely related to the Rayleigh dissipation function.[4] The kinematic terms $\mathcal{Z}_{qa}$ are also dependent on the external forces. The set of differential equations (Eq. (19.5)–(19.9)), which describe macroscopic variables in manifold $\mathfrak{M}$ is called generalized hydrodynamic equations, for they not only reduce to the classical hydrodynamic equations of Navier and Stokes, as the system tends to equilibrium from a state far removed from equilibrium, but are also consistent with the laws of thermodynamics.

Phenomenologically, the precise forms for the constitutive equations represented by Eq. (19.9) depend on the nature and level of description for the process of interest, as all constitutive equations are in macroscopic thermal physics. For example, if we are interested in phenomena occurring near equilibrium Eq. (19.9) may be approximated by linear equations and, furthermore, if the processes are steady, then the time derivative term on the left can be omitted to a good approximation. Such an approximation results in linear constitutive equations. Typical examples are Fourier's law of heat conduction, Newton's law of viscosity, Fick's law of diffusion, and so

---

[4]Lord Rayleigh, *Theory of Sound* (Dover, New York, 1949).

on. Some specific forms for the constitutive equations will be given when we discuss examples for application of the formalism presented here.

Furthermore, if the external forces are electromagnetic, then Maxwell's electrodynamic equations must be appended[5] to the constitutive equations (Eq. (19.9)). In this regard, it is helpful to recall that the Poisson–Boltzmann equation enters the theory of electrolytes in this context when we discuss the thermodynamics of strong electrolytes; see Chapter 16.

The extended Gibbs relation (Eq. (19.4)) implies that the tangents (in manifold $\mathfrak{T}$) to the calortropy surface $\widehat{\Psi}$ in $\mathfrak{M}$ are given by

$$T^{-1} = \left(\frac{\partial \widehat{\Psi}}{\partial \mathcal{E}} \right)_{v,c,\widehat{\Phi}}, \tag{19.10}$$

$$p = T \left(\frac{\partial \widehat{\Psi}}{\partial v} \right)_{T,c,\widehat{\Phi}}, \tag{19.11}$$

$$\mu_a = -T \left(\frac{\partial \widehat{\Psi}}{\partial c_a} \right)_{T,v,c',\widehat{\Phi}}, \tag{19.12}$$

$$X_{qa} = T \left(\frac{\partial \widehat{\Psi}}{\partial \widehat{\Phi}_{qa}} \right)_{T,v,c,\widehat{\Phi}'}. \tag{19.13}$$

The prime on c in the subscript means the exclusion of c_a, and similarly for $\widehat{\Phi}'$. Phenomenologically, the left-hand sides of these derivatives are constitutive relations, such as the caloric equation of state, the equation of state, the chemical potentials, and the generalized potentials which may be given in terms of the variables spanning manifolds $\mathfrak{M}$ and $\mathfrak{T}$. Development of such constitutive relations is one of the principal tasks in the phenomenological theory of irreversible thermodynamics. Integrating these constitutive relations in $\mathfrak{M}$, the calortropy density surface $\widehat{\Psi}$ is obtained as a storage of constitutive information of the substance in nonequilibrium. In irreversible thermodynamics, such a task necessarily involves solutions of the coupled set of partial differential equations (Eqs. (19.5)–(19.9)), which constitute the generalized hydrodynamic equations mentioned earlier. It should be remarked that the generalized hydrodynamic equations reduce to the classical hydrodynamic equations—the Navier–Stokes–Fourier–Fick equations

---

[5]There is a question regarding the form for the electromagnetic momentum balance equation. See B. C. Eu, *Phys. Rev. A* **33**, 4121 (1986) and also B. C. Eu, *Kinetic Theory and Irreversible Thermodynamics* (Wiley, New York, 1992).

in classical hydrodynamics, if the irreversible processes occur near equilibrium, so that evolution equations for non-conserved variables are linearized with respect to the thermodynamic forces and fluxes.

19.1.2 *Variables in the Tangent Manifold*

To develop and investigate the constitutive relations for the variables spanning the tangent manifold $\mathfrak{T}$ it is convenient to transform the internal energy to another kind as we usually have done in equilibrium thermodynamics. Thus we define the extended enthalpy by

$$\mathcal{H} = \mathcal{E} + pv \tag{19.14}$$

and the extended Gibbs free energy by

$$\mathcal{G} = \mathcal{H} - T\widehat{\Psi}. \tag{19.15}$$

Then the extended Gibbs relations for $\mathcal{H}$ and $\mathcal{G}$ are given by

$$d\mathcal{H} = T\,d\widehat{\Psi} + v\,dp + \sum_{a=1}^{c} \widehat{\mu}_a\,dc_a - \sum_{q\geq 1}\sum_{a=1}^{c} X_{qa}\,d\widehat{\Phi}_{qa} \tag{19.16}$$

and

$$d\mathcal{G} = -\widehat{\Psi}\,dT + v\,dp + \sum_{a=1}^{c} \widehat{\mu}_a\,dc_a - \sum_{q\geq 1}\sum_{a=1}^{c} X_{qa}\,d\widehat{\Phi}_{qa}. \tag{19.17}$$

These differential forms imply that, for example,

$$T = \left(\frac{\partial \mathcal{H}}{\partial \widehat{\Psi}}\right)_{p,c,\widehat{\Phi}}, \tag{19.18}$$

$$v = \left(\frac{\partial \mathcal{H}}{\partial p}\right)_{T,c\widehat{\Phi}}, \tag{19.19}$$

$$\widehat{\mu}_a = \left(\frac{\partial \mathcal{H}}{\partial c_a}\right)_{T,p,c',\widehat{\Phi}}, \tag{19.20}$$

$$X_{qa} = -\left(\frac{\partial \mathcal{H}}{\partial \widehat{\Phi}_{qa}}\right)_{T,p,c,\widehat{\Phi}'}, \tag{19.21}$$

which give rise to the extended Maxwell relations:

$$\left(\frac{\partial T}{\partial \widehat{\Phi}_{qa}}\right)_{\widehat{\Psi},p,c,\widehat{\Phi}'} = -\left(\frac{\partial X_{qa}}{\partial \widehat{\Psi}}\right)_{p,c,\widehat{\Phi}} = -\frac{T}{c_{p\Phi}}\left(\frac{\partial X_{qa}}{\partial T}\right)_{p,c,\widehat{\Phi}}, \tag{19.22}$$

$$\left(\frac{\partial v}{\partial \widehat{\Phi}_{qa}}\right)_{\widehat{\Psi},p,c,\widehat{\Phi}'} = \left(\frac{\partial X_{qa}}{\partial p}\right)_{T,c,\widehat{\Phi}}, \tag{19.23}$$

$$\left(\frac{\partial \widehat{\mu}_a}{\partial \widehat{\Phi}_{qa}}\right)_{\widehat{\Psi},p,c,\widehat{\Phi}'} = -\left(\frac{\partial X_{qa}}{\partial c_a}\right)_{T,p,c',\Phi},$$

(19.24)

$$\left(\frac{\partial X_{qa}}{\partial \widehat{\Phi}_{rb}}\right)_{\widehat{\Psi},p,c,\widehat{\Phi}'} = \left(\frac{\partial X_{rb}}{\partial \widehat{\Phi}_{qa}}\right)_{T,p,c,\widehat{\Phi}'},$$

(19.25)

where $c_{p\Phi}$ is the extended isobaric heat capacity at constant $\widehat{\Phi} = \{\widehat{\Phi}_{qa}\}$:

$$c_{p\Phi} = T\left(\frac{\partial \widehat{\Psi}}{\partial T}\right)_{p,c,\widehat{\Phi}}.$$

(19.26)

These relations can be employed to calculate the $\widehat{\Phi}$ dependence of the constitutive relations for the variables in $\mathfrak{L}$. For this purpose it is convenient to express the generalized potentials in the form

$$X_{qa} = -g_{qa}(\widehat{\Phi})\widehat{\Phi}_{qa},$$

(19.27)

where $g_{qa}(\widehat{\Phi})$ is a scalar-valued function of $\widehat{\Phi}$ as well as other variables of $\mathfrak{M}$. For processes near equilibrium g_{qa} may be approximated by a function independent of $\widehat{\Phi}_{qa}$.

19.1.2.1 *Nonequilibrium Effect on Temperature*

We calculate the dependence of temperature on the non-conserved variables $\widehat{\Phi}_{qa}$ in nonequilibrium fluids. On integrating Eq. (19.22) over $\widehat{\Phi}_{qa}$ along the path of constant p and $\{c_a\}$, we obtain the $\widehat{\Phi}_{qa}$ dependence of temperature

$$T = T_0 + \sum_{q\geq 1}\sum_{a=1}^{c} \int_0^{\widehat{\Phi}_{qa}} d\widehat{\Phi}_{qa} \frac{T}{c_{p\Phi}} \left(\frac{\partial g_{qa}}{\partial T}\right)_{p,c,\widehat{\Phi}} \widehat{\Phi}_{qa},$$

(19.28)

where T_0 is the temperature of the system at $\widehat{\Phi}_{qa} = 0$, namely, at equilibrium where the generalized potentials vanish. Thus in the linear regime where g_{qa} is independent of $\widehat{\Phi}_{qa}$ and also $c_{p\Phi}$ is approximately constant, namely, $c_{p\Phi} \simeq c_p$, T is quadratic with respect to $\widehat{\Phi}_{qa}$:

$$\Delta T = T - T_0 \approx \sum_{q\geq 1}\sum_{a=1}^{c} \frac{1}{2c_p}\left(\frac{\partial g_{qa}}{\partial T}\right)_{p,c} \widehat{\Phi}_{qa}^2.$$

(19.29)

If $(\partial g_{qa}/\partial T)_{p,c}$ is positive, the nonequilibrium processes generally tend to increase temperature with increasing $\widehat{\Phi}_{qa}$: $\Delta T \geq 0$.

19.1.2.2 *Nonequilibrium Effect on Pressure*

The effect on pressure can be calculated similarly to the effect on temperature of irreversible processes. For this purpose it is convenient to define the extended Helmholtz free energy:

$$\mathcal{A} = \mathcal{E} - T\widehat{\Psi}, \tag{19.30}$$

which gives rise to the differential one-form

$$d\mathcal{A} = -\widehat{\Psi}\, dT - p\, dv + \sum_{a=1}^{c} \widehat{\mu}_a\, dc_a - \sum_{q \geq 1}\sum_{a=1}^{c} X_{qa}\, d\widehat{\Phi}_{qa}. \tag{19.31}$$

From this follows the extended Maxwell relation

$$\left(\frac{\partial p}{\partial \widehat{\Phi}_{qa}} \right)_{T,v,c,\widehat{\Phi}'} = \left(\frac{\partial X_{qa}}{\partial v} \right)_{T,c,\widehat{\Phi}}. \tag{19.32}$$

Integration of this equation over $\widehat{\Phi}_{qa}$ along a path of constant $T, v, c \equiv \{c_a\}$ yields

$$p = p_0 - \sum_{q \geq 1}\sum_{a=1}^{c} \int_0^{\widehat{\Phi}_{qa}} \left(\frac{\partial g_{ab}}{\partial v} \right)_{T,c,\widehat{\Phi}} \widehat{\Phi}_{qa}\, d\widehat{\Phi}_{qa}, \tag{19.33}$$

where $p_0 = p(\widehat{\Phi} = 0)$, the pressure at equilibrium. In the linear regime of irreversible processes, since g_{ab} is independent of $\widehat{\Phi}_{qa}$ it follows that the nonequilibrium correction is quadratic with respect to $\widehat{\Phi}_{qa}$:

$$p = p_0 - \frac{1}{2}\sum_{q \geq 1}\sum_{a=1}^{c} \left(\frac{\partial g_{ab}}{\partial v} \right)_{T,c,\widehat{\Phi}} \widehat{\Phi}_{qa}^2. \tag{19.34}$$

Other derivatives (Maxwell relations) can be similarly integrated to obtain the constitutive relations in the nonequilibrium regime.

19.1.2.3 *Nonequilibrium Effect on Chemical Potentials*

The differential one-form for $\mathcal{G}$, Eq. (19.17), implies that chemical potentials are the extended partial molar Gibbs free energy

$$\widehat{\mu}_a = \left(\frac{\partial \mathcal{G}}{\partial c_a} \right)_{T,p,c',\widehat{\Phi}}, \tag{19.35}$$

which in turn means that $\mathcal{G}$ is a first degree homogeneous function of $\{c_a : 1 \leq a \leq c\}$. That is,

$$\mathcal{G} = \sum_{a=1}^{c} \widehat{\mu}_a c_a. \tag{19.36}$$

Upon substitution into Eq. (19.17), it gives rise to the extended Gibbs–Duhem equation

$$\sum_{a=1}^{c} c_a \, d\widehat{\mu}_a = -\widehat{\Psi} \, dT + v \, dp - \sum_{q \geq 1} \sum_{a=1}^{c} X_{qa} \, d\widehat{\Phi}_{qa}, \tag{19.37}$$

which is the integrability condition[6] of the differential one-form (Eq. (19.17)). It also suggests that the chemical potentials may be regarded as functions of $(T, p, c_a, \widehat{\Phi}_{qa})$. Therefore we find

$$d\widehat{\mu}_a = -\overline{\Psi}_a \, dT + \overline{v}_a \, dp + \sum_{a=1}^{c} \sum_{b=1}^{c-1} \mu_{ab} \, dc_b - \sum_{q \geq 1} \sum_{b=1}^{c} \overline{X}_{qb;a} \, d\widehat{\Phi}_{qb}, \tag{19.38}$$

where

$$\overline{\Psi}_a = -\left(\frac{\partial \widehat{\mu}_a}{\partial T}\right)_{p,c,\widehat{\Phi}} = -\left(\frac{\partial \widehat{\Psi}}{\partial c_a}\right)_{T,p,c',\widehat{\Phi}}, \tag{19.39}$$

$$\overline{v}_a = \left(\frac{\partial \widehat{\mu}_a}{\partial p}\right)_{T,c,\widehat{\Phi}} = \left(\frac{\partial v}{\partial c_a}\right)_{T,p,c',\widehat{\Phi}}, \tag{19.40}$$

$$\overline{X}_{qb;a} = -\left(\frac{\partial \widehat{\mu}_a}{\partial \widehat{\Phi}_{qb}}\right)_{T,p,c,\widehat{\Phi}'} = \left(\frac{\partial X_{qb}}{\partial c_a}\right)_{T,p,c',\widehat{\Phi}}, \tag{19.41}$$

$$\mu_{ab} = \left(\frac{\partial \widehat{\mu}_a}{\partial c_b}\right)_{T,p,c',\widehat{\Phi}}. \tag{19.42}$$

Therefore $\overline{\Psi}_a$, $\overline{v}_a$, and $\overline{X}_{qa}$ are the partial molar calortropy, volume, and generalized potential, respectively. Therefore integrating Eq. (19.38) over $\widehat{\Phi}_{qb}$ along a path of constant T, p, and $\{c_a\}$ we obtain chemical potentials

$$\widehat{\mu}_a = \widehat{\mu}_a^{eq} - \sum_{q \geq 1} \sum_{b=1}^{c} \int_0^{\widehat{\Phi}_{qb}} \overline{X}_{qb;a} \, d\widehat{\Phi}_{qb}, \tag{19.43}$$

where $\widehat{\mu}_a^{eq} = \widehat{\mu}_a(\widehat{\Phi}_{qa} = 0)$. Since it is possible to write

$$\overline{X}_{qb;a} = -\overline{g}_{qb;a} \widehat{\Phi}_{qb}, \tag{19.44}$$

---

[6]See Footnote 1 and also M. Chen and B. C. Eu, *J. Math. Phys.* **34**, 3012 (1993).

where $\overline{g}_{qb;a}$ is the partial molar g_{qb}, in the linear regime where g_{qa} is independent of $\widehat{\Phi}_{qa}$ the chemical potentials are quadratic with respect to $\widehat{\Phi}_{qa}$:

$$\widehat{\mu}_a = \widehat{\mu}_a^{eq} + \frac{1}{2} \sum_{q \geq 1} \sum_{b=1}^{c} \overline{g}_{qb;a} \widehat{\Phi}_{qb}^2. \tag{19.45}$$

By using the formalism presented here, it is possible to develop and investigate the thermodynamics of irreversible processes in nonequilibrium fluids in a manner parallel to equilibrium thermodynamics discussed in the previous chapters in this work.

19.1.2.4 *Nonequilibrium Effect on Equilibrium Constants*

Since we have excluded the possibility of chemical reactions, the present formalism, strictly speaking, does not apply to the case of chemical reactions. However, chemical reactions appear only in the mass fraction balance equation as a source term. Moreover, chemical reactions generally occur on a faster time scale than the hydrodynamic processes underlying the non-conserved variables. Therefore, even though there are irreversible processes with regard to the non-conserved variables the chemical equilibrium condition is still given by

$$\sum_{a=1}^{r} \omega_a \widehat{\mu}_a = 0, \tag{19.46}$$

where ω_a are the stoichiometric coefficients of the chemical reaction. Here we are assuming there is a single chemical reaction.

Assuming that irreversible processes are occurring near equilibrium, we insert Eq. (19.45) into Eq. (19.46) to obtain

$$\sum_{a=1}^{r} \omega_a \left[\widehat{\mu}_a^0 + (RT/\rho) \ln(x_a f_a) \right] + \frac{1}{2} \sum_{q \geq 1} \sum_{b=1}^{c} \Delta g_{qb} \widehat{\Phi}_{qb}^2 = 0, \tag{19.47}$$

where we have used the phenomenological formula

$$\widehat{\mu}_a^{eq} = \widehat{\mu}_a^0 + (RT/\rho) \ln(x_a f_a) \quad (\rho = \text{density}) \tag{19.48}$$

and Δg_{qb} is the change in g_{qb} arising from the chemical reaction:

$$\Delta g_{qb} = \sum_{a=1}^{r} \overline{g}_{qb;a} \omega_a. \tag{19.49}$$

Therefore the equilibrium constant is given by

$$K(T) = K^0(T) \exp\left[-\frac{1}{2}\sum_{q\geq 1}\sum_{b=1}^{c}\Delta g_{qb}\widehat{\Phi}_{qb}^2\right],\qquad (19.50)$$

where $K^0(T)$ is the equilibrium constant in the absence of $\widehat{\Phi}_{qb}$:

$$K^0(T) = \exp\left[-\Delta G^0/RT\right],\qquad (19.51)$$

$$\Delta G^0 = \sum_{a=1}^{r}\omega_a\rho\widehat{\mu}_a^0 = \sum_{a=1}^{r}\omega_a\mu_a^0,\qquad (19.52)$$

where ΔG^0 is the standard Gibbs free energy change for the reaction. This result indicates that the equilibrium constant is modified by the presence of non-conserved variables undergoing irreversible processes in the fluid; for example, when the fluid is sheared, or compressed by a shock wave, or subjected to a temperature gradient or concentration gradients giving rise to diffusion, etc.

19.2   Flow of a Non-Newtonian Liquid

The viscosity of a fluid, among other properties of the fluid, is essential for understanding the flow phenomena in macroscopic matter, and study of viscosity in fluid flow has a long history in science tracing back to Isaac Newton. The subject field, known as rheology[7] at present, is important in physical chemistry, engineering, and physiology among many other disciplines in science and engineering. We discuss this topic as an example of irreversible phenomena under the purview of the theory of irreversible thermodynamics presented.

Viscous flow phenomena occurring near equilibrium is adequately described by the linear viscosity, namely, the Newtonian viscosity, which is the limiting viscosity at a sufficiently small shear rate. It is well known that description of flow with the Newtonian viscosity is inadequate as the shear rate is increased and the fluid thus gets increasingly removed from equilibrium. It is generally found that in such a state the viscosity of the fluid becomes dependent on the shear rate in contrast to the Newtonian viscosity. The viscosity is then said to be non-Newtonian. Non-Newtonian

---

[7]See, for example, H. A. Barnes, J. F. Hutton, and K. Walters, *An Introduction to Rheology* (Elsevier, Amsterdam, 1989).

viscosities of such fluids are generally nonlinear with respect to the shear rate. In rheology, we mostly deal with non-Newtonian flow phenomena. In this section, we show how this flow phenomena can be treated from the phenomenological theory viewpoint by employing the theory of irreversible processes described earlier.

19.2.1   *Velocity Profile of Flow in a Rectangular Channel*

To make the discussion as simple as possible we will assume that there is present only a shearing perturbation in a single-component simple liquid, which does not support a bulk viscosity; in other words, the excess normal stress Δ is equal to zero. Neither is there heat flow present. Then the only relevant nonequilibrium variable is the traceless symmetric part of the stress tensor, and its evolution equation is that of the shear stress $\widehat{\mathbf{\Pi}}$. It takes the form

$$\rho d_t \widehat{\mathbf{\Pi}} = -2p[\nabla \mathbf{u}]^{(2)} - 2[\mathbf{\Pi} \cdot \nabla \mathbf{u}]^{(2)} - \frac{p}{\eta_0}\mathbf{\Pi}q(\kappa), \qquad (19.53)$$

for which the pressure tensor $\mathbf{P}$ is decomposed into components

$$\mathbf{\Pi} = \mathbf{P} - p\boldsymbol{\delta} \qquad (19.54)$$

with p denoting the hydrostatic pressure, and $\boldsymbol{\delta}$ the unit second rank tensor, and $\mathbf{\Pi}$ denotes the traceless symmetric part of $\mathbf{P}$ defined by

$$\mathbf{\Pi} = \rho\widehat{\mathbf{\Pi}} = [\mathbf{P}]^{(2)} = \frac{1}{2}(\mathbf{P} + \mathbf{P}^t) - \frac{1}{3}\boldsymbol{\delta}\mathrm{Tr}\mathbf{P} \qquad (19.55)$$

with the superscript t denoting the transpose. The symbol $[\mathbf{A}]^{(2)}$ denotes the traceless symmetric part of second rank tensor $\mathbf{A}$ as in Eq. (19.55). Other symbols in the equations are as follows: η_0 is the Newtonian shear viscosity; ρ is the mass density; $q(\kappa)$ is the nonlinear factor defined by

$$q(\kappa) = \frac{\sinh \kappa}{\kappa}, \qquad (19.56)$$

$$\kappa = \frac{\tau}{2\eta_0}(\mathbf{\Pi}:\mathbf{\Pi})^{1/2}, \qquad (19.57)$$

$$\tau = \frac{(mk_BT)^{1/4}\sqrt{\eta_0}}{p\sigma}. \qquad (19.58)$$

In these expressions m is the reduced mass, σ is the size parameter of the molecule, k_B is the Boltzmann constant, and κ is the Rayleigh dissipation function for shearing. For Eq. (19.53) the flux of stress tensor, $\boldsymbol{\psi}^{(p)}$, is set equal to zero because of the assumption that only the shear stress is present.

Setting $\psi^{(p)} = 0$, is in fact, is equivalent to the closure of the hierarchy of evolution equations for the non-conserved variables. Similarly, $\mathbf{V}^{(1)}$ that originally appears in the full evolution equation for the stress tensor is also ignored because it is a second-rank tensor given in terms of intermolecular forces and peculiar velocities of molecules that belong to the moment other than the shear stress. This type of closure has been found adequate for the description of shock wave phenomena[8] and other flow processes in simple fluids. Therefore we will also use it in this discussion. Since $\mathbf{V}^{(1)}$ is associated with the velocity times the virial associated with the flow of the fluid, it does not make a contribution in the order of approximation we are interested in for the stress tensor in this discussion.

Flow of a fluid that occurs confined between two plates aligned in parallel to, say, the xz plane in a suitably fixed laboratory coordinate system is called plane Couette flow. On the basis of the evolution equation or the constitutive equation for flow, Eq. (19.53), we consider a plane Couette flow in a flow configuration.[9] The plates are separated by distance L and positioned at $y = L/2$ and $y = -L/2$, respectively, while moving in opposite directions at speed $u_d/2$ and $-u_d/2$, respectively. In the flow configuration defined, the flow is in the direction of x, but neutral in the z direction whereas the velocity gradient of the laminar flow is present in the direction of the y axis. The problem is simplified if the flow is assumed to be incompressible. Since it is also sufficient to consider the case of the normal stress differences being equal to zero because the normal stress differences are of higher order than the shear stress, we neglect the normal stress differences.

The steady state constitutive equation for the shear stress tensor is then obtained from Eq. (19.53) in the form

$$\Pi q(\kappa) = -2\eta_0 \gamma, \tag{19.59}$$

where $\Pi \equiv \Pi_{xy} = \Pi_{yx}$ is the shear stress and γ is the shear rate (velocity gradient) defined by

$$\gamma = \frac{1}{2} \frac{\partial u_x}{\partial y} \tag{19.60}$$

with u_x denoting the x component of the velocity $\mathbf{u}$. If the channel is sufficiently long in the flow (x) direction, translational invariance holds to

---

[8]It is discussed in B. C. Eu, *Nonequilibrium Statistical Mechanics* (Kluwer, Dordrecht, 1998). See also the reference cited in Footnote 1.

[9]L. D. Landau and E. M. Lifshitz, *Fluid Mechanics* (Pergamon, Oxford, 1958).

a good approximation, making u_x independent of x. But u_x would remain a function of y. It should be noted that if the nonlinear factor $q(\kappa)$ is set equal to unity then Eq. (19.59) becomes the Newtonian law of viscosity in the classical hydrodynamics of Navier and Stokes: $\Pi = -2\eta_0\gamma$ for the present flow problem.

The Rayleigh dissipation function κ under the conditions mentioned earlier is given by

$$\kappa = \frac{\tau}{\sqrt{2}\eta_0}|\Pi|. \tag{19.61}$$

The steady state momentum balance equation consistent with the shear stress equation (Eq. (19.59)) is given by

$$\frac{\partial}{\partial y}\Pi = -p_x, \tag{19.62}$$

where

$$p_x = \frac{\partial p}{\partial x}.$$

This equation is coupled to Eq. (19.59). This pair of equations is solved for the velocity profile $u_x(y)$ for the plane Couette flow. The following two cases are possible for the flow configuration.

19.2.1.1 *The Case of $p_x = 0$*

For this case, Eqs. (19.59) and (19.62) with $p_x = 0$ are solved subject to the boundary conditions

$$u_x\left(\pm\frac{1}{2}L\right) = \pm\frac{1}{2}u_d, \tag{19.63}$$

where $\pm\frac{1}{2}u_d$ are the speeds of the plates moving in opposite directions along the x axis. This is called stick boundary conditions because the fluid sticks to the boundary walls. This sticking tendency of the fluid creates a velocity profile in the channel. We are interested in the velocity profile.

The solutions subject to the boundary conditions are easily obtained:

$$u_x = \frac{u_d}{L}y, \tag{19.64}$$

$$\Pi = -\frac{\sqrt{2}\eta_0}{\tau}\ln\left(\frac{\tau u_d}{\sqrt{2}L} + \sqrt{1 + \left(\frac{\tau u_d}{\sqrt{2}L}\right)^2}\right). \tag{19.65}$$

The negative branch is chosen for Π. Notice that the shear stress tensor (i.e., xy component) Π is a constant independent of the coordinates; it depends on the shear rate u_d/L only and the velocity profile is linear with respect to y. This conclusion does not remain true in different flow configurations.

Since in rheology the non-Newtonian viscosity η is defined by the constitutive relation

$$\Pi = -2\eta\gamma, \tag{19.66}$$

it follows from the solution (Eq. (19.65)) obtained that the non-Newtonian viscosity η of the fluid is given by

$$\eta = \eta_0 \frac{\sinh^{-1}\left(\frac{\tau u_d}{\sqrt{2}L}\right)}{\left(\frac{\tau u_d}{\sqrt{2}L}\right)}. \tag{19.67}$$

It should be noted that the shear rate γ is now given by

$$\gamma = \frac{u_d}{2L}. \tag{19.68}$$

Therefore the non-Newtonian viscosity obtained is dependent on the shear rate according to the formula (19.67). Notice that u_d, and hence γ, is an experimental input and a constant independent of density and temperature. However, the density and temperature dependences of η are vested in the Newtonian viscosity, which should generally depend on them. The prediction of γ dependence by this formula for η has been found in agreement with molecular dynamics simulation results for Lennard–Jones fluids in the same flow configuration as described. We will return to this question later.

19.2.1.2   *The Case of $p_x \neq 0$*

It is possible to obtain simple analytic solutions for the case of flow when p_x is not equal to zero, but a constant. We consider a flow subject to the boundary conditions

$$u(y)|_{y=\pm L/2} = 0. \tag{19.69}$$

In other words, a non-Newtonian fluid flows in the channel under a constant pressure gradient in the x direction. Therefore in this case, the boundary walls do not move, but the fluid sticks at the walls.

Since the shear stress at $y = 0$ (along the axis of the channel) should be equal to zero, integration of Eq. (19.62) yields

$$\Pi(y) = -p_x y. \tag{19.70}$$

Substitution of this result into Eq. (19.59) yields the differential equation for u_x:

$$\frac{\partial u_x}{\partial y} = \frac{2}{\tau} \sinh\left(\frac{\tau p_x}{2\eta_0} y\right). \tag{19.71}$$

The solution of this differential equation is

$$u_x = \alpha u_{\max}\left[\cosh\left(\frac{\delta}{2}\right) - \cosh\left(\frac{\delta y}{L}\right)\right], \tag{19.72}$$

where $u_{\max} = u_x(y = 0)$ and

$$\alpha = -\frac{4\eta_0}{\tau^2 p_x u_{\max}}, \qquad \delta = \frac{\tau p_x L}{2\eta_0}. \tag{19.73}$$

On expanding the hyperbolic functions in power series of the arguments in the limit of small δ, this velocity profile reduces to the well-known parabolic Poiseuille profile. This is the limit of either small η_0/τ or small $p_x L$.

It is possible to obtain analytic solutions of the governing equations for the case of flow in which the boundary walls move at $\pm u_d/2$ and hence the boundary conditions are given by Eq. (19.63). We leave this problem as an exercise for the reader.

The flow rate can be calculated with the velocity profile (Eq. (19.72)) obtained. Let the mass flow rate through the channel of cross section $A = L^2$ under the pressure gradient p_x be denoted by Q_v:

$$Q_v = \int_{-L/2}^{L/2} dz \int_{-L/2}^{L/2} dy \rho u_x(y). \tag{19.74}$$

The integrand is the volume flow per unit volume per unit time. Assuming the fluid is incompressible and using $u_x(y)$ in Eq. (19.72) we obtain

$$Q_v = \frac{\rho L^4}{12\eta_0} \frac{\Delta p}{\Delta x} F(\delta), \tag{19.75}$$

where

$$F(\delta) = \frac{12}{\delta}\left[\cosh\left(\frac{\delta}{2}\right) - \frac{2}{\delta}\sinh\left(\frac{\delta}{2}\right)\right] \tag{19.76}$$

and Δp is the pressure difference between the head and tail ends of the rectangular tube and Δx is the length of the rectangular tube. As δ tends to zero, $F(\delta) \to 1$ and Q_v reduces to the well-known Hagen–Poiseuille volume flow rate of a Newtonian fluid.

19.2.2 Non-Newtonian Viscosity

To assess the quality of the non-Newtonian viscosity formula (Eq. (19.67)) obtained for flow of a simple fluid we compare the present theoretical prediction with the computer simulation results performed on a Lennard–Jones fluid. To this end it is convenient to make use of dimensionless variables reduced with respect to parameters associated with the model fluid. The relevant variables in reduced units are:

$$
\begin{aligned}
T^* &= k_B T/\epsilon, \qquad \rho^* = n\sigma^3, \\
\eta^* &= \eta\sigma^2/(m\epsilon)^{1/2}, \qquad \eta_0^* = \eta_0\sigma^2/(m\epsilon)^{1/2}, \\
\gamma^* &= \gamma\sigma(m/\epsilon)^{1/2} = \frac{u_d}{2L}\sigma\sqrt{\frac{m}{\epsilon}}, \\
\tau^* &= \tau/\sigma(m/\epsilon)^{1/2} = \frac{\sqrt{\eta_0^*}}{\rho^* T^{*3/4}}, \\
\tau_e^* &= \sqrt{2}\tau^*.
\end{aligned}
\tag{19.77}
$$

Here ϵ is the well depth of the LJ potential, σ is its contact diameter, m is the mass of the particle, and n is the number density.

Then the non-Newtonian shear viscosity formula is expressible in terms of a universal function of the reduced variable product

$$
\tau_e^*\gamma^* = \frac{\sqrt{2\eta_0^*}\,\gamma^*}{\rho^* T^{*3/4}},
\tag{19.78}
$$

where τ_e^* may be regarded as the reduced stress relaxation time. Therefore if the non-Newtonian viscosity is scaled by the Newtonian viscosity then the scaled non-Newtonian viscosity becomes a universal function of $\tau_e^*\gamma^*$ (reduced shear rate):

$$
\frac{\eta^*}{\eta_0^*}(\rho^*, T^*, \gamma^*) = \frac{\sinh^{-1}(\tau_e^*\gamma^*)}{\tau_e^*\gamma^*},
\tag{19.79}
$$

which is independent of material parameters and thus indicates that there are rheological corresponding states of ρ^*, T^*, and γ^* since η_0^* is a function of ρ^* and T^* for the Lennard–Jones liquid. The non-Newtonian viscosity formula (Eq. (19.67)) is known as the Ree–Eyring formula in the rheology literature, which was obtained on the basis of the absolute reaction rate theory of Eyring. The original Ree–Eyring formula, however, contains a number of empirical parameters. In contrast to their semiempirical formula the present non-Newtonian viscosity formula is derived from the kinetic theory of dense fluids and the generalized hydrodynamics derived therefrom and hence, in principle, does not contain empirical parameters at all. Therefore the present formula provides the molecular theory foundation for the Ree–Eyring formula, at least, for simple fluids, if the Newtonian viscosity η_0^* is calculated by means of a molecular theory.

The scaled variable $\tau_e^* \gamma^*$ indicates that it will be possible to predict the non-Newtonian viscosity η^* for all reduced shear rates at any density and temperature. We will see that this indeed is the case when all the available molecular dynamics simulation data by various authors are assembled and compared with the prediction by the reduced formula (19.79), as shown Fig. 19.1, for which η_0^* is calculated by using the density fluctuation theory formula for η_0^*. In the density fluctuation theory[10] the modified free volume theory for self-diffusion and the generic van der Waals equation of state (i.e., the canonical equation of state described in Chapter 10) are combined to obtain a molecular theory expression for η_0^*. In Fig. 19.1, the reduced shear rate dependence of the non-Newtonian viscosity thus computed (solid

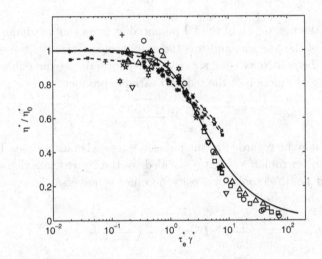

Fig. 19.1   Scaled non-Newtonian shear viscosity (η^*/η_0^*) plotted against the scaled reduced shear rate ($\tau^* \gamma^*$) for all available molecular simulation data and compared with the universal formula given in Eq. (19.79) at various reduced temperatures and reduced densities. The symbols and filled circles connected by broken lines are MD simulation data reported in the literature whereas the solid curve is the corresponding state (universal) non-Newtonian viscosity predicted by the present theory. Reproduced with permission from R. Laghaei, A. E. Nasrabad, B. C. Eu, *J. Chem. Phys.* **123**, 234507 (2005) © 2005 American Institute of Physics.

---

[10]See, B. C. Eu, *Transport Coefficients of Fluids* (Springer, Heidelberg, 2006).

curve) is compared with the results of molecular dynamics (MD) simulation data by Ashurst *et al.*,[11] Heyes *et al.*,[12] and Evans.[13] Various symbols in the figures represent the MD simulation data by various authors, which are reduced with η_0^* predicted by the theory of Newtonian viscosity, which is a statistical mechanical theory developed by Eu and collaborators; see Footnote 10. The meanings of the symbols are: $\bigcirc$ at $T^* = 2.5$, $\rho^* = 0.01$; $\square$ at $T^* = 2.5$, $\rho^* = 0.05$; $\triangle$ at $T^* = 2.5$, $\rho^* = 0.1$; $\triangledown$ at $T^* = 1.5$, $\rho^* = 0.05$; ($\bigstar$) at $T^* = 3.0$, $\rho^* = 0.05$; ($\varhexstar$) at $T^* = 0.85$, $\rho^* = 0.76$. For the triple point data at $T^* = 0.722$, $\rho^* = 0.8449$ the crosses ($+$) are by Heyes *et al.*, asterisks ($*$) are by Evans, and the crosses ($\times$) are by Ashurst *et al.*, all of which are scaled by $\eta_0^* = 2.89$ estimated from the simulation data. The left-pointed open triangles ($\triangleleft$) are the data of Heyes and Szczepanski at $T^* = 5.0$, $\rho^* = 0.1$, which are scaled with $\eta_0^* = 0.45$ reported in the NIST reference data base. The broken curve with filled circles ($\bullet$) is the simulation results of Evans *et al.* scaled with the Newtonian viscosity computed by Meier *et al.*[14] $\eta_0^* = 3.24$, and the broken curve with filled squares ($\blacksquare$) is the simulation results of Evans *et al.* scaled with their own simulation result for the Newtonian viscosity $\eta_0^* = 3.41$. The value of 3.24 for η_0^* given by Meier *et al.* and Ferrario *et al.*[15] appears to yield better zero shear rate non-Newtonian viscosity, suggesting that the result of Meier *et al.* is more accurate than the older value by Evans *et al.*

In the reduced plot presented in Fig. 19.1, the simulation data are considerably scattered around the theoretical solid curve, suggesting the delicate task of evaluating the non-Newtonian viscosity by molecular dynamics simulation methods. Nevertheless, the simulation data of various authors are still around the universal non-Newtonian viscosity curve (solid curve) predicted by the present theory, and it indicates that the non-Newtonian viscosity of the LJ fluid may indeed have a scaling function that is universal for all temperatures and densities.

[11]W. T. Ashurst and W. G. Hoover, *Phys. Rev. A* **11**, 658 (1975); W. G. Hoover, D. J. Evans, R. B. Hickman, A. J. C. Ladd, W. T. Ashurst, and B. Moran, *Phys. Rev. A* **22**, 1690 (1980).

[12]D. M. Heyes, J. J. Kim, C. J. Montrose, and T. A. Litovitz, *J. Chem. Phys.* **73**, 3987 (1980); D. M. Heyes, *Physica* **133A**, 473 (1985); D. M. Heyes and R. Szczepanski, *J. Chem. Soc. Faraday Trans.* II **83**, 319 (1987).

[13]D. J. Evans, *Phys. Rev. A* **23**, 1988 (1981); .D. J. Evans, G. P. Morriss, and L. M. Hood, *Mol. Phys.* **68**, 637 (1989).

[14]K. Meier, A. Laesecke, and S. Kabelac, *J. Chem. Phys.* **121**, 3671 (2004); *ibid.* **121**, 9526 (2004).

[15]M. Ferrario, G. Ciccotti, L. Holian, and J. P. Ryckaert, *Phys. Rev. A* **44**, 6936 (1991).

19.3   Chemical Oscillations and Pattern Formation

We have studied the thermodynamics of chemical equilibria and learned
how we may calculate the equilibrium constants in terms of thermodynamic quantities. Chemical reactions at the equilibrium level, however, do
not reveal dynamics of chemical reactions. We can learn about the dynamical aspects, only if the kinetic evolution of reactions is studied. Such
kinetic evolutions can be studied from the standpoint of irreversible thermodynamics, a general outline of which is presented in this chapter, with
a particular example that emphasizes the type of wave equations involved
in dynamical study of chemical reactions. Other concurrently occurring
hydrodynamic processes in the system may interfere with the chemical evolution, and many experiments in recent years show that some fascinating
phenomena can be observed.

Chemical reactions can happen when molecules come sufficiently close
to each other in space within the range of chemical forces on the order
of a few bond lengths at most. They therefore occur at shorter spatial
scales than the usual hydrodynamic scales. And moreover, they are usually
faster than most of hydrodynamic processes such as flow of matter arising
from the stress and heat flow present in the fluid, but can occur roughly
on the same time scale as diffusion of molecules, and consequently their
evolutions can get coupled with each other. Such coupling phenomena
have been experimentally observed in connection with chemical oscillations,
initially through the famous Belusov–Zhaboutinsky (BZ) reaction,[16] which
harbingered numerous subsequent studies of oscillatory chemical reactions
and pattern formations in connection with chemistry, biochemistry, and
biology.

The BZ reaction was initially examined from the irreversible thermodynamic framework by I. Prigogine and his collaborators[17] by using the
local equilibrium hypothesis and linear irreversible thermodynamics formalism. This formalism utilizes steady-state constitutive equations for diffusion fluxes, which are linear with respect to the diffusion fluxes. When
used in the concentration evolution equations for components undergoing
chemical reactions, the system of equations necessarily gives rise to a system

[16]B. P. Belusov, *Sborn referat. radiat. meditsin za 1958*: Collection of Abstracts on
Radiation Medicine (Medgiz, Moscow, 1959), p. 145; A. M. Zhaboutinsky, *Biofizika* **2**,
306 (1964); *Russ. J. Phys. Chem.* **42**, 1649 (1968).
[17]P. Glansdorff and I. Prigogine, *Thermodynamic Theory of Structure, Stability and
Fluctuations* (Wiley-Interscience, New York, 1971).

of reaction–diffusion equations. More generally than the BZ reaction mentioned, reaction–diffusion reactions occur in a model for morphogenesis according to A. M. Turing's theory.[18] In this theory, reaction–diffusion equations appear giving rise to forms, and forms occur beyond the regime of stability of a set of coupled chemical reactions. For this reason alone there is a considerable interest studying reaction–diffusion equations in connection with biological phenomena.

The reaction–diffusion equations appearing in the theories of Prigogine and his school and Turing in morphogenesis are all nonlinear parabolic[19] partial differential equations, which are expected to describe wave phenomena associated with the chemical waves accompanying the chemical reactions. One characteristic of parabolic partial differential equations is that the speed of wave propagation is infinite, suggesting that a wave initiating at a point in space instantly reaches points away from the source and hence the boundaries. This generally contradicts the experience with the chemical waves studied in the laboratory. Such parabolic equations arise because of using the linear theory of irreversible thermodynamics. In this section, we will show that according the generalized hydrodynamic theory presented earlier in this chapter the governing equations for the chemical wave propagation is hyperbolic, which means that the speed of wave propagation is finite.

The subject matter discussed in this subsection ranges over such a wide area that a subsection in a book would not do justice to the whole subject area that can take up a monograph. It is therefore all the more important to provide the basic key idea of how the dynamical theory incorporating the principles of irreversible thermodynamics can be formulated. Our aim here is just to introduce the reader to the notion that the subject matters can come under the purview of irreversible thermodynamics principles and indeed can be studied with the thermodynamics formalism described earlier in this chapter. However, the associated wave phenomena can be studied more appropriately by using hyperbolic reaction–diffusion equations than the conventionally employed parabolic reaction–

---

[18] A. M. Turing, *Phil. Trans. Roy. Soc. London B* **237**, 37 (1952).

[19] For example, the time-dependent Schrödinger equation, a wave equation that has a first-order time derivative term, is a parabolic partial differential equation, whereas a classical wave equation, which has a second-order time derivative term in addition to a second-order spatial derivative term, is a hyperbolic partial differential equation. For more precise mathematical definitions, see P. M. Morse and H. Feshbach, *Methods of Theoretical Physics* (McGraw-Hill, New York, 1953), Vol. I, Chapter 6.

diffusion equations, which necessarily predict an infinite speed of propagation of disturbances caused in the fluid by chemical reactions occurring locally.

If there are chemical reactions undergoing in the fluid, the mass fraction evolution equations (Eq. (19.6)) should be modified by adding the reaction rate term. Thus if we denote the reaction rate for species a by

$$\Lambda_a^{(r)} = \sum_{k=1}^{m} \nu_{ak} R_k, \tag{19.80}$$

where R_k is the reaction rate of the kth reaction and ν_{ak} is stoichiometric coefficients times the mass of species a of the reaction, then the mass fraction evolution equation should read

$$\rho \frac{dc_a}{dt} = -\nabla \cdot \mathbf{J}_a + \Lambda_a^{(r)}. \tag{19.81}$$

In the sense that the rate term $\Lambda_a^{(r)}$ appears as a source term in the mass fraction balance equation, c_a is no longer a conserved quantity if a chemical reaction is present in the system. Now, if linear constitutive equations are assumed, as is conventionally done, for diffusion fluxes

$$\sum_{b=1}^{m} L_{ab} \mathbf{J}_b + \frac{k_B T}{m_a} \nabla \rho_b = 0, \tag{19.82}$$

where ρ_b is the mass density of b and L_{ab} are phenomenological coefficients obeying the Onsager reciprocal relations $L_{ab} = L_{ba}$, and if they are then substituted into Eq. (19.81), a coupled set of parabolic partial differential equations arise for the evolution equations for mass fractions c_a. Such parabolic wave equations predict an infinite speed of wave propagation as mentioned earlier. In other words, in the context of the chemical reactions considered, the effects of chemical reactions occurring at a point in a local volume will instantly propagate throughout the fluid. This intuitively does not appear to be realistic, and indeed experimental evidence does not support it. Nevertheless, such parabolic reaction–diffusion equations have been conventionally employed to reaction–diffusion phenomena. In this section, we show that the thermodynamically consistent generalized hydrodynamic theory introduced earlier produces hyperbolic reaction–diffusion equations in contrast to the conventional approach to the reaction–diffusion phenomena. It opens up a new vista to the study of reaction–diffusion phenomena and, perhaps, morphogenesis which chemical reactions and pattern formation underlie according to the concept of Turing.

To be specific, we will take an example of typical coupled chemical reactions involving a feedback loop: A modified Selkov model consisting of three steps of chemical reactions[20]

$$A \underset{k_{-1}}{\overset{k_1}{\rightleftarrows}} S,$$

$$S + 2P \underset{k_{-2}}{\overset{k_2}{\rightleftarrows}} 3P,$$

$$P \underset{k_{-3}}{\overset{k_3}{\rightleftarrows}} B.$$

These reactions have been studied for some aspects of glycolysis. In these reactions, A and B are kept at fixed concentrations and the intermediates S and P change in time. According to the mass action law, the reaction rates for the two intermediate species S and P are given by

$$R_S = k_1\rho_A - k_{-1}\rho_S - k_2\rho_S\rho_P^2 + k_{-2}\rho_P^3, \tag{19.83}$$

$$R_P = k_2\rho_S\rho_P^2 - k_{-2}\rho_P^3 - k_3\rho_P + k_{-3}\rho_B, \tag{19.84}$$

where k_i and k_{-i} are the reaction rate constants for the forward and reverse reactions in the ith step in the coupled chemical reactions under consideration. To make the equations simpler we assume that the phenomenological coefficients $L_{SP} = L_{PS}$ are equal to zero. Therefore there are only self-diffusion processes taken into account for species P and S. This assumption is generally not valid for real systems, but it is taken here to make the equations as simple as possible. This assumption is, in fact, taken in virtually all works on chemical waves in the literature. Since it is convenient to work with dimensionless equations, the following reduced variables are introduced:

$$\begin{aligned}
\tau &= k_3 t, & \xi &= \mathbf{r}/L, \\
\mathbf{u} &= \mathbf{J}_P/Lk_3\sqrt{k_3/k_2}, & \mathbf{v} &= \mathbf{J}_S/Lk_3\sqrt{k_3/k_2}, \\
X &= \rho_P/\sqrt{k_3/k_2}, & Y &= \rho_S/\sqrt{k_3/k_2}, \\
A &= (k_1/k_3\sqrt{k_3/k_2})\rho_A, & B &= (k_{-3}/k_3\sqrt{k_3/k_2})\rho_B, \\
K &= k_{-2}/k_2, & R &= k_{-1}/k_3.
\end{aligned}$$

[20] Y. Termonia and J. Ross, *Proc. Nat. Acad. Sci. USA* **78**, 2952, 3563 (1982); P. Richter, P. Rehmus, and J. Ross, *Prog. Theor. Phys.* **66**, 385 (1981).

It is also convenient to define the dimensionless reaction–diffusion number defined by[21]

$$N_{\rm rd} = \frac{k_B T}{\sqrt{m_P m_S}} k_3^{-1} \left(D_S D_P\right)^{-1/2}, \tag{19.85}$$

where D_S and D_P are essentially the self-diffusion coefficients defined by the diagonal parts of the diffusion constant matrix in the S and P species space. The reaction–diffusion number is the ratio of reaction time × diffusion time to the time scale of hydrodynamic speed squared. We also define the additional dimensionless parameters

$$f = \sqrt{\frac{m_S D_S}{m_P D_P}}, \quad \widehat{D}_X = \frac{D_P}{k_3 L^2}, \quad \widehat{D}_Y = \frac{D_S}{k_3 L^2}. \tag{19.86}$$

With these reduced variables, the evolution equations take the forms

$$\frac{\partial X}{\partial \tau} = -\frac{\partial}{\partial \xi} \cdot \mathbf{u} + B - X + X^2 Y - K X^3, \tag{19.87}$$

$$\frac{\partial Y}{\partial \tau} = -\frac{\partial}{\partial \xi} \cdot \mathbf{v} + A - R Y - X^2 Y + K X^3, \tag{19.88}$$

$$\frac{\partial \mathbf{u}}{\partial \tau} = -N_{\rm rd} f \left(\widehat{D}_X \frac{\partial X}{\partial \xi} + \mathbf{u} \right), \tag{19.89}$$

$$\frac{\partial \mathbf{v}}{\partial \tau} = -N_{\rm rd} f^{-1} \left(\widehat{D}_Y \frac{\partial Y}{\partial \xi} + \mathbf{v} \right). \tag{19.90}$$

Therefore, put in the context of the present reactions, the reaction–diffusion number $N_{\rm rd}$ gives a measure of relative time scales of the chemical evolution of material species X and Y to the ratio of the hydrodynamic time squared to the diffusion time, when the reduced time is reckoned in the relative scale of the rate constant k_3 and the mean diffusion constant to the sound speed.

To determine the character of this set of differential equations the eigenvalues of the characteristic matrix of this system are sought. They are determined by the characteristic determinant

$$\begin{vmatrix} -\lambda & 0 & 1 & 0 \\ 0 & -\lambda & 0 & 1 \\ N_{\rm rd}\widehat{D}_X f & 0 & -\lambda & 0 \\ 0 & N_{\rm rd}\widehat{D}_Y/f & 0 & -\lambda \end{vmatrix} = 0. \tag{19.91}$$

---

[21]The reaction–diffusion number should rank among various non-dimensional fluid dynamic numbers such as Mach, Reynolds, Prandtl numbers, etc. It characterizes the relative time and spatial scales of a chemical reaction and hydrodynamic processes.

The eigenvalues of this determinant are:

$$\lambda_1, \lambda_2, \lambda_3, \lambda_4 = \pm\sqrt{N_{rd}\widehat{D}_X f}, \quad \pm\sqrt{N_{rd}\widehat{D}_Y/f}. \tag{19.92}$$

Since the eigenvalues are all real, the system of Eqs. (19.87)–(19.90) is seen to be hyperbolic. It would be interesting to see in which regime the hyperbolic behavior occurs. For the purpose we estimate N_{rd} for typical values for the parameters involved. If the mean diffusion coefficient is of the order of $10^{-9}\,\mathrm{m^2 s^{-1}}$ and the rate constant is of the order of $10^{12}\,\mathrm{s^{-1}}$, then N_{rd} is of the order of 10^2 at the room temperature. In this case, in the regime of $N_{rd} \lesssim 10^2$ the evolution must be hyperbolic, while in the longtime regime the system behaves as if it is parabolic. Thus in the transient regime of evolution and below it, the dynamic evolution of the system must be described by a hyperbolic system of evolution equations.

It is helpful to put the equations into a form of wave equation that will enable us to compare with the classical wave equations and thereby give us better insight. Eliminating the variables $\mathbf{u}$ and $\mathbf{v}$ from the set of Eqs. (19.87)–(19.90) we obtain the coupled second-order partial differential equations

$$\frac{\partial^2 \mathbf{Z}}{\partial \tau^2} + \overline{N}_{rd}\mathbf{H}\frac{\partial \mathbf{Z}}{\partial \tau} - \overline{N}_{rd}\mathbf{D}\frac{\partial^2 \mathbf{Z}}{\partial \xi^2} = \overline{N}_{rd}\mathbf{R}(X,Y), \tag{19.93}$$

where

$$\mathbf{Z} = \begin{pmatrix} X \\ Y \end{pmatrix}, \tag{19.94}$$

$$\mathbf{R} = \begin{pmatrix} B - X + X^2 Y - KX^2 \\ (A - RY - X^2 Y + KX^3)\,f^{-2} \end{pmatrix}$$

$$= \begin{pmatrix} 1 & 0 \\ 0 & f^{-2} \end{pmatrix} \begin{pmatrix} B - X + X^2 Y - KX^2 \\ A - RY - X^2 Y + KX^3 \end{pmatrix} \equiv \mathbf{f}_2 \mathbf{R}', \tag{19.95}$$

$$\mathbf{D} = \begin{pmatrix} \widehat{D}_X & 0 \\ 0 & \widehat{D}_Y f^{-2} \end{pmatrix} = \mathbf{f}_2\widehat{\mathbf{D}}, \tag{19.96}$$

$$\mathbf{H} = \begin{pmatrix} H_{xx} & H_{xy} \\ H_{yx} & H_{yy} \end{pmatrix}, \tag{19.97}$$

with the definitions

$$H_{xx} = 1 + (1 - 2XY + 3KX^2)/\overline{N}_{rd},$$
$$H_{xy} = -X^2/\overline{N}_{rd},$$
$$H_{yx} = (2XY - 3KX^2)/\overline{N}_{rd},$$
$$H_{yy} = f^{-2} + (R + X^2)/\overline{N}_{rd}. \tag{19.98}$$

The matrix nonlinear wave equation (Eq. (19.93)) is a pair of coupled nonlinear telegraphist equations, which are hyperbolic partial differential equations. Telegraphist equations are well known in the field of wave equations. Roughly speaking, the inverses of the eigenvalues of the coefficient matrix to the second derivative term in the wave equation (Eq. (19.93)) give the wave velocities (group velocity, more precisely), which are proportional to $1/\sqrt{N_{\mathrm{rd}}D_X}$ or $1/\sqrt{N_{\mathrm{rd}}D_Y}$. Therefore in the limit of $N_{\mathrm{rd}}D_X$, $N_{\mathrm{rd}}D_Y \to \infty$, the wave equations become parabolic. Such limits are precisely the reaction–diffusion equations used in the conventional approach to chemical oscillation phenomena. Therefore, if $N_{\mathrm{rd}}D_X$ is finite, the equations are hyperbolic, and the modes of evolution of the waves predicted by the hyperbolic wave equations are expected to be different from those in the limit of $N_{\mathrm{rd}}D_X \to \infty$, namely, parabolic wave equations.

It is not possible to obtain analytic solutions in closed forms for the nonlinear wave equations (Eq. (19.93)). The numerical solutions have been studied by Al-Ghoul and Eu.[22] Numerical solutions reveal rather rich structures, which can range over a widely different spectrum of patterns and modes of oscillation different from those observed with parabolic wave equations.

Before implementing numerical solutions and a general procedure of a solution method, it is preferable to examine the steady state of the equations and the stability of the steady state.

If the system is spatially homogeneous then the second spatial derivative term in Eq. (19.93) vanishes and the homogeneous steady state is determined by the equation

$$\mathbf{R}(X_0, Y_0) = 0, \tag{19.99}$$

where X_0 and Y_0 are the steady state. More explicitly, they are the solutions of the algebraic equations

$$A - RY_0 - X_0^2 Y_0 + KX_0^3 = 0,$$
$$B - X_0 + X_0^2 Y_0 - KX_0^2 = 0.$$

Eliminating Y_0 from these coupled equations yields a cubic equation for X_0. It is well known that a cubic equation can have three types of solutions depending on the sign of the discriminant, which may be written as

$$\Delta(B, C) = 4\alpha^{-2} \left[B^2 + 4\lambda B/\alpha + 4 \left(\lambda^2 - q^3 \right)/\alpha^2 \right], \tag{19.100}$$

---

[22]See M. Al-Ghoul and B. C. Eu, *Physica D* **90**, 119, (1996); *ibid.* **97**, 531 (1996).

where

$$\alpha = R/(1 - KR),$$
$$\lambda = -\alpha^2 C/6R + \alpha^3 C^3/27R^3, \qquad (19.101)$$
$$q = \frac{1}{3}\alpha(\alpha C^2/3R^2 - 1).$$

There are three real roots if $\Delta(B,C) < 0$ in which case there holds the condition

$$-2\left(\lambda + q^{3/2}\right)/\alpha < B < -2\left(\lambda - q^{3/2}\right)/\alpha. \qquad (19.102)$$

There are three real roots with two of them degenerate if $f(B,C) = 0$, which is satisfied if

$$B = -2\left(\lambda \pm q^{3/2}\right)/\alpha, \qquad (19.103)$$

whereas there is only one real root and two complex conjugate roots if $\Delta(B,C) > 0$, which means

$$B + 2\left(\lambda \pm q^{3/2}\right)/\alpha > 0. \qquad (19.104)$$

To see the distinction with regard to stability of the steady state between hyperbolic and parabolic wave equations, we consider inhomogeneous steady states, which are displaced from the homogeneous steady state. For the purpose we first linearize Eq. (19.93) with respect to (X_0, Y_0). With the definitions of displacements from (X_0, Y_0)

$$x = X - X_0, \quad y = Y - Y_0 \qquad (19.105)$$

we obtain the linearized wave equations

$$\frac{\partial^2 \mathbf{Z}_l}{\partial \tau^2} + \overline{N}_{\rm rd}\mathbf{H}_0\frac{\partial \mathbf{Z}_l}{\partial \tau} - \overline{N}_{\rm rd}\mathbf{D}\frac{\partial^2 \mathbf{Z}_l}{\partial \xi^2} = -\overline{N}_{\rm rd}\mathbf{DMZ}_l, \qquad (19.106)$$

where

$$\mathbf{Z}_l = \begin{pmatrix} x \\ y \end{pmatrix},$$

$$\mathbf{H}^0 = \mathbf{H}(X_0, Y_0), \qquad (19.107)$$

$$\mathbf{M} = -\mathbf{D}^{-1}\begin{pmatrix} -1 + 2X_0Y_0 - 3KX_0^2 & X_0^2 \\ (-2X_0Y_0 + 3KX_0^2)\,f^{-2} & -(R - X_0^2)f^{-2} \end{pmatrix}.$$

This set of linearized wave equations can be easily solved and the dispersion relations of waves can be calculated. Solving Eq. (19.106) by the Fourier transform

$$\mathbf{Z}_l = \sum_\omega \sum_\mathbf{k} \mathbf{\Phi}(\mathbf{k}, \omega) \exp[i(\mathbf{k} \cdot \boldsymbol{\xi} - \omega\tau)], \tag{19.108}$$

we obtain the linear algebraic set

$$\left(-\omega^2 \mathbf{I} - i\omega \overline{N}_{\mathrm{rd}} \mathbf{H}^0 + k^2 \overline{N}_{\mathrm{rd}} \mathbf{D} + \overline{N}_{\mathrm{rd}} \mathbf{DM}\right) \mathbf{\Phi}(\mathbf{k}, \omega) = 0. \tag{19.109}$$

Here $\mathbf{I}$ is the 2×2 unit matrix. The dispersion relations are given by the secular determinant

$$\det\left|-\omega^2 \mathbf{I} - i\omega \overline{N}_{\mathrm{rd}} \mathbf{H}_0 + k^2 \overline{N}_{\mathrm{rd}} \mathbf{D} + \overline{N}_{\mathrm{rd}} \mathbf{DM}\right| = 0. \tag{19.110}$$

It is a quartic equation of $z = -i\omega$ and k, which may be written as

$$d_4(z, k) = z^4 + Pz^3 + Qz^2 + Tz + U = 0, \tag{19.111}$$

where the coefficients are defined by

$$K_{xx} = k^2 D_{xx} + (\mathbf{DM})_{xx},$$
$$K_{yy} = k^2 D_{yy} + (\mathbf{DM})_{yy},$$
$$P = \overline{N}_{\mathrm{rd}}(H_{xx}^0 + H_{yy}^0),$$
$$Q = \overline{N}_{\mathrm{rd}}(K_{xx} + K_{yy}) + \overline{N}_{\mathrm{rd}}^2(H_{xx}^0 H_{yy}^0 - H_{xy}^0 H_{yx}^0),$$
$$T = \overline{N}_{\mathrm{rd}}^2\left[K_{xx}H_{yy}^0 + K_{yy}H_{xx}^0 - H_{xy}^0(\mathbf{DM})_{yx} - H_{yx}^0\right],$$
$$U = \overline{N}_{\mathrm{rd}}^2\left[K_{xx}K_{yy} - (\mathbf{DM})_{xy}(\mathbf{DM})_{yx}\right]. \tag{19.112}$$

For the polynomials to have all the roots with the negative imaginary part the Hurwitz conditions[23] must be satisfied:

$$P > 0, \quad PQ - T > 0, \quad PQT - T^2 - P^2U > 0. \tag{19.113}$$

The first condition P is independent of k, while the second condition $f_2(B, C) = PQ - T$, and the third condition $f_3(B, C) = PQT - T^2 - P^2U$ are even functions of k. Since $d_4(z, k)$ is real it has two pairs of complex conjugate roots. The level curves of the Hurwitz conditions are given by $P = 0$, $f_2(B, C) = 0$, and $f_3(B, C) = 0$. On the level curves the real parts of the roots vanish and the solution of the wave equations (Eq. (19.106)) becomes oscillatory in time. The stability phase diagram can be constructed

---

[23] For Hurwitz conditions, see F. R. Gantmacher, *Theories des Matrices* (Dunod, Paris, 1966), Vol. 2.

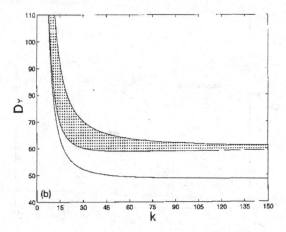

Fig. 19.2 Stability phase diagram in $(k, \widehat{D}_Y)$ plane for the hyperbolic wave equations. The ordinate is in the units of 10^{-4}. The shaded region is for the stable phase and the rest is for the unstable phases. Reproduced with permission from M. Al-Ghoul and B. C. Eu, *Physica* D **90**, 119 (1996) © 1996 Published by Elsevier Science B.V.

by using the Hurwtiz conditions. In Fig. 19.2, the stability phase diagram is plotted in $(k, \widehat{D}_Y)$ plane for the case of $R = 0.1$, $K = 1$, $f = 1$, $\widehat{D}_X = 0.006$, $B = 0.09$. The first level curve $P = 0$ does not appear in this figure because it does not depend on k. According to Turing, new structures can occur as the steady state solutions bifurcate as the parameters change into those of the unstable phase from those of the stable phase. This stability diagram therefore suggests such new structures can get organized for wide ranges of k values. This is in contrast to the case of the parabolic wave equations for which there is only one level curve possible, and when the stability phase diagram is constructed, it is as in Fig. 19.3. Since one of Hurwitz conditions for the parabolic wave equations is always positive there appears only one level curve in Fig. 19.3. Since the unstable region is confined to the region bound by the parabola the Turing bifurcations are possible only in the confined region bound by $k_{\max} = \max(k)$. This difference is probably one of the most significant differences between the hyperbolic and parabolic wave equations besides the difference in the speed of wave propagation for the system under examination.

We give a short summary of numerical solution results for the reaction–diffusion equations. If there are no diffusion fluxes, the one-dimensional version of Eq. (19.93) predicts a limit cycle in (X, Y) plane if the initial concentrations are appropriate. As the diffusion is turned on, N_{rd}

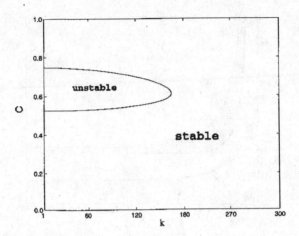

Fig. 19.3 Stability phase diagram in (k, C) plane for the parabolic wave equations. The parameter values are the same as Fig. 19.2 for the hyperbolic wave equations. Reproduced with permission from M. Al-Ghoul and B. C. Eu, *Physica D* **90**, 119 (1996) © 1996 Published by Elsevier Science B.V.

becomes finite and the limit cycle turns quasiperiodic; for example, at $N_{\mathrm{rd}} = 0.1$ the power spectrum for the hyperbolic system shows there are three fundamental frequencies, whereas there are two fundamental frequencies in the case of the parabolic system, indicating a period doubling. As N_{rd} decreases further, the power spectrum of the hyperbolic system indicates a chaotic motion. The traveling wave patterns also differ in that a sharp wave front appears for the hyperbolic system, whereas there does not appear such a sharp front, but a diffuse one stretching over all space, indicating an infinite speed of propagation in the case of the parabolic system.

For a two-dimensional version of Eq. (19.93) the numerical solutions yields a very rich collection of structures: hexagonal structures, stripes, maze structures, chaotic structures, etc., depending on the values of N_{rd} and the initial and boundary conditions. Random initial conditions give rise to patterns which evolve to break symmetry and form a new series of patterns as N_{rd} varies. Solitary waves are also observed. The hyperbolic wave equations therefore are observed to produce much richer patterns and structures than the parabolic wave equations. This probably is owing to the wider domain of instability of the hyperbolic wave equations than the parabolic wave equations, as is evident from the two phase diagrams presented in Figs. 19.2 and 19.3.

Because of space limitation it is not possible to discuss further details of the numerical study made. What is briefly mentioned here allows us only a glimpse into the chemical reaction dynamics. There remain wide subject areas of dynamics and irreversible thermodynamics which are little explored as yet in the present approach described. The nonlinear telegraph equations such as Eq. (19.93) can serve as the starting point of such a study.

Appendix A

Local Form of Energy Conservation Law

The first law of thermodynamics may be formulated in local form by using the formalism of continuum mechanics. This form is not used in thermodynamics, but it gives us a more precise description of heat and work in local form. This form of conservation law will be useful for fluid dynamic considerations.

We consider a fluid composed of r chemically inert components. The components, denoted by $i = 1, 2, \ldots, r$, are subject to external conservative potentials ψ_i on components $i = 1, 2, \ldots, r$. We denote the mass density of component i by ρ_i and the total mass density by ρ:

$$\rho = \sum_{i=1}^{r} \rho_i. \tag{A.1}$$

The total energy density (specific energy) will be denoted by E. The total energy $\mathbb{E}$ of the fluid enclosed in volume V is then given by the integral

$$\mathbb{E} = \int_V dV \, \rho E(\mathbf{r}, t). \tag{A.2}$$

The rate of change in energy is

$$\frac{d}{dt} \int_V dV \, \rho E(\mathbf{r}, t) = \int_V dV \frac{\partial}{\partial t} \, \rho E(\mathbf{r}, t). \tag{A.3}$$

Let us denote the energy flux by $\mathbf{J}_e$. This vector is parallel to the unit vector normal to the surface of the volume whose outward direction is taken positive. Thus, when the energy flows out of the system, the sign of the flux is positive. Since the energy is conserved, the energy change within the volume must be balanced by the inflow of energy from the outside.

Therefore we find

$$\int_V dV \frac{\partial}{\partial t} \rho E(\mathbf{r}, t) = -\int_\Omega d\mathbf{\Omega} \cdot \mathbf{J}_e, \qquad (A.4)$$

where $d\mathbf{\Omega}$ denotes the surface element and the surface integral is over the entire surface Ω of volume V. By the Gauss theorem, Eq. (A.4) may be written as

$$\int_V dV \left[\frac{\partial}{\partial t} \rho E(\mathbf{r}, t) + \nabla \cdot \mathbf{J}_e \right] = 0. \qquad (A.5)$$

Since the volume is arbitrary, Eq. (A.5) implies the following local equation

$$\frac{\partial}{\partial t} \rho E(\mathbf{r}, t) + \nabla \cdot \mathbf{J}_e = 0 \qquad (A.6)$$

for the local expression of the energy conservation law. Since the fluid may flow with a velocity $\mathbf{u}$, its kinetic energy density is $\frac{1}{2} u^2$, and if it is subject to an external field with a potential, its potential energy density is

$$\psi = \rho^{-1} \sum_{i=1}^{r} \psi_i \rho_i, \qquad (A.7)$$

where ψ_i is the potential energy density of species i. If we denote the internal energy density by $\mathcal{E}$ then the total energy density is

$$E = \tfrac{1}{2} u^2 + \psi + \mathcal{E}. \qquad (A.8)$$

The energy flux also consists of various components. They are the convective total energy flow $\rho E \mathbf{u}$ arising from the convective motion of the fluid, an energy flow $\mathbf{P} \cdot \mathbf{u}$ stemming from the mechanical work done on the system where $\mathbf{P}$ is the stress (pressure) tensor, a potential energy flow $\sum_i \psi_i \mathbf{J}_i$ arising from the diffusion of various components relative to the fluid motion in the field of force, where $\mathbf{J}_i$ is the diffusion flow of component i, and the heat flow $\mathbf{J}_h$:

$$\mathbf{J}_e = \rho E \mathbf{u} + \mathbf{P} \cdot \mathbf{u} + \sum_{i=1}^{r} \psi_i \mathbf{J}_i + \mathbf{J}_h. \qquad (A.9)$$

The diffusion flow $\mathbf{J}_i$ is defined by

$$\mathbf{J}_i = \rho_i (\mathbf{u}_i - \mathbf{u}), \qquad (A.10)$$

where $\mathbf{u}_i$ is the velocity of component i. It is important to recognize that the meaning of heat flow $\mathbf{J}_h$ is given by Eq. (A.9). That is, given the meaning of the first three terms on the right-hand side making up the energy flux in Eq. (A.9), the heat flow is the rest of the energy flow so that the energy conservation law as stated in Eq. (A.5) holds valid. Therefore, the meaning of heat flow will change if there are additional terms that should be taken into account to make up $\mathbf{J}_e$. In this sense, our understanding of $\mathbf{J}_h$ can evolve as our understanding of the energy conservation law broadens over time. Therefore thermodynamics may be said to be anthropomorphic.

Since the mass must conserve, the equation of continuity holds

$$\frac{\partial}{\partial t}\rho = -\nabla \cdot (\rho \mathbf{u}) \tag{A.11}$$

and the momentum balance equation is

$$\frac{\partial}{\partial t}\rho \mathbf{u} = -\nabla \cdot (\mathbf{P} + \rho \mathbf{uu}) + \sum_{i=1}^{r} \rho_i \mathbf{F}_i, \tag{A.12}$$

where $\mathbf{F}_i$ is the force density on component i:

$$\mathbf{F}_i = -\nabla \psi_i. \tag{A.13}$$

It can be shown with Eqs. (A.11) and (A.12) that the kinetic and potential energy densities together obey the balance equation

$$\frac{\partial}{\partial t}\rho \left(\tfrac{1}{2}u^2 + \psi\right) = -\nabla \cdot \left(\tfrac{1}{2}\rho u^2 \mathbf{u} + \rho \psi \mathbf{u} + \mathbf{P} \cdot \mathbf{u} + \sum_{i=1}^{r} \psi_i \mathbf{J}_i\right)$$
$$+ \mathbf{P}{:}\nabla \mathbf{u} - \sum_{i=1}^{r} \mathbf{J}_i \cdot \mathbf{F}_i. \tag{A.14}$$

When Eqs. (A.8) and (A.9) are substituted into Eq. (A.6) and the terms are rearranged with the help of Eq. (A.14), the energy conservation law may be cast in the form

$$\rho \frac{d\mathcal{E}}{dt} = -\nabla \cdot \mathbf{J}_h - p\nabla \cdot \mathbf{u} - \mathbf{\Pi}{:}\nabla \mathbf{u} + \sum_{i=1}^{r} \mathbf{J}_i \cdot \mathbf{F}_i \tag{A.15}$$

where d/dt means the substantial time derivative

$$\frac{d}{dt} = \frac{\partial}{\partial t} + \mathbf{u} \cdot \nabla$$

and we have split the pressure tensor $\mathbf{P}$ into the hydrostatic pressure p and a symmetric tensor[1] $\mathbf{\Pi}$:

$$\mathbf{P} = p\boldsymbol{\delta} + \mathbf{\Pi} \tag{A.16}$$

---

[1] It is not the same tensor as the traceless symmetric tensor $\mathbf{\Pi}$ used in Chapter 19, despite the same symbol.

with $\mathbf{U}$ denoting the unit tensor

$$\delta = \begin{pmatrix} 1 & 0 & 0 \\ 0 & 1 & 0 \\ 0 & 0 & 1 \end{pmatrix}. \tag{A.17}$$

With the definition of the specific volume by $v = \rho^{-1}$ the equation of continuity may be written as

$$\rho \frac{dv}{dt} = \nabla \cdot \mathbf{u}. \tag{A.18}$$

We may also write

$$\rho \frac{dQ}{dt} = -\nabla \cdot \mathbf{J}_h \tag{A.19}$$

where dQ is the differential heat added to the unit mass of the fluid. With Eqs. (A.18) and (A.19) we may recast Eq. (A.15) in the form

$$\rho \frac{d\mathcal{E}}{dt} = \rho \frac{dQ}{dt} - \rho p \frac{dv}{dt} - \mathbf{\Pi}{:}\nabla\mathbf{u} + \sum_{i=1}^{r} \mathbf{J}_i \cdot \mathbf{F}_i. \tag{A.20}$$

The last three terms on the right-hand side of Eq. (A.20) represent the work done on the system per unit time; the third term on the right is the viscous heating term, and the last term the work per unit time arising from diffusion against the external force. Therefore we will combine them to write

$$\rho \frac{dW}{dt} = -\rho p \frac{dv}{dt} - \mathbf{\Pi}{:}\nabla\mathbf{u} + \sum_{i=1}^{r} \mathbf{J}_i \cdot \mathbf{F}_i, \tag{A.21}$$

so that

$$\frac{d\mathcal{E}}{dt} = \frac{dQ}{dt} + \frac{dW}{dt} \tag{A.22}$$

This is the mathematical statement of the first law of thermodynamics we wished to formulate in local form. Note that the last two terms on the right-hand side of Eq. (A.21) represent the work arising from the dissipative effects arising from viscous frictions and mass diffusion caused by the external force, respectively. In the case of electrically charged fluids subject to an external electric field the last term is related to the Joule heating. When these dissipative terms vanish, the work per unit time is then given by the pressure–volume work per unit time alone:

$$\frac{dW}{dt} = -p \frac{dv}{dt}. \tag{A.23}$$

More precisely, it is the pressure–volume work performed in infinitesimal timo interval dt. Note that $dW = (dW/dt)\,dt$ and similarly for heat. It is useful to remember that when work is expressed without taking into account the internal work—namely, the second and third terms—in Eq. (A.21), the first law of thermodynamics is considered in the case of non-dissipative processes.

Finally, we note that the internal energy conservation law considered here excludes the effect of radiative heating. If R denotes the radiative heating the internal energy conservation law then should read

$$\rho\frac{d\mathcal{E}}{dt} = \rho\frac{dQ}{dt} - \rho p\frac{dv}{dt} - \boldsymbol{\Pi}{:}\nabla\mathbf{u} + \sum_{i=1}^{r}\mathbf{J}_i \cdot \mathbf{F}_i + R. \qquad (A.24)$$

This equation, Eqs. (A.11), and (A.12) are coupled to constitutive equations for $\boldsymbol{\Pi}$, $\mathbf{J}_i$, radiation intensity, and so on, which must be supplied, if a full description of the system is desired.

Appendix B

Various Coefficients Used in Chapter 10

In this Appendix, we have collected various coefficients appearing in expansions employed in the formulation of the algorithms to calculate the subcritical thermodynamic properties of fluids in Chapter 10. With these coefficients used in various expressions, it is possible to calculate exactly and numerically various quantities developed for the subcritical thermodynamic properties with the models for the GvdW parameters in the canonical equation of state.

B.1 Coefficients of $\Pi^{(i)}(x_{sk}, t)$

To begin with, we observe $\Pi(x_{sk}, t) \equiv \Pi^{(0)}(x_{sk}, t)$ and the derivatives $\Pi^{(i)}(x_{sk}, t)$ of Π are decomposable into cubic polynomials in x_{sk}:

$$\Pi^{(i)}(x_{sk}, t) = \Pi_0^{(i)} + \Pi_1^{(i)} x_{sk} + \Pi_2^{(i)} x_{sk}^2 + \Pi_3^{(i)} x_{sk}^3 \quad (0 \leq i \leq 4). \quad (B.1)$$

The coefficients in this polynomial consist of t-independent and t-dependent parts as follows:

$$\Pi_j^{(i)} = P_{ij} + \widehat{\Pi}_j^{(i)}(t), \quad (B.2)$$

where P_{ij} is the t-independent part made up of the parameters determining the critical point and $\widehat{\Pi}_j^{(i)}(t)$ is the t-dependent part determined by the t-dependent part of the GvdW parameters, namely, $\widehat{a}_i(t)$ and $\widehat{b}_i(t)$. These coefficients, although somewhat complicated but straightforward to obtain from the definition of $\Pi^{(i)}(x, t)$, are listed for $i = 0, \ldots, 4$ in the following section.

B.2   Coefficients P_{ij}

$P_{00} = P = 0,$

$P_{10} = \tau + \zeta(\nu - 1)(a_1 + 2) + \nu(\zeta + 1)(b_1 + 1) = P_1 = 0,$

$P_{11} = 2\zeta(3\nu - 1) + \zeta(3\nu - 2)a_1 + \nu(1 + 3\zeta)b_1,$

$P_{12} = \zeta(3\nu - 1)a_1 + 3\zeta\nu(b_1 + 1),$

$P_{13} = \zeta\nu(a_1 + b_1),$

$P_{20} = 2\zeta(\nu - 1)(2a_1 + a_2 + 1) + 2\nu(b_1 + b_2)(\zeta + 1)$

$\qquad + 2\zeta\nu(a_1 + 2)(b_1 + 1) = 2!\, P_2 = 0,$

$P_{21} = 6\zeta\nu + 4\zeta(3\nu - 1)a_1 + 8\zeta\nu b_1 + 4\zeta\nu a_1 b_1$

$\qquad + 2\zeta(3\nu - 2)a_2 + 2\nu(3\zeta + 1)b_2,$

$P_{22} = 10\zeta\nu a_1 + 6\zeta\nu b_1 + 6\zeta\nu a_1 b_1 + 2\zeta(5\nu - 1)a_2 + 6\zeta\nu b_2,$

$P_{23} = 2\zeta\nu(b_2 + a_1 b_1 + a_2),$

$P_{30} = 6\zeta(\nu - 1)(a_1 + 2a_2 + a_3) + 6\nu(\zeta + 1)(b_2 + b_3)$

$\qquad + 6\zeta\nu(a_1 + 2)(b_1 + b_2) + 6\zeta\nu(2a_1 + a_2 + 1)(b_1 + 1) = 3!\, P_3 = 0,$

$P_{31} = 18\zeta\nu a_1 + 12\zeta(3\nu - 1)a_2 + 6\zeta(3\nu - 2)a_3 + 18\zeta\nu b_1 + 36\zeta\nu b_2$

$\qquad + 6(3\zeta + 1)\nu b_3 + 18\zeta\nu(2a_1 b_1 + a_1 b_2 + a_2 b_1),$

$P_{32} = 6\zeta(3\nu - 1)a_3 + 18\zeta\nu b_3 + 18\zeta\nu(a_2 + b_2 + a_1 b_1 + a_1 b_2 + a_2 b_1),$

$P_{33} = 6\zeta\nu(a_1 b_2 + a_2 b_1) + 6\zeta\nu(a_3 + b_3),$

$P_{40} = 24\nu(\zeta + 1)(b_3 + b_4) + 24\zeta(\nu - 1)(a_2 + 2a_3 + a_4)$

$\qquad + 24\zeta\nu(2a_1 + a_2 + 1)(b_1 + b_2) + 24\zeta\nu(2 + a_1)(b_2 + b_3)$

$\qquad + 24\zeta\nu(a_1 + 2a_2 + a_3)(b_1 + 1) = 4!\, P_0 = 0,$

$P_{41} = 72\zeta\nu(b_1 + 2b_2 + b_3)a_1 + 72\zeta\nu(2b_1 + b_2)a_2 + 72\zeta\nu b_1 a_3 + 72\zeta\nu b_2$

$\qquad + 72\zeta\nu a_2 + 48\zeta(3\nu - 1)a_3 + 24\zeta(2\nu - 1)a_4 + 144\zeta\nu b_3$

$\qquad + 24(3\zeta + 1)\nu b_4,$

$P_{42} = 24\zeta a_4(\nu - 1) + 48\zeta\nu(a_3 + a_4) + 48\zeta\nu b_4 + 24\zeta\nu(b_3 + b_4)$

$\qquad + 48\zeta\nu b_1(a_2 + a_3) + 24\zeta\nu a_3(1 + b_1) + 24\zeta\nu a_1 b_3 + 24\zeta\nu b_3(2 + a_1)$

$\qquad + 24\zeta\nu a_1(b_2 + b_3) + 48\zeta\nu b_2(a_1 + a_2) + 24\zeta\nu a_2(b_1 + b_2),$

$P_{43} = 24\zeta\nu a_1 b_3 + 24\zeta\nu a_2 b_2 + 24\zeta\nu a_3 b_1 + 24\zeta\nu b_4 + 24\zeta\nu a_4.$

Note that $\widehat{a}_i^{(k)} = \widehat{a}_i^{(k)}(t)$ and $\widehat{b}_i^{(k)} = \widehat{b}_i^{(k)}(t)$, the t-dependent part of $a_i^{(k)}(t)$ and $b_i^{(k)}(t)$. The t-dependence of $\widehat{a}_i^{(k)}(t)$, etc. are henceforth suppressed for brevity of notation. It should be recalled that $P_{i0} \equiv P_i = 0$ by virtue of Proposition 1 at the critical point. These five conditions reduce the seven parameters in the quadratic model to two free parameters, which must also satisfy the stability condition, that is, the fifth density derivative being negative if the critical state is to be stable.

B.3   Coefficients $\widehat{\Pi}_j^{(i)}(t)$

$$\widehat{\Pi}_0^{(0)} = \zeta(\nu - 1)\widehat{a}_0^{(k)} + \nu(1 + \zeta)\widehat{b}_0^{(k)} + \zeta\nu\widehat{a}_0^{(k)}\widehat{b}_0^{(k)},$$

$$\widehat{\Pi}_1^{(0)} = \zeta(3\nu - 2)\widehat{a}_0^{(k)} + (1 + 3\zeta)\nu\widehat{b}_0^{(k)} + 3\zeta\nu\widehat{a}_0^{(k)}\widehat{b}_0^{(k)},$$

$$\widehat{\Pi}_2^{(0)} = (3\nu - 1)\zeta\widehat{a}_0^{(k)} + 3\zeta\nu\widehat{b}_0^{(k)} + 3\zeta\nu\widehat{a}_0^{(k)}\widehat{b}_0^{(k)},$$

$$\widehat{\Pi}_3^{(0)} = \zeta\nu\widehat{a}_0^{(k)} + \zeta\nu\widehat{b}_0^{(k)} + \zeta\nu\widehat{a}_0^{(k)}\widehat{b}_0^{(k)},$$

$$\widehat{\Pi}_0^{(1)} = \zeta(\nu - 1)\left(2\widehat{a}_0^{(k)} + \widehat{a}_1^{(k)}\right) + \zeta\nu(a_1 + 2)\widehat{b}_0^{(k)} + \nu(\zeta + 1)\left(\widehat{b}_0^{(k)} + \widehat{b}_1^{(k)}\right)$$
$$+ \zeta\nu(b_1 + 1)\widehat{a}_0^{(k)} + \zeta\nu\widehat{a}_0^{(k)}\left(\widehat{b}_0^{(k)} + \widehat{b}_1^{(k)}\right) + \zeta\nu\widehat{b}_0^{(k)}\left(2\widehat{a}_0^{(k)} + \widehat{a}_1^{(k)}\right) + \tau t,$$

$$\widehat{\Pi}_1^{(1)} = 2\zeta(3\nu - 1)\widehat{a}_0^{(k)} + 6\zeta\nu\widehat{b}_0^{(k)} + \zeta(3\nu - 2)\widehat{a}_1^{(k)} + \nu(1 + 3\zeta)\widehat{b}_1^{(k)}$$
$$+ 3\zeta\nu\left(\widehat{b}_0^{(k)}a_1 + \widehat{a}_0^{(k)}b_1\right) + 3\zeta\nu\left(2\widehat{a}_0^{(k)}\widehat{b}_0^{(k)} + \widehat{a}_0^{(k)}\widehat{b}_1^{(k)} + \widehat{a}_1^{(k)}\widehat{b}_0^{(k)}\right),$$

$$\widehat{\Pi}_2^{(1)} = 3\zeta\nu\left(\widehat{a}_0^{(k)} + \widehat{b}_0^{(k)}\right) + \zeta(3\nu - 1)\widehat{a}_1^{(k)} + 3\zeta\nu\widehat{b}_1^{(k)} + 3\zeta\nu\left(\widehat{a}_0^{(k)}b_1 + \widehat{b}_0^{(k)}a_1\right)$$
$$+ 3\zeta\nu\left(\widehat{a}_0^{(k)}\widehat{b}_0^{(k)} + \widehat{a}_0^{(k)}\widehat{b}_1^{(k)} + \widehat{a}_1^{(k)}\widehat{b}_0^{(k)}\right),$$

$$\widehat{\Pi}_3^{(1)} = \zeta\nu\left(\widehat{a}_1^{(k)} + \widehat{b}_1^{(k)}\right) + \zeta\nu\left(\widehat{b}_0^{(k)}a_1 + \widehat{a}_0^{(k)}b_1\right) + \zeta\nu\left(\widehat{a}_0^{(k)}\widehat{b}_1^{(k)} + \widehat{a}_1^{(k)}\widehat{b}_0^{(k)}\right),$$

$$\widehat{\Pi}_0^{(2)} = 2\zeta\left[\nu - 1 + \nu(2 + 3b_1 + b_2)\right]\widehat{a}_0^{(k)} + 2\zeta\nu(3 + 3a_1 + a_2)\widehat{b}_0^{(k)}$$
$$+ 2\zeta(3\nu + \nu b_1 - 2)\widehat{a}_1^{(k)} + 2\nu(1 + 3\zeta + \zeta a_1)\widehat{b}_1^{(k)}$$
$$+ 2\zeta\nu\widehat{a}_0^{(k)}\left(\widehat{b}_1^{(k)} + \widehat{b}_2^{(k)}\right) + 2\zeta(\nu - 1)\widehat{a}_2^{(k)} + 2\nu(\zeta + 1)\widehat{b}_2^{(k)}$$
$$+ 2\zeta\nu\left(\widehat{b}_0^{(k)} + \widehat{b}_1^{(k)}\right)\left(2\widehat{a}_0^{(k)} + \widehat{a}_1^{(k)}\right) + 2\zeta\nu\widehat{b}_0^{(k)}\left(\widehat{a}_0^{(k)} + 2\widehat{a}_1^{(k)} + \widehat{a}_2^{(k)}\right),$$

$$\widehat{\Pi}_1^{(2)} = 2\zeta\nu(3 + 4b_1 + 3b_2)\widehat{a}_0^{(k)} + 6\zeta\nu(1 + 2a_1 + a_2)\widehat{b}_0^{(k)}$$
$$+ 4\zeta(3\nu + \nu b_1 - 1)\widehat{a}_1^{(k)} + 6\zeta\nu(2 + a_1)\widehat{b}_1^{(k)} + 2\zeta(3\nu - 2)\widehat{a}_2^{(k)}$$

$$+2\nu(3\zeta+1)\widehat{b}_2^{(k)} + 6\zeta\nu a\widehat{b}_0^{(k)} + 12\zeta\nu\big(\widehat{a}_0^{(k)}\widehat{b}_1^{(k)} + \widehat{a}_1^{(k)}\widehat{b}_0^{(k)}\big)$$

$$+6\zeta\nu\big(\widehat{a}_0^{(k)}\widehat{b}_2^{(k)} + \widehat{a}_1^{(k)}\widehat{b}_1^{(k)} + \widehat{b}_0^{(k)}\widehat{a}_2^{(k)}\big),$$

$$\widehat{\Pi}_2^{(2)} = 6\zeta\nu(b_1+b_2)\widehat{a}_0^{(k)} + 10\zeta\nu(a_1+a_2)\widehat{b}_0^{(k)} + 2\zeta\nu\big(5+3b_1\big)\widehat{a}_1^{(k)}$$

$$+6\zeta\nu\big(1+a_1\big)\widehat{b}_1^{(k)} + 2\zeta\big(5\nu-1\big)\widehat{a}_2^{(k)} + 6\zeta\nu\widehat{b}_2^{(k)} + 6\zeta\nu\widehat{a}_1^{(k)}\widehat{b}_1^{(k)}$$

$$+10\zeta\nu\widehat{b}_0^{(k)}\widehat{a}_2^{(k)} + 6\zeta\nu\widehat{a}_0^{(k)}\widehat{b}_2^{(k)} + 2\zeta\nu\big(3\widehat{a}_0^{(k)}\widehat{b}_1^{(k)} + 5\widehat{a}_1^{(k)}\widehat{b}_0^{(k)}\big),$$

$$\widehat{\Pi}_3^{(2)} = 2\zeta\nu b_2\widehat{a}_0^{(k)} + 2\zeta\nu a_2\widehat{b}_0^{(k)} + 2\zeta\nu b_1\widehat{a}_1^{(k)} + 2\zeta\nu a_1\widehat{b}_1^{(k)} + 2\zeta\nu\big(\widehat{a}_2^{(k)} + \widehat{b}_2^{(k)}\big)$$

$$+2\zeta\nu\big(\widehat{a}_0^{(k)}\widehat{b}_2^{(k)} + \widehat{a}_1^{(k)}\widehat{b}_1^{(k)} + \widehat{b}_0^{(k)}\widehat{a}_2^{(k)}\big),$$

$$\widehat{\Pi}_0^{(3)} = 6\zeta\nu\big(1+3b_1+3b_2+b_3\big)\widehat{a}_0^{(k)} + 6\zeta\nu\big(1+3a_1+3a_2+a_3\big)\widehat{b}_0^{(k)}$$

$$+6\zeta\big[\nu-1+\nu(b_1+b_2)+2\nu(b_1+1)\big]\widehat{a}_1^{(k)} + 6\zeta\nu\big(3+3a_1+a_2\big)\widehat{b}_1^{(k)}$$

$$+\big[2(\nu-1)+\nu(b_1+1)\big]\widehat{a}_2^{(k)} + 6\nu\big[\zeta+1+\zeta(a_1+2)\big]\widehat{b}_2^{(k)}$$

$$+6\zeta(\nu-1)\widehat{a}_3^{(k)} + 6\nu(\zeta+1)\widehat{b}_3^{(k)} + 6\zeta\nu\widehat{a}_0^{(k)}\big(\widehat{b}_0^{(k)} + \widehat{b}_1^{(k)} + \widehat{b}_2^{(k)} + \widehat{b}_3^{(k)}\big)$$

$$+6\zeta\nu\big(\widehat{b}_1^{(k)} + \widehat{b}_2^{(k)}\big)\big(2\widehat{a}_0^{(k)} + \widehat{a}_1^{(k)}\big) + 6\zeta\nu\big(\widehat{b}_0^{(k)} + \widehat{b}_1^{(k)}\big)\big(2\widehat{a}_1^{(k)} + \widehat{a}_2^{(k)}\big)$$

$$+6\zeta\nu\widehat{b}_0^{(k)}\big(\widehat{a}_1^{(k)} + 2\widehat{a}_2^{(k)} + \widehat{a}_3^{(k)}\big),$$

$$\widehat{\Pi}_1^{(3)} = 18\zeta\nu\big(b_1+2b_2+b_3\big)\widehat{a}_0^{(k)} + 18\zeta\nu\big(a_1+2a_2+a_3\big)\widehat{b}_0^{(k)}$$

$$+18\zeta\nu\big(2b_1+b_2+1\big)\widehat{a}_1^{(k)} + 18\zeta\nu\big(2a_1+a_2+1\big)\widehat{b}_1^{(k)}$$

$$+6\zeta\big(6\nu+3\nu b_1-2\big)\widehat{a}_2^{(k)} + 18\zeta\nu\big(a_1+2\big)\widehat{b}_2^{(k)} + 6\zeta(3\nu-2)\widehat{a}_3^{(k)}$$

$$+6\nu(3\zeta+1)\widehat{b}_3^{(k)} + 18\zeta\nu\widehat{a}_0^{(k)}\widehat{b}_1^{(k)} + 18\zeta\nu\widehat{a}_1^{(k)}\widehat{b}_0^{(k)} + 36\zeta\nu\widehat{a}_0^{(k)}\widehat{b}_2^{(k)}$$

$$+36\zeta\nu\widehat{a}_1^{(k)}\widehat{b}_1^{(k)} + 36\zeta\nu\widehat{b}_0^{(k)}\widehat{a}_2^{(k)} + 18\zeta\nu\widehat{a}_0^{(k)}\widehat{b}_3^{(k)} + 18\zeta\nu\widehat{a}_1^{(k)}\widehat{b}_2^{(k)}$$

$$+18\zeta\nu\widehat{b}_0^{(k)}\widehat{a}_3^{(k)} + 18\zeta\nu\widehat{a}_2^{(k)}\widehat{b}_1^{(k)},$$

$$\widehat{\Pi}_2^{(3)} = 18\zeta\nu\big(b_2+b_3\big)\widehat{a}_0^{(k)} + 18\zeta\nu\big(a_2+a_3\big)\widehat{b}_0^{(k)} + 18\zeta\nu\big(b_1+b_2\big)\widehat{a}_1^{(k)}$$

$$+18\zeta\nu\big(a_1+a_2\big)\widehat{b}_1^{(k)} + 18\zeta\nu\big(1+b_1\big)\widehat{a}_2^{(k)} + 18\zeta\nu\big(a_1+1\big)\widehat{b}_2^{(k)}$$

$$+18\zeta\nu\widehat{a}_0^{(k)}\widehat{b}_2^{(k)} + 18\zeta\nu\widehat{b}_0^{(k)}\widehat{a}_2^{(k)} - 6\zeta\big(1-3\nu\big)a_3^{(k)} + 18\zeta\nu\widehat{b}_3^{(k)}$$

$$+18\zeta\nu\widehat{a}_1^{(k)}\widehat{b}_1^{(k)} + 18\zeta\nu\widehat{a}_1^{(k)}\widehat{b}_2^{(k)} + 18\zeta\nu\widehat{a}_2^{(k)}\widehat{b}_1^{(k)}$$

$$+18\zeta\nu\widehat{a}_0^{(k)}\widehat{b}_3^{(k)} + 18\zeta\nu\widehat{b}_0^{(k)}\widehat{a}_3^{(k)},$$

$$\widehat{\Pi}_3^{(3)} = 6\zeta\nu\big(b_3\widehat{a}_0^{(k)} + b_2\widehat{a}_1^{(k)} + b_1\widehat{a}_2^{(k)} + a_3\widehat{b}_0^{(k)} + a_2\widehat{b}_1^{(k)} + a_1\widehat{b}_2^{(k)}\big)$$
$$+ 6\zeta\nu\big(\widehat{a}_1^{(k)}\widehat{b}_2^{(k)} + \widehat{a}_2^{(k)}\widehat{b}_1^{(k)}\big) + 6\zeta\nu\big(\widehat{a}_0^{(k)}\widehat{b}_3^{(k)} + \widehat{b}_0^{(k)}\widehat{a}_3^{(k)}\big)$$
$$+ 6\zeta\nu\big(\widehat{a}_3^{(k)} + \widehat{b}_3^{(k)}\big),$$

$$\widehat{\Pi}_0^{(4)} = 24\zeta\nu\big(b_1 + 3b_2 + 3b_3 + b_4\big)\widehat{a}_0^{(k)} + 24\zeta\nu\big(3b_1 + 3b_2 + b_3 + 1\big)\widehat{a}_1^{(k)}$$
$$+ 24\zeta\big(3\nu + 3\nu b_1 + \nu b_2 - 1\big)\widehat{a}_2^{(k)} + 24\zeta\big(3\nu + \nu b_1 - 2\big)\widehat{a}_3^{(k)}$$
$$+ 24\zeta(\nu - 1)\widehat{a}_4^{(k)} + 24\zeta\nu\big(a_1 + 3a_2 + 3a_3 + a_4\big)\widehat{b}_0^{(k)}$$
$$+ 24\zeta\nu\big(3a_1 + 3a_2 + a_3 + 1\big)\widehat{b}_1^{(k)} + 24\zeta\nu\big(3a_1 + 3a_2 + a_3 + 1\big)\widehat{b}_2^{(k)}$$
$$+ 24\nu\big(3\zeta a_0 + \zeta a_1 + 1\big)\widehat{b}_3^{(k)} + 24\nu\big(za_0 + 1\big)\widehat{b}_4^{(k)} + 24\zeta\nu\widehat{a}_0^{(k)}\big(\widehat{b}_3^{(k)} + \widehat{b}_4^{(k)}\big)$$
$$+ 24\zeta\nu\big(\widehat{b}_2^{(k)} + \widehat{b}_3^{(k)}\big)\big(2\widehat{a}_0^{(k)} + \widehat{a}_1^{(k)}\big)$$
$$+ 24\zeta\nu\big(\widehat{b}_1^{(k)} + \widehat{b}_2^{(k)}\big)\big(\widehat{a}_0^{(k)} + 2\widehat{a}_1^{(k)} + \widehat{a}_2^{(k)}\big)$$
$$+ 24\zeta\nu\big(\widehat{b}_0^{(k)} + \widehat{b}_1^{(k)}\big)\big(\widehat{a}_1^{(k)} + 2\widehat{a}_2^{(k)} + \widehat{a}_3^{(k)}\big)$$
$$+ 24\zeta\nu\widehat{b}_0^{(k)}\big(\widehat{a}_2^{(k)} + 2\widehat{a}_3^{(k)} + \widehat{a}_4^{(k)}\big),$$

$$\widehat{\Pi}_1^{(4)} = 72\zeta\nu\big(b_2 + 2b_3 + b_4\big)\widehat{a}_0^{(k)} + 24\zeta\nu\big(3a_2 + 6a_3 + 2a_4\big)\widehat{b}_0^{(k)}$$
$$+ 72\zeta\nu\big(b_1 + 2b_2 + b_3\big)\widehat{a}_1^{(k)} + 72\zeta\nu\big(a_1 + 2a_2 + a_3\big)\widehat{b}_1^{(k)} + 72\zeta\nu\widehat{a}_1^{(k)}\widehat{b}_1^{(k)}$$
$$+ 72\zeta\nu\big(2b_1 + b_2 + 1\big)\widehat{a}_2^{(k)} + 72\zeta\nu\big(2a_1 + a_2 + 1\big)\widehat{b}_2^{(k)} + 72\zeta\nu\widehat{a}_0^{(k)}\widehat{b}_2^{(k)}$$
$$+ 72\zeta\nu\widehat{b}_0^{(k)}\widehat{a}_2^{(k)} + 24\zeta\big(6\nu + 3\nu b_1 - 2\big)\widehat{a}_3^{(k)} + 24\zeta\big(2\nu - 1\big)\widehat{a}_4^{(k)}$$
$$+ 72\zeta\nu\big(a_1 + 2\big)\widehat{b}_3^{(k)} + 24\nu\big(3\zeta + 1\big)\widehat{b}_4^{(k)} + 144\zeta\nu\widehat{a}_0^{(k)}\widehat{b}_3^{(k)}$$
$$+ 144\zeta\nu\widehat{a}_1^{(k)}\widehat{b}_2^{(k)} + 144\zeta\nu\widehat{b}_0^{(k)}\widehat{a}_3^{(k)} + 144\zeta\nu\widehat{a}_2^{(k)}\widehat{b}_1^{(k)} + 72\zeta\nu\widehat{a}_0^{(k)}\widehat{b}_4^{(k)}$$
$$+ 72\zeta\nu\widehat{a}_1^{(k)}\widehat{b}_3^{(k)} + 48\zeta\nu\widehat{b}_0^{(k)}\widehat{a}_4^{(k)} + 72\zeta\nu\widehat{a}_2^{(k)}\widehat{b}_2^{(k)} + 72\zeta\nu\widehat{b}_1^{(k)}\widehat{a}_3^{(k)},$$

$$\widehat{\Pi}_2^{(4)} = 72\zeta\nu(b_3 + b_4)\widehat{a}_0^{(k)} + 72\zeta\nu(b_2 + b_3)\widehat{a}_1^{(k)} + 72\zeta\nu(b_1 + b_2)\widehat{a}_2^{(k)}$$
$$+ 72\zeta\nu\big(b_1 + 1\big)\widehat{a}_3^{(k)} + 24\zeta\big(3\nu - 1\big)\widehat{a}_4^{(k)} + 72\zeta\nu\big(a_3 + a_4\big)\widehat{b}_0^{(k)}$$
$$+ 72\zeta\nu\big(a_2 + a_3\big)\widehat{b}_1^{(k)} + 72\zeta\nu\big(a_1 + a_2\big)\widehat{b}_2^{(k)} + 72\zeta\nu\big(a_1 + 1\big)\widehat{b}_3^{(k)}$$
$$+ 72\zeta\nu\widehat{b}_4^{(k)} + 72\zeta\nu\big(\widehat{b}_3^{(k)} + \widehat{b}_4^{(k)}\big)\widehat{a}_0^{(k)} + 72\zeta\nu\big(\widehat{b}_2^{(k)} + \widehat{b}_3^{(k)}\big)\widehat{a}_1^{(k)}$$
$$+ 72\zeta\nu\big(\widehat{b}_1^{(k)} + \widehat{b}_2^{(k)}\big)\widehat{a}_2^{(k)} + 72\zeta\nu\big(\widehat{b}_0^{(k)} + \widehat{b}_1^{(k)}\big)\widehat{a}_3^{(k)} + 72\zeta\nu\widehat{b}_0^{(k)}\widehat{a}_4^{(k)},$$

$$\widehat{\Pi}_3^{(4)} = 24\zeta\nu\big(b_4\widehat{a}_0^{(k)} + b_3\widehat{a}_1^{(k)} + b_2\widehat{a}_2^{(k)} + b_1\widehat{a}_3^{(k)} + \widehat{a}_4^{(k)}\big)$$

$$+ 24\zeta\nu\big(a_4\widehat{b}_0^{(k)} + a_3\widehat{b}_1^{(k)} + a_2\widehat{b}_2^{(k)} + a_1\widehat{b}_3^{(k)} + \widehat{b}_4^{(k)}\big)$$

$$+ 24\zeta\nu\big(\widehat{a}_0^{(k)}\widehat{b}_4^{(k)} + \widehat{a}_1^{(k)}\widehat{b}_3^{(k)} + \widehat{a}_2^{(k)}\widehat{b}_2^{(k)} + \widehat{b}_1^{(k)}\widehat{a}_3^{(k)} + \widehat{b}_0^{(k)}\widehat{a}_4^{(k)}\big).$$

B.4　Coefficients of the Spinodal Equations

With P_{ij} and $\widehat{\Pi}_j^{(i)}$ presented earlier, it is now possible to show the coefficients of the spinodal equations (Eqs. (10.79) and (10.81)). In the model represented by Eqs. (10.18) and (10.20) they are given as follows:

$$\varphi_{10} = \nu\big(1 + b_1 + \widehat{b}_0^{(k)} + \widehat{b}_1^{(k)}\big)\Pi_0^{(0)} + \Pi_0^{(1)}\big(1 - \nu - \nu\widehat{b}_0^{(k)}\big),$$

$$\varphi_{11} = \nu\Pi_0^{(0)}\big(\widehat{b}_1^{(k)} + b_1\big) + \nu\Pi_1^{(0)}\big(\widehat{b}_0^{(k)} + \widehat{b}_1^{(k)} + b_1 + 1\big)$$

$$+ \Pi_1^{(1)}\big(1 - \nu\widehat{b}_0^{(k)} - \nu\big) - \nu\Pi_0^{(1)}\big(\widehat{b}_0^{(k)} + 1\big),$$

$$\varphi_{12} = \nu\Pi_1^{(0)}\big(\widehat{b}_1^{(k)} + b_1\big) + \nu\Pi_2^{(0)}\big(\widehat{b}_0^{(k)} + \widehat{b}_1^{(k)} + b_1 + 1\big)$$

$$+ \Pi_2^{(1)}\big(1 - \nu\widehat{b}_0^{(k)} - \nu\big) - \nu\Pi_1^{(1)}\big(\widehat{b}_0^{(k)} + 1\big),$$

$$\varphi_{13} = \nu\Pi_2^{(0)}\big(\widehat{b}_1^{(k)} + b_1\big) + \nu\Pi_3^{(0)}\big(\widehat{b}_0^{(k)} + \widehat{b}_1^{(k)} + b_1 + 1\big)$$

$$+ \Pi_3^{(1)}\big(1 - \nu\widehat{b}_0^{(k)} - \nu\big) - \nu\Pi_2^{(1)}\big(\widehat{b}_0^{(k)} + 1\big)$$

$$\varphi_{14} = \nu\Pi_3^{(0)}\big(b_1 + \widehat{b}_1^{(k)}\big) - \nu\Pi_3^{(1)}\big(1 + \widehat{b}_0^{(k)}\big),$$

$$\varphi_{20} = 2\nu\Pi_0^{(0)}\big(b_1 + b_2 + \widehat{b}_1^{(k)} + \widehat{b}_2^{(k)}\big) + \Pi_0^{(2)}\big(1 - \nu - \nu\widehat{b}_0^{(k)}\big),$$

$$\varphi_{21} = 2\nu\Pi_0^{(0)}\big(b_2 + \widehat{b}_2^{(k)}\big) + 2\nu\Pi_1^{(0)}\big(b_1 + b_2 + \widehat{b}_1^{(k)} + \widehat{b}_2^{(k)}\big)$$

$$+ \Pi_1^{(2)}\big(1 - \nu - \nu\widehat{b}_0^{(k)}\big) - \nu\Pi_0^{(2)}\big(\widehat{b}_0^{(k)} + 1\big),$$

$$\varphi_{22} = 2\nu\Pi_1^{(0)}\big(b_2 + \widehat{b}_2^{(k)}\big) + 2\nu\Pi_2^{(0)}\big(b_1 + b_2 + \widehat{b}_1^{(k)} + \widehat{b}_2^{(k)}\big)$$

$$+ \Pi_2^{(2)}\big(1 - \nu - \nu\widehat{b}_1^{(k)}\big) - \nu\Pi_1^{(2)}\big(\widehat{b}_0^{(k)} + 1\big),$$

$$\varphi_{23} = 2\nu\Pi_2^{(0)}\big(b_2 + \widehat{b}_2^{(k)}\big) + 2\nu\Pi_3^{(0)}\big(b_1 + b_2 + \widehat{b}_1^{(k)} + \widehat{b}_2^{(k)}\big)$$

$$+ \Pi_3^{(2)}\big(1 - \nu - \nu\widehat{b}_0^{(k)}\big) - \nu\Pi_2^{(2)}\big(\widehat{b}_0^{(k)} + 1\big),$$

$$\varphi_{24} = 2\nu\Pi_3^{(0)}\big(b_2 + \widehat{b}_2^{(k)}\big) - \nu\Pi_3^{(2)}\big(\widehat{b}_0^{(k)} + 1\big),$$

$$\varphi_{30} = 6\nu\Pi_0^{(0)}\big(\widehat{b}_2^{(k)} + \widehat{b}_3^{(k)} + b_2 + b_3\big) - \Pi_0^{(3)}\big(\nu\widehat{b}_0^{(k)} + \nu - 1\big),$$

$$\varphi_{31} = 6\nu\Pi_1^{(0)}\left(\widehat{b}_2^{(k)} + \widehat{b}_3^{(k)} + b_2 + b_3\right) + 6\nu\Pi_0^{(0)}\left(\widehat{b}_3^{(k)} + b_3\right)$$
$$- \Pi_1^{(3)}\left[\nu\left(\widehat{b}_0^{(k)} + 1\right) - 1\right] - \nu\Pi_0^{(3)}\left(\widehat{b}_0^{(k)} + 1\right),$$

$$\varphi_{32} = 6\nu\Pi_2^{(0)}\left(\widehat{b}_2^{(k)} + \widehat{b}_3^{(k)} + b_2 + b_3\right) + 6\nu\Pi_1^{(0)}\left(\widehat{b}_3^{(k)} + b_3\right)$$
$$- \Pi_2^{(3)}\left[\nu\left(\widehat{b}_0^{(k)} + 1\right) - 1\right] - \nu\Pi_1^{(3)}\left(\widehat{b}_0^{(k)} + 1\right),$$

$$\varphi_{33} = \nu\left(\Pi_3^{(0)}\left(6\widehat{b}_2^{(k)} + 6\widehat{b}_3^{(k)} + 6b_2 + 6b_3\right) + \Pi_2^{(0)}\left(6\widehat{b}_3^{(k)} + 6b_3\right)\right)\Pi_2^{(3)}$$
$$- \Pi_3^{(3)}\left(\nu\left(\widehat{b}_0^{(k)} + 1\right) - 1\right) - \nu\left(\widehat{b}_0^{(k)} + 1\right),$$

$$\varphi_{34} = 6\nu\Pi_3^{(0)}\left(\widehat{b}_3^{(k)} + b_3\right) - \nu\Pi_3^{(3)}\left(\widehat{b}_0^{(k)} + 1\right),$$

$$\varphi_{40} = 24\nu\Pi_0^{(0)}\left(\widehat{b}_3^{(k)} + \widehat{b}_4^{(k)} + b_3 + b_4\right) - \Pi_0^{(4)}\left(\nu\widehat{b}_0^{(k)} + \nu - 1\right),$$

$$\varphi_{41} = 24\nu\Pi_1^{(0)}\left(\widehat{b}_3^{(k)} + \widehat{b}_4^{(k)} + b_3 + b_4\right) + 24\nu\Pi_0^{(0)}\left(\widehat{b}_4^{(k)} + b_4\right)$$
$$- \Pi_1^{(4)}\left[\nu\left(\widehat{b}_0^{(k)} + 1\right) - 1\right] - \nu\Pi_0^{(4)}\left(\widehat{b}_0^{(k)} + 1\right),$$

$$\varphi_{42} = 24\nu\Pi_2^{(0)}\left(\widehat{b}_3^{(k)} + \widehat{b}_4^{(k)} + b_3 + b_4\right) + 24\nu\Pi_1^{(0)}\left(\widehat{b}_4^{(k)} + b_4\right)$$
$$- \Pi_2^{(4)}\left[\nu\left(\widehat{b}_0^{(k)} + 1\right) - 1\right] - \nu\Pi_1^{(4)}\left(\widehat{b}_0^{(k)} + 1\right),$$

$$\varphi_{43} = 24\nu\Pi_3^{(0)}\left(\widehat{b}_3^{(k)} + \widehat{b}_4^{(k)} + b_3 + b_4\right) + 24\nu\Pi_2^{(0)}\left(\widehat{b}_4^{(k)} + b_4\right)$$
$$- \Pi_3^{(4)}\left[\nu\left(\widehat{b}_0^{(k)} + 1\right) - 1\right] - \nu\Pi_2^{(4)}\left(\widehat{b}_0^{(k)} + 1\right),$$

$$\varphi_{44} = 24\nu\Pi_3^{(0)}\left(\widehat{b}_4^{(k)} + b_4\right) - \nu\Pi_3^{(4)}\left(\widehat{b}_0^{(k)} + 1\right).$$

It is then useful to decompose these coefficients into t-independent and t-dependent parts

$$\varphi_{ij} = \omega_{ij} + \widehat{\varphi}_{ij}(t),$$

where

$$\omega_{i0} = (1 - \nu)P_{i0} = 0 \quad (i = 1, \ldots, 4),$$

whereas the rest of ω_{ij} are given by the following:

$$\omega_{11} = \nu b_1 + \nu(b_1 + 1)P_{01} - \nu P_{10} + (1 - \nu)P_{11},$$

$$\omega_{12} = \nu b_1 P_{01} + \nu(b_1 + 1)P_{02} - \nu P_{11} + (1 - \nu)P_{12},$$

$$\omega_{13} = \nu b_1 P_{02} + \nu(1 + b_1)P_{03} - \nu P_{12} + (1 - \nu)P_{13},$$

$$\omega_{14} = \nu(b_1 P_{03} - P_{13}),$$

$$\omega_{21} = 2\nu(b_1 + b_2)P_{01} + (1 - \nu)P_{21} - \nu P_{20},$$

$$\omega_{22} = 2\nu b_2 P_{01} + 2\nu(b_1 + b_2)P_{02} + (1 - \nu)P_{22} - \nu P_{21},$$

$$\omega_{23} = 2\nu b_2 P_{02} + 2\nu(b_1 + b_2)P_{03} + (1 - \nu)P_{23} - \nu P_{22},$$

$$\omega_{24} = 2\nu b_2 P_{03} - \nu P_{23},$$

$$\omega_{31} = 6\nu(b_2 + b_3)P_{01} - (\nu - 1)P_{31} - \nu P_{30},$$

$$\omega_{32} = 6\nu(b_2 + b_3)P_{02} + 6\nu b_3 P_{01} - (\nu - 1)P_{32} - \nu P_{31},$$

$$\omega_{33} = 6\nu(b_2 + b_3)P_{03} + 6\nu b_3 P_{02} - (\nu - 1)P_{33} - \nu P_{32},$$

$$\omega_{34} = 6\nu b_3 P_{03} - \nu P_{33},$$

$$\omega_{41} = 24\nu(b_3 + b_4)P_{01} - \nu P_{40} - (\nu - 1)P_{41},$$

$$\omega_{42} = 24\nu(b_3 + b_4)P_{02} + 24\nu b_4 P_{01} - (\nu - 1)P_{42} - \nu P_{41},$$

$$\omega_{43} = 24\nu(b_3 + b_4)P_{03} + 24\nu b_4 P_{02} - (\nu - 1)P_{43} - \nu P_{42},$$

$$\omega_{44} = 24\nu b_4 P_{03} - \nu P_{43}.$$

Then $\widehat{\varphi}_{ij}(t)$ can be calculated from the formulas presented for $\varphi_{ij}(t)$ by subtracting ω_{ij}. They are quite bulky, but straightforward to calculate. Here we list the expressions for $\widehat{\varphi}_{i0}(t)$ since they determine the t dependence of the spinodal curve in the leading order.

$$\widehat{\varphi}_{10} = \nu(1 + b_1)\widehat{\Pi}_0^{(0)} + \nu(\widehat{b}_0^{(k)} + \widehat{b}_1^{(k)})\widehat{\Pi}_0^{(0)} + (1 - \nu)\widehat{\Pi}_0^{(1)} - \nu\widehat{b}_0^{(k)}\widehat{\Pi}_0^{(1)},$$

$$\widehat{\varphi}_{20} = 2\nu(b_1 + b_2)\widehat{\Pi}_0^{(0)} + 2\nu\widehat{\Pi}_0^{(0)}(\widehat{b}_1^{(k)} + \widehat{b}_2^{(k)}) + (1 - \nu)\widehat{\Pi}_0^{(2)} - \nu\widehat{b}_0^{(k)}\widehat{\Pi}_0^{(2)},$$

$$\widehat{\varphi}_{30} = (1 - \nu)\widehat{\Pi}_0^{(3)} - \nu\widehat{b}_0^{(k)}\widehat{\Pi}_0^{(3)},$$

$$\widehat{\varphi}_{40} = 24\nu\widehat{\Pi}_0^{(0)}(b_3 + b_4 + \widehat{b}_3^{(k)} + \widehat{b}_4^{(k)}) + (1 - \nu)\widehat{\Pi}_0^{(4)} - \nu\widehat{b}_0^{(k)}\widehat{\Pi}_0^{(4)}.$$

Since $\omega_{i0} = 0$ we find

$$\varphi_{ij}(t) = \widehat{\varphi}_{i0}(t).$$

In the quadratic model we find $\widehat{\varphi}_{i0}(t)$ as follows:

$$\widehat{\varphi}_{10} = \zeta(\nu - 1)(\nu b_1 - \nu + 2)\widehat{a}_0^{(k)} - [\nu(\nu - 2) + \zeta\nu(3\nu - 4 + \nu a_1 - a_1)]\widehat{b}_0^{(k)}$$

$$- \zeta(1 - \nu)^2\widehat{a}_1^{(k)} + (\zeta + 1)\nu(1 - \nu)\widehat{b}_1^{(k)} + \text{second order terms},$$

$$\widehat{\varphi}_{20} = 2\zeta(1 - \nu)(3\nu + 2\nu b_1 - 1)\widehat{a}_0^{(k)}$$

$$+ 2\nu[\zeta(3a_1 + a_2 + \nu b_1 + \nu b_2 + 3) + \nu(b_1 + b_2)]\widehat{b}_0^{(k)}$$

$$+ 2\zeta(1 - \nu)(3\nu + \nu b_1 - 2)\widehat{a}_1^{(k)} + 2\nu(1 - \nu)(1 + 3\zeta + \zeta a_1)\widehat{b}_1^{(k)}$$

$$- 2\zeta(\nu - 1)^2\widehat{a}_2^{(k)} + 2\nu(\zeta + 1)(1 - \nu)\widehat{b}_2^{(k)} + \text{second order terms},$$

$$\widehat{\varphi}_{30} = 6\zeta\nu(1-\nu)(1+3b_1+3b_2)\widehat{a}_0^{(k)} + 6\zeta(1-\nu)\left[3\nu-1+\nu(3b_1+b_2)\right]\widehat{a}_1^{(k)}$$
$$+ (1-\nu)(3\nu-2+\nu b_1)\mathring{a}_2^{(k)} + 6\zeta\nu(1-\nu)(1+3a_1+3a_2)\widehat{b}_0^{(k)}$$
$$+ 6\zeta\nu(1-\nu)(3+3a_1+a_2)\widehat{b}_1^{(k)} + 6\nu(1-\nu)(3\zeta+1+\zeta a_1)\widehat{b}_2^{(k)}$$
$$+ \text{second order terms.}$$

Index